DIAPHRAGM WALLS & ANCHORAGES

DIAPHRAGM WALLS & ANCHORAGES

Proceedings of the Conference organized by the Institution of Civil Engineers and held in London, 18–20 September 1974

Institution of Civil Engineers, London, 1975

ORGANIZING COMMITTEE

M. W. Leonard (Chairman)
P. J. C. Buckley
R. W. Cooke
T. H. Hanna
C. K. Haswell
A. S. West

EDITOR

Thelma J. Darwent

© The Institution of Civil Engineers, 1975.

ISBN: 0 7277 0005 7

Published by Thomas Telford Ltd for the Institution of Civil Engineers at 26-34 Old Street, London EC1V 9AD

Made and printed in Great Britain by William Clowes & Sons, Limited, London, Beccles and Colchester.

Contents

Opening address

Sir Kirby Laing, JP, MA(Cantab)
President of the Institution of Civil Engineers

I think the organizing committee is to be congratulated on the work it has done. From the support that it has received from all over the world it is obvious that a conference of this sort on this particular subject is most apposite at the present time. Certainly the support of so many authors of world-wide renown is most encouraging.

While diaphragm walling in the sense of this Conference is in fact new in the UK (the first application was the construction of the Hyde Park underpass in 1967) we must remember that the use of sheet piling, which in effect is diaphragm walling, started almost beyond living memory.

It is not enough for a conference of this sort to report on past achievements. It should look forward. I am therefore glad to see references to methods of improving surface finishes, but I am sorry not to see any comments about the noise of machines which are used in the construction of diaphragm walls, nor about conditions prevailing on sites where bentonite is in use. Also, why no reference to the conditions prevailing on surrounding roads when this type of construction is used in confined sites in city centres?

I think those of us involved in the construction of diaphragm walls have a responsibility to our neighbours, and in this way we civil engineers once again become involved in the affairs of the environment.

I have much pleasure in declaring the Conference open and trust that all participants will benefit from the discussions that will take place.

Practical considerations affecting the construction of diaphragm walls

Z. SLIWINSKI, BSc, MICE, *Consultant, Cementation Piling and Foundations Ltd*
W. G. K. FLEMING, BSc, PhD, MICE, *Projects Manager (Technical), Cementation Piling and Foundations Ltd*

This Paper traces briefly the development of cast in place concrete diaphragm walls and sets out to draw attention to some practical points which need consideration, discussion or further research. The practical limitations of the method, in relation to such items as the concentration of the suspension necessary to avoid excessive fluid loss, the stability of trenches in soft ground, the methods of excavation which can be used in hard strata, and the problems which occur when the standing groundwater level is high, are discussed and some guidance is given. The behaviour and control of bentonite suspensions are considered in relation to the cycle of bentonite use and the process of contamination of the fluid, including possible countermeasures which can be taken to reduce contamination. The problems of control on site using the available parameters are enumerated, and the effects of allowing the properties of the fluid to go outside certain defined safe limits are stated. The desirable properties of concrete for placing by tremie are reviewed and details of a suitable basic mix are given. From the point of view of design, consideration is given to the use of reinforcement, and to the development of shaft friction on the walls of a trench and end bearing at its base. The factors influencing choice of panel dimensions are defined, and some suggestions are made about detailing. The permeability of walls is examined, taking into account both the concrete of the panels and the performance of the joints. Tolerances are also discussed with reference to verticality, the finished wall face, the top surface of the concrete as cast and the positioning of recesses. Finally, some of the risks associated with the process of diaphragm wall construction are examined, including those associated with stability of the trench and possible settlement of the adjacent ground.

INTRODUCTION

Bentonite suspensions have found a substantial and continuing use in oil well drilling. They allow deep boreholes to be stabilized when lining is impractical and they also serve to convey bored materials to the surface. For such applications they have been subject to detailed study. In the field of soil mechanics, bentonite has long been of interest because of its curious behaviour in forming thixotropic suspensions.

2. The first use of the material in the construction industry to form retaining walls appears to have been in boring contiguous piles by direct or reverse circulation methods. The real breakthrough in the development of the technique came when it was found possible to use grabs for direct excavation, using the fluid only to support the sides of a trench or pile. This gave rise to a simplifica-

tion of equipment and the method rapidly became economical.

3. The first slot excavations, combined with reinforced concrete structures, were used only about 20 years ago. A present assessment of the total area of such diaphragm walls in the world stands at some 6 to 7 million square metres.

4. The spectacular growth of the method may be attributed to certain factors. The first is the availability of bentonite, produced from montmorillonite clays, and refined and converted to sodium montmorillonite for commercial use. Secondly, the technique is well suited to a wide range of application. This includes cut-off walls under water retaining structures such as dams, weirs and reservoirs. In such cases the wall may be rigid (concrete) or plastic (concrete and bentonite mixture) and load carrying capacity is not required. The range also includes load bearing and earth-retaining walls for a variety of applications such as underpasses, deep basements, underground stations, tunnels, docks and pump houses. The third factor contributing to growth of the method is the development of trenching equipment, which frequently takes the form of heavy clamshell grabs which may be hydraulically operated, and which can be attached to standard crawler crane units. Also relevant is the development of plant for processing bentonite suspensions, now often including equipment for efficient mixing and for desanding. Finally there is the economy of the method as compared with more traditional methods, often saving time because of relative simplicity and saving cost because of the incorporation of diaphragm walls into permanent structures.

5. However, it must be recognized that there is still much to learn and, although the views of specialists are broadly in accord, the technology is still developing. In some directions, where we would wish to depend on established facts, only enlightened opinion is available. The object of this Paper is to raise some of the practical problems which are in need of further research or exchange of views between engineers and contracting specialists.

PRACTICAL LIMITATIONS OF THE METHOD

Loss of bentonite

6. One of the basic functions of bentonite suspension is to form a filter cake skin on the surface of the soil exposed by excavation. The skin will form only if the fluid pressure in the trench exceeds any external groundwater

pressure and if the permeability of the soil is within certain limits.

7. A normal 4–6% concentration of suspension will be retained by ground having a permeability k up to about 10^{-1} or 10^{-2} cm/s as defined by Darcy's law. If the ground is more open than this a denser suspension in the range up to about 12% may prove satisfactory. In extraordinary cases, where even this dense fluid is not retained, a number of additives may be tried as indicated in Table 1. None of these methods is likely to be very effective if permeability exceeds about 5 cm/s.

8. As filter cake formation is dependent on fluid loss is is not normally possible to find any significant filter cake formation in low permeability soils such as clay. Pieces of clay from the faces of diaphragm walls, examined during the subsequent main excavation, show little or no trace of bentonite as a rule.

Table 1. Treatment for loss of bentonite

Treatment method	Materials used	Function	Comment
Add intermediate size particles to suspension	Clay Silt Sand	Particles carried into pores or fissures causing blockage	Not suitable if pores are very large
Chemical additive treatment	Potassium aluminate Aluminium chloride Calcium	Flocculates bentonite gel, flocs block pores	Once seal is made the fluid may have to be renewed
Cement/bentonite mixture	Cement added to fluid at mixing	Allowed to penetrate areas of loss and then set	Mixture usually supplied only to area of loss, and area subsequently re-excavated under normal fluid
Fibrous materials (used in oil drilling work)	Graded fibrous or flake materials, e.g. plant fibres, glass, rayon, cellophane flakes, mica, ground rubber tyres, ground plastic, perlite, nut shells, etc.	A suitable mixture will bridge fissures and fill a wide range of pore sizes effectively	Trial and error method with wide selection range; many materials available under trade names from oil drilling sources

Very soft soils

9. Experience shows that, in soils of very low strength, it may not be possible to form diaphragm walls by conventional methods, because the internal bentonite pressure in the trench may be less than the active pressure of the soil. However, the examination of stability is not straightforward since panel length, shape and arching effects modify the normal two-dimensional considerations. Any soil exhibiting strength parameters ϕ (angle of internal friction) $= 0$ and c (cohesion) < 10 kN/m^2 should be considered with caution; panels should be kept short, and without complications such as arise with the use of counterforts. Conditions of instability are most prevalent in soft marine clays and in fresh hydraulic fills.

Hard strata

10. Occasionally it is specified that a diaphragm wall shall penetrate some distance into rock or semi-rock soils, and this can pose serious excavation problems. A standard clamshell grab of the type used by many companies cannot penetrate hard rock and even in soft rock its efficiency is considerably reduced. The penetration of the grab depends on its design and construction, and more specifically on its closing force and weight (Fig. 1). For occasional hard layers, heavy chisels are used to break the rock into pieces which the grab can collect.

11. For penetration into hard rock the standard equipment may be replaced by a special percussive reverse circulation machine, but its output can be painfully slow. Figures of 0·1–0·5 m/h are sometimes quoted.

12. For semi-rock materials such as marl, Cementation use the clamshell grab, assisted by an auxiliary rotary boring machine of the type used in large diameter piling (Fig. 2).

High water table

13. A high groundwater table may be a substantial impediment to the construction of diaphragm walls. It hampers the casting of guide walls and reduces the differen-

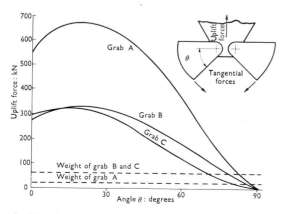

Fig. 1. Uplift force on grab during closing cycle

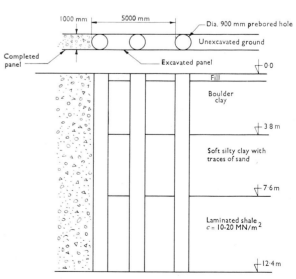

Fig. 2. Example of excavation of a diaphragm wall panel using preboring to overcome difficulty in grabbing through hard shale. (Plant used for grabbing: Cemtrencha with hydraulically operated grab 1000 mm size)

tial between the pressure inside the trench and the active pressure of the ground outside it. This has a detrimental effect on the formation of filter cake and on the general stability conditions. In addition the unavoidable fluctuation of the bentonite level in the trench during digging must be taken into account.

14. In general the level of bentonite suspension in the excavation must be at least 1·25 m above the groundwater level in the surrounding ground, and if this cannot be realized it may be necessary either to raise the site level or to lower the groundwater using well points.

BEHAVIOUR AND CONTROL OF BENTONITE SUSPENSION

The cycle of bentonite use

15. The bentonite suspension is prepared by adding a measured quantity of the powder to clean fresh water. It is mixed and allowed to hydrate and is then introduced to the excavation. When the excavation is complete and the reinforcement inserted, the fluid is displaced by concrete and pumped away for disposal or for treatment, storage, and further use. Figure 3 shows a typical site mixing and handling arrangement.

16. The time taken for hydration, after which the bentonite suspension has attained the desired rheological properties, depends on the method of mixing. A strong shearing action gives good performance and rapid hydration. It is technically and economically advantageous to use a fully hydrated suspension. Figures 4 and 5 show the rate of hydration for a colloidal mixer.

17. During excavation and concreting the bentonite fluid changes in colour and general appearance. It becomes more viscous and dense, sometimes partly loses its thixotropic properties, and it may be difficult to pump. The causes of these changes are complex, but in general they are due to the acceptance of soil and cement particles into the suspension. This contamination has both physical and chemical effects. The physical effects are that inert soil particles increase the fluid density and viscosity and enter into filter cake as it is formed, making it more pervious. This means that the thickness of cake builds up more rapidly and that the cake is softer.

18. Chemical contamination is in most cases by calcium, and the reaction consists in an exchange of calcium for sodium ions. Viscosity and gel strength increase rapidly, causing a quicker build-up of filter cake, which is less easy to displace with concrete than the sodium-based variety. Calcium in soil can come from shelly deposits, gypsum or anhydrite, but a more common source of contamination is the cement used in the concrete.

19. These changes happen on every site and it is evident that certain properties must be controlled within limits if the process of construction is to be satisfactory.

20. As mentioned above, bentonite from an excavation may be pumped away either for disposal or for cleaning and re-use. Reconditioning consists essentially of desanding by screening or by separating out the heavier particles in the suspension using a hydrocyclone. Both methods are effective and prolong the working life of the fluid, but it is not yet possible to avoid the disposal of a certain residual quantity. Even if the very fine contaminating particles could be removed, the process would be expected to involve considerable expense.

21. There has been some debate regarding the general ideas of reconditioning bentonite, and there is a school of thought which seems to prefer a single use followed by disposal. However, the argument is somewhat academic because the same standards of control have to be applied to each situation. The high cost of disposal, and the environmental considerations linked with finding acceptable places for dumping, will no doubt eventually lead to the general use of efficient reconditioning plant.

Control of bentonite suspension on sites

22. In current practice the control of bentonite varies

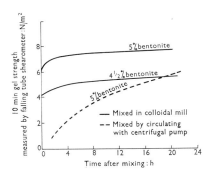

Fig. 4. Gel strength/time relationship

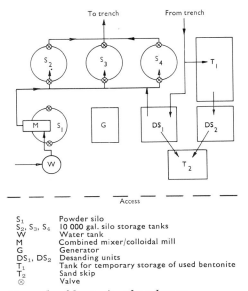

S$_1$ Powder silo
S$_2$, S$_3$, S$_4$ 10 000 gal. silo storage tanks
W   Water tank
M   Combined mixer/colloidal mill
G   Generator
DS$_1$, DS$_2$ Desanding units
T$_1$ Tank for temporary storage of used bentonite
T$_2$ Sand skip
⊗   Valve

Fig. 3. Example of bentonite plant layout

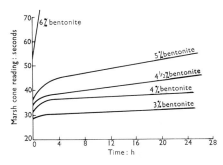

Fig. 5. Build-up of Marsh cone viscosity when colloidal mixer is used, temperature 20–22°C

within very wide limits. Visual inspection and the judgement of an experienced foreman, who may try the properties of bentonite between his fingers, is acceptable at one extreme while at the other extreme we may find a complex specification with many laboratory tests adopted from the oil industry or bentonite suppliers.

23. The first approach can of course be over-simplified, and visual inspection only can be deceptive. The second approach can produce a mass of data which is difficult to interpret in relation to the process of diaphragm walling, because the measurements are often interrelated in quite a complicated way; over-stringent limits are sometimes set and the execution of the work can become very costly. It seems common sense to assume that the extremes should be ruled out and that control should be based on measurable parameters which can be related directly to the construction process.

24. Table 2 shows a wide range of parameters which are measurable in a laboratory. The problem of efficient site control consists of selecting the more useful of these parameters and correlating them to the actual performance of the method. Some of the tests are not suitable for measuring the properties of bentonite from a panel, because of suspended solids. The Fan viscometer, for instance, is applicable only to checks on supplied bentonite materials. The recently published FPS specification[1] strikes a reasonably balanced view on the use of a few of the most important measurements.

Table 2. Bentonite quality parameters

Parameter	Test method
Viscosity	Marsh cone
Rheology	
(a) Plastic viscosity	
(b) Apparent viscosity	Fan viscometer
(c) Bingham yield strength	
(d) 10 min shear strength	
10 min gel strength	Shearometer
Density	Mud balance
pH	Electrical meter or paper strips
Fluid loss	Filter press
Sand content	Wet sieving through 200 BS mesh

Control limits

25. Experience shows that the use of bad suspensions may cause immediate or potential troubles, and the following are some points which have to be taken into account.

26. Excessive overbreak and cavities may occur below guide walls. This may be caused by shallow guide walls, a high water table, or an insufficient head of bentonite, but it can be seriously aggravated by a low viscosity suspension with a low gel strength. The wash from the moving grab may penetrate behind the guide wall and when the fluid drains back it can bring soil with it.

27. The bentonite filter cake may build up quickly and become thick and porous. This is well known in oil drilling where the rapid building of filter cake can obstruct the operation. In diaphragm walling and piling the excavation tools are not obstructed, but there could be a local influence on the friction between a wall and the soil.

28. Excessive loss of bentonite may occur in pervious strata, causing filter cake formation to be poor or non-existent. Loss of bentonite may be due to very permeable soil, but it can also be due to viscosity and gel strength being too low. If the bentonite suspension is not fully hydrated, these properties may not be adequately developed.

29. The concrete may not fully displace the bentonite. The finished wall must be sound without any bentonite inclusions. To ensure this the suspension must displace easily from the bottom of the trench, from around inserts, boxes, reinforcement and panel joints, and clean contact surfaces must be left. Much remains to be learned about the mechanism of displacement and Cementation Research Limited has put considerable effort into a study of the relationship between the properties of concrete and bentonite. So far it has been confirmed that the density of bentonite is a major factor, and that viscosity and gel strength are also involved.

30. Excessive loss of bentonite may take place because of its removal with materials in the grab. The concentration of bentonite must meet stability requirements in the trench, but an excessively concentrated fluid is lost too quickly in the spoil and represents unnecessary waste.

31. Difficulties in handling the bentonite suspension may occur. This problem is mostly due to contamination and is well known on sites.

32. From these considerations it is evident that the bentonite suspension has to fulfil many conditions, and that the requirements are often contradictory. Table 3

Table 3. Required properties of bentonite

Function of suspension	Parameter				
	Viscosity	Shear strength	Density	Fluid loss	pH
Form filtercake and stabilize bore by hydrostatic pressure application	Moderate to high	Moderate to high	High	Moderate to low	
Reduce cavitation caused by tool disturbance	High	High	—	Moderate	Low
Minimize loss of fluid in pervious strata	High	High	—	—	Low
Minimize loss of fluid in excavation spoil	Low	Low	—	—	—
Prevent accumulation of dense particles at base of excavation prior to concreting	High	High	High	—	—
Ensure free flow of concrete from tremie and easy displacement of bentonite from excavation and reinforcement	Low	Low	Low	—	Low
Allow easy pumping of bentonite fluid	Low	Low	Low	—	—
Prevent sedimentation in pipes and tanks	Moderate	High	High	—	—

Table 4. FPS specification for bentonite supplied to trench

Item	Range of results at 20°C	Test method
Density	Less than 1·10 g/ml	Mud density balance
Viscosity	30–90 s	Marsh cone method
Shear strength (10 min gel strength)	1·4–10 N/m²	Shearometer
pH	9·5–12	pH indicator paper strips

Table 5. Suggested concrete mix

Slump	150–200 mm
Water/cement ratio	Below 0·6
Aggregate type	Natural rounded stone if possible, 20 mm max. size
Sand type	Natural sand complying with zone 2 or 3 grading
Sand content	35–45% of the total aggregate weight
Cement	Not less than 400 kg/m³ of OPC or SR cement
Additives	The use of plasticisers is to be recommended. These can include air entraining agents, which prevent early stiffening of the mix and can improve durability of an exposed face. Retarders may be used when necessary.

illustrates some of the contradictions. Blank spaces in this table mean that the effect of the parameter as normally measured is likely to be secondary rather than completely insignificant.

33. The range of properties set out by the FPS specification,[1] as shown in Table 4, appears to offer a reasonable compromise for most circumstances.

34. The same specification suggests that the density of bentonite suspension taken from near the base of the trench before concrete is placed should be not greater than 1·3 g/ml.

35. It is the Authors' experience that improvements in standards of control not only reduce troubles from adverse soil conditions but also effect an improvement in the standard of the finished product.

CONCRETE SPECIFICATION

36. The specification for concrete in a cast in place diaphragm wall must fulfil the normal criteria relating to strength and durability. In addition, impermeability is an important feature of such walls.

37. The conditions of placing concrete in an excavated trench under bentonite are completely different from those applying to normal structural work where the concrete can be compacted into forms and where supervision is relatively easy. Concrete has to be poured into a diaphragm wall panel through a tremie pipe, displacing the fluid completely, and the operation is more akin to grouting than structural concreting.

38. The requirements to ensure displacement of bentonite have already been considered (§ 29 and Table 3). The requirements for concrete which can be tremied successfully can be stated as follows. The mix must be of a plastic consistency, behaving as a heavy fluid: if the concrete mix has a significant shear strength, the active pressure which it exerts as it flows may be insufficient to displace the bentonite suspension with which it is in contact. The mix must be cohesive: a mix which bleeds or disintegrates under the pressure of self weight can block the tremie or can accept bentonite. The mix must not set nor stiffen too quickly, but should remain workable until the panel is completed.

39. The consistency of concrete depends on the quantity of water included in it in relation to the fines, including cement. In order to make a flowable mix the water has to be trapped in the fines fraction, and since the particles opposing water flow are mainly those below 0·5 mm size these must be present in sufficient quantity.

40. Suggestions for a suitable mix are given in Table 5.

41. The control of concrete is of great importance. Although the slump test is crude it is normally adequate provided that at the same time the concrete quality is observed by a person of some experience. Segregation and bleeding are usually noticeable to a trained eye.

PRACTICAL CONSIDERATIONS AFFECTING DESIGN

42. The achievement of a full displacement of bentonite is often discussed in relation to reinforcement, wall friction and bearing capacity.

Reinforcement design and detail

43. The effect of bentonite on the bond between reinforcement and concrete has received some attention in the past. Although the mechanism of displacement is not fully understood, it seems likely that any adhesion between the steel and bentonite is of the order of the gel strength of the fluid, which in turn is comparatively small in relation to the shearing stress caused by the rising concrete. A CIRIA report[2] presented a series of tests on bond, using plain and deformed bars.

44. Taking all the current evidence into account it seems reasonable to accept the recommendations given in the FPS specification.[1] It would also seem to be desirable to design reinforcing cages so as to ensure that transverse reinforcement in the horizontal direction, which could lead to trapping of bentonite around intersections, is kept to a minimum. The concept of how concrete is going to flow upward through a reinforcing cage needs to be borne in mind when preparing a cage design, and particular attention should also be paid to the effect of insert boxes which are subsequently to form recesses in the wall.

The effect of bentonite on wall friction

45. A failure to displace bentonite from the trench walls could adversely affect the skin friction which might be taken into account in design. The suspicion that bentonite always has a detrimental influence appears to originate in the use of bentonite as a lubricant for caisson and tunnel work. However, the conditions in which walls or piles are formed are very different to those applying when a free bentonite layer is deliberately introduced. The adhesion between the bentonite gel and the sides of an excavation is of the order of the gel strength, which in this application seldom exceeds 15×10^{-6} N/mm². The rising concrete displaces the bentonite and exercises a sweeping action on the excavation walls. The shear force at the wall, resulting from cohesion and friction of the advancing concrete, is several times greater than the adhesion with the bentonite, and so virtually all free bentonite is removed.

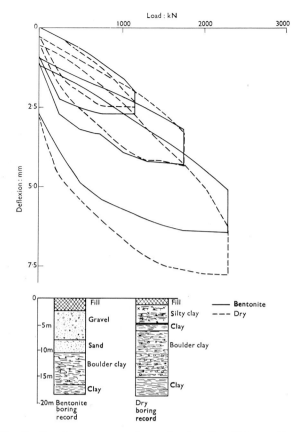

Fig. 6. Load test comparison of two piles 600 mm dia. on same site, under bentonite and dry

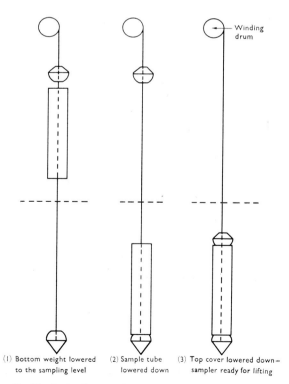

Fig. 7. Method of operation of patented Cementation mud sampler

46. In impervious clays, where bentonite does not penetrate nor filter cake form, full displacement is to be expected and a perfect soil to concrete contact is attained. In pervious layers of sand or gravel, all the free bentonite is displaced but the advancing concrete cannot displace the layers of soil impregnated with bentonite, neither is harder bentonite filter cake removed, nor even soft filter cake in cavities. These latter deposits have substantial strength, and evidence that significant friction may be taken into account is gradually accumulating from pile tests.

47. A summary of the Authors' views could be expressed as follows. In impervious soils such as clay the bentonite has no effect on friction, and no reduction of the normal factors for deriving adhesion is justified. There may indeed be a slight increase of adhesion. In granular soils the bentonite has some effect. The reduction of friction, calculated from effective lateral pressure and a coefficient of friction,[3] should be considered in the region 10–30%.

48. These views are supported by the knowledge that many thousands of piles have been constructed under bentonite, and they are performing satisfactorily. Also many pile tests have been carried out, in both cohesive and cohesionless soils, confirming safe load assessments based on calculations including allowance for friction or adhesion. Load tests on three instrumented deep piles at Bidston Moss have indicated high load transfer in both granular and clay soils. Special tests on two piles, one constructed under bentonite, and the other constructed in the dry, on the same site and in conditions as shown in Fig. 6, indicate no practical difference in load/deformation performance of the piles. A model pile test in sand, carried out in Cementation's research laboratory, indicated that a 5 mm thick filter cake around the pile caused a reduction of load transfer of up to about 10%.

49. In addition there are several publications[4−7] which confirm that friction is not significantly reduced when piles or walls are constructed under bentonite suspension.

Bearing at the trench base

50. Incomplete displacement from the base of a trench could affect the end bearing capacity of panels but there is no evidence that this situation arises. On the contrary, experience of the use of diaphragm wall panels for heavily loaded foundations seems to indicate that displacement is successful.

51. The practice of using diaphragm wall panels as load bearing elements appears to be generally accepted in continental practice.

52. It is nevertheless important to sample and check the bentonite at the bottom of an excavation, and if its density is too high then any slurry must be removed and replaced by fresh fluid. Fig. 7 shows a simple and reliable sampler developed by Cementation. Use of this tool has allowed the relationship between depth and bentonite density to be examined, and it has been observed that when good mixing and control are exercised the bentonite density varies only slightly with depth in a trench.

JOINTS AND PANEL LENGTHS

53. A concrete diaphragm wall cannot be formed as a continuous monolithic construction at this stage. Vertical joints are used to divide the wall into panels, and hori-

zontal reinforcement is not continuous from one panel to the next. Each panel acts as an independent unit except in rare cases where shear connectors are introduced.

54. A normal joint is formed by inserting a round stop-end pipe at the end of the excavation, and this is withdrawn when concrete has set, forming a semi-circular joint against which the concrete of the next panel is poured. This can be regarded as a standard joint. Other ingenious stop-end devices have been used to meet particular requirements but the circular pipe is the most popular. The following paragraphs relate to standard circular stop ends.

Choice of panel dimensions
55. The choice of panel dimensions depends on various considerations as follows.

(*a*) Stability. In general short panels are more stable than long panels, and in soft ground short and uncomplicated panels are necessary.
(*b*) Concreting. It must be possible to place the total volume of concrete for a panel before setting or significant stiffening takes place. In practice this usually means 3–3½ hours. Also, the distance from the tremie pipe to the end of the excavation should be limited to about 4 m. This distance may have to be reduced to about 2 m if the wall is to be thin and the reinforcement heavy.
(*c*) Excavation equipment. The most important limitation is that the panel cannot be shorter than a single bite of the grab. For longer panels there are no special limits, but it is desirable that the grab works with symmetrical load applied to its jaws, especially in hard soils. For economy the length should be sufficient to assure full bites of the grab, although this is not essential.
(*d*) Number of stop ends. The placing and extraction of stop ends is costly, and also each joint is a potential source of leakage, so it is desirable that the number of panel joints should be small. However, panel volumes should be of reasonable size and moderation in length is advised because of the possible shrinkage of panels.
(*e*) Reinforcement. The size and weight of each reinforcement cage must allow easy handling with normal site equipment.
(*f*) Anchor capacities and position. If anchors have to be used then the panel length has to be selected to make an economical arrangement.
(*g*) Programme requirements. The choice of panel length should allow for the completion of a whole number of panels per day. This is both technically and economically advantageous.

56. In practice an approximate panel size should be selected on the basis of items (*a*) and (*b*). This should be analysed for condition (*c*) and then the remaining items should be considered. The final checks should be made on the corner and the closing panels.

Suggested method of detailing
57. The design of reinforcement involves a consideration of the use of stop ends, and since some panels have no stop end, some have one and others have two, this can lead to complications in detailing.
58. It is suggested that for simplicity the stop end or ends should always be considered to lie outside the panel length L as indicated in Fig. 8. Assuming that the cover

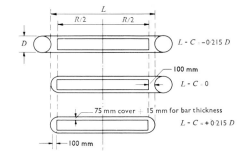

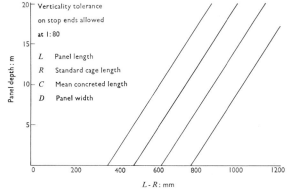

Fig. 8. Relationships for panel dimensions

to main steel is 90 mm, measured from the face of the panel to the centre of the main steel, and that the tolerances on verticality of stop ends are as given in the FPS specification, the relationships between wall thickness, panel length, reinforcing cage length and equivalent concrete panel length are as shown.

59. In detailing it is essential to show clearly the orientation of reinforcement cages, and there should be no possibility of wrong identification of the earth and excavated faces.

PERMEABILITY OF A DIAPHRAGM WALL

60. Since diaphragm walls are frequently required to resist both earth and water pressures, the means by which this is achieved warrants some examination, and consideration has to be given both to the panels of wall and to the joints between panels.

61. The permeability of concrete is expressed in general by the use of a coefficient k in accordance with Darcy's law. Hardened cement pastes have been studied in the USA by Powers *et al.*,[8,9] who examined the relationship between water/cement ratio and permeability, and reached the conclusion that for ratios up to 0·6 the permeability k was less than 10^{-11} cm/s. For wetter mixes the permeability increased rapidly with water content. Other work on cement pastes has shown permeabilities of the order of 10^{-10} cm/s for a water/cement ratio 0·7.

62. The permeability of various types of aggregate has been studied by Valenta[10] and, for common unweathered rocks ranging from sandstone to basalt, permeabilities in the range 10^{-9}–10^{-13} cm/s have been found.

63. The permeability of concrete has also been studied by Valenta[10] and, as with cement paste, once the water/cement ratio exceeds a critical value of about 0·6 a rapid

increase of permeability takes place. For a dense mix with 20 mm maximum aggregate and a water/cement ratio of up to 0·6, permeability is not likely to exceed 10^{-10} cm/s according to this research. Values of the same order have been found from samples taken from large diameter piles formed under bentonite.

64. If one considers a 600 mm thick wall retaining a 10 m head of water, taking permeability 10^{-10} cm/s and porosity 15%, and taking into account a suction pressure of one atmosphere assisting water flow, the quantity passing into the wall would be in the order of 0·3 litres for 1000 m² in 24 hours.

65. It is therefore evident that the concrete used for diaphragm walls can for practical purposes be considered impermeable and that the risk of sulphate attack is low.

66. However, in practice the permeability of a panel must also depend on the formation of cracks and on any local defects in the concrete such as may result from segregation. Due to the humid curing conditions, drying shrinkage cracks should be almost eliminated. Since panel lengths are in general relatively short any remaining shrinkage, whether due to drying or temperature effects, appears in practice to be taken up by movement at vertical joints, and intermediate cracks do not develop. The possibility of defects in concrete caused by segregation of the mix at placing cannot be completely excluded but, with normal concrete control and care, such occurrences should be limited to isolated cases.

67. The simple butt joint between panels cannot be claimed to be proof against water entry but significant leakages are rare, probably due to the presence of soil impregnated with bentonite gel behind the joint, and to the presence of some thin layer of contaminated bentonite at the edges of the joint. Where leaks occur they can usually be ascribed to differential deflexions between wall panels, and these differentials are at their worst near corners. Fig. 9 shows a diagram of a London site where some leaks occurred, and it was notable that the maximum measured deflexion differentials and percolation occurred in the area of a re-entrant corner.

68. The whole matter of deflexion differentials between wall panels depends on panel shape in plan, wall height, the use of anchors, excavation procedure and other factors. The use of shear transfer devices between panels, in order to reduce differential deflexions, would seem to demand very robust connexions. The resulting risk of causing cracks in the wall elsewhere away from joints, where they could be much more troublesome, would seem in most cases to be unjustified. The present practice for dealing with damp joints is to allow the leak to appear, and the differential wall deflexion for the most part to take place, and then to inject cement or chemical grout into the soil at the back of the joint, either vertically or horizontally through drillings depending on access. Alternatively a steel or other suitable plate, bedded on epoxy-resin mortar, can be bolted to the concrete over the internal face of the joint.

VERTICALITY CONTROL AND TOLERANCES
Verticality
69. The generally accepted tolerance on verticality is 1:80, but the control which can be achieved depends on the soil, the plant and the early detection of any deviation. Boulders, obstructions or inclined hard strata significantly increase the difficulties, and correction usually involves the use of long heavy chisels.

70. Occasionally concrete which has bypassed the stop end, possibly via a cavity, can form an obstruction to the excavation of the adjacent panel, particularly if there is an interval of several days between completion of the first panel and excavation of the second. Figure 10 illustrates how this may occur.[11]

Finished wall face
71. The FPS specification[1] recommends a tolerance of 100 mm for protrusions on the finished face of a wall. Such protrusions are caused by irregularities in the face of the excavation which are filled by the concrete. In homogeneous clay or dense sand and gravel the tolerance value given can usually be achieved. However, if there are underground obstructions or layers of soft ground interbedded with harder soils, it is almost inevitable that there will be some extension of irregularities outside this limit and allowance needs to be made for trimming.

72. If the soil upon which the guide wall has been cast is not compact, one of the areas where protrusions are likely to occur is just below the guide wall base level. Should this difficulty be foreseen, it may be prudent to take special measures to discourage the wash out of soil from beneath the wall as suggested in Fig. 11.

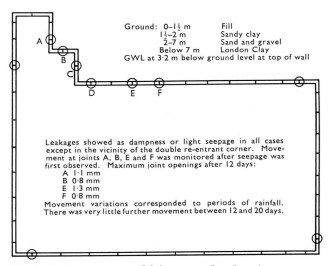

Ground: 0–1½ m   Fill
1½–2 m   Sandy clay
2–7 m   Sand and gravel
Below 7 m   London Clay
GWL at 3·2 m below ground level at top of wall

Leakages showed as dampness or light seepage in all cases except in the vicinity of the double re-entrant corner. Movement at joints A, B, E and F was monitored after seepage was first observed. Maximum joint openings after 12 days:
A 1·1 mm
B 0·8 mm
E 1·3 mm
F 0·8 mm
Movement variations corresponded to periods of rainfall. There was very little further movement between 12 and 20 days.

Fig. 9. Leakages at panel joints on a London site

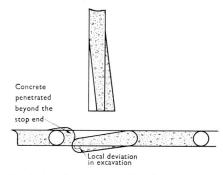

Concrete penetrated beyond the stop end

Local deviation in excavation

Fig. 10. Deviation from verticality due to concrete penetrating beyond stop end

Top surface of concrete as cast

73. The top surface of a concreted wall panel always needs some trimming. The interface between the upper layer of concrete, with some contamination in it, and the sound concrete underneath is not a sharply defined horizontal plane. The thickness of the contaminated concrete depends on a variety of factors such as the soil conditions, the panel length, width and depth, the workability of the concrete and the time taken to complete the pour. The practical way of establishing its approximate depth is to examine the first panels at an early stage in the work.

74. The top surface of rising concrete is not level, and this makes an additional complication in defining tolerances. With the tremie position central in a panel, the highest point of the rising surface is close to the tremie. The surface of the concrete then slopes away gently, the difference in level between the centre and end of the panel depending on the panel length and the submerged depth of the tremie. Fig. 12 shows a case study in which measurements of the rising concrete profile were made during concreting.

Positioning of recesses

75. Frequently recesses are required in a wall corresponding to the levels of future floors, and reinforcing bars, which can be bent out to form a mechanical connexion, are placed behind the recesses.

76. The insert boxes which are used must be kept to reasonable dimensions if they are not to obstruct the flow of concrete and endanger the integrity of a panel.

77. Theoretically there should be no difficulty in positioning insert boxes if they are robust and firmly attached to the reinforcement cage, and if the cages are accurately positioned and firmly held. Occasionally an insert box is displaced on the reinforcing cage and this may be due to the box being caught on a small ridge formed where the ground changes from a soft or loose state to a relatively stiff state. Cages are usually heavy to handle and a brief resistance to motion during insertion may not register with the crane operator. Care needs to be exercised that there is an adequate clearance for the insert box, taking trench width, spacer and cage dimensions into account.

THE RISK ELEMENT IN THE CONSTRUCTION OF DIAPHRAGM WALLS

78. The main risks attached to diaphragm wall construction are associated with effects on adjacent structures and services. General stability of the trench, deflexion of the ground during excavation and deflexion of the completed wall have to be taken into account.

79. General stability can be analysed in an approximate manner and, short of a sudden loss of bentonite from a trench to an old pipe or underground cavity, it is possible to identify cases of risk. Precautionary measures may vary from limiting the length of trench open at any given time to injection of the ground under adjacent footings. Sudden losses of bentonite have to be dealt with quickly by backfilling.

80. Deflexion of the ground during excavation is much more difficult to assess and it depends mainly on the compaction or stiffness of the soil. In general, compact ground

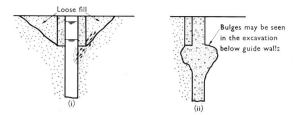

Fig. 11 (a). Formation of bulges below guide walls: (i) variation of bentonite level in the trench induces flow from outside which creates cavities; (ii) shallow guide walls and loose fill behind high water table induces formation of bulges

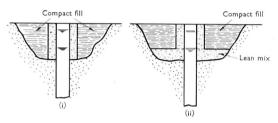

Fig. 11(b). Examples of preventive action: (i) deep guide wall with compact fill behind; (ii) lean mix base to guide wall

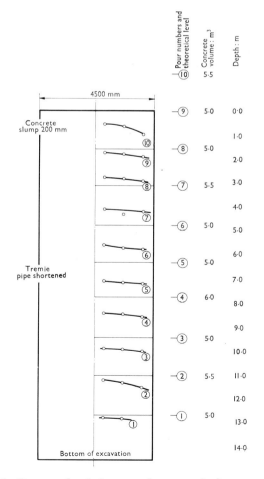

Fig. 12. Concrete levels in a panel measured after successive pours

exerts less active pressure against the bentonite in a trench than loose soil, and its movement is commensurately less. There is some evidence to suggest that settlement of the ground surface immediately adjacent to an open bentonite filled trench could be in the region of 1/1000–1/500 of the excavated depth, depending on the state of the ground and any surcharge pressure. This movement will diminish with increasing distance from the wall, and whether or not it is of a serious character depends on the use and sensitivity of adjacent buildings. In most cases where the soil is moderately compact and normal panel lengths of up to about 5 m are used, associated settlement of adjacent structures is not significant.

81. The completed diaphragm retaining wall is subject to further deflexion as the main excavation encompassed by it takes place, and both vertical and horizontal movements of the retained soil close to the wall occur. The analysis of this condition is again very difficult, and is complicated by the use of ground anchors. Measurements of deflexion have been made in several cases and in general it would appear that the movements which commonly take place are insufficient to cause concern for adjacent buildings. Similar movements are in any event bound to happen, almost irrespective of the chosen method of retaining the soil, unless specific steps are taken to reduce them. However, where an adjacent building is founded on shallow footings which bear on soft soil the risk of damage is not negligible and this should be taken into account at the design stage. Arrangements can be made for additional underpinning or alternatively for monitoring of movements and application of specified remedies if pre-set deflexions are exceeded.

REFERENCES

1. FEDERATION OF PILING SPECIALISTS. *Specification for cast in place concrete diaphragm walling.* Federation of Piling Specialists, London, 1973.
2. CONSTRUCTION INDUSTRY RESEARCH AND INFORMATION ASSOCIATION. *The effect of bentonite on the bond between steel reinforcement and concrete.* Construction Industry Research and Information Association, London, 1967, Interim Research Report No. 9.
3. MEYERHOF G. G. The ultimate bearing capacity of foundations. *Géotechnique*, 1951, **2**, No. 4, 313.
4. CHADEISSON R. Influence of boring methods on the behaviour of cast in place piles. *Proc. 5th Int. Conf. Soil Mech., Paris*, 1961, **2**, 27.
5. BURLAND J. B. Conference discussion. *Grouts and drilling muds in engineering practice*, Butterworths, London, 1963, 223.
6. REESE L. C. *et al.* Bored piles installed by slurry displacement. *Proc. 8th Int. Conf. Soil Mech., Moscow*, 1973, **2**, Part 1, 203.
7. GEFFEN S. A. and AMIR J. M. Effect of construction procedure on load-carrying behaviour of single piles and piers. *Proc. 4th Asian Regional Conf. on Soil Mechanics and Foundations.* Asian Institute of Technology, Bangkok, 1971, **1**, 263.
8. POWERS T. C., COPELAND L. E. and MANN H. M. *J. Portland Cement Assn Res. Dev. Labs*, 1959, **1**, 38.
9. TAYLOR W. H. *Concrete technology and practice.* Angus and Robertson, Sidney, 1965, 189.
10. VALENTA O. Durability of concrete. *Proc. 5th Int. Symp. on the Chemistry of Cement, Tokyo*, 1968, Part III, 193.
11. SCHNEEBELI G. *Les parois moulées dans le sol.* Editions Eyrolles, Paris, 1972.

Diaphragm wall projects at Seaforth, Redcar, Bristol and Harrow

F. A. FISHER, BSc, LLB, FICE, *Partner, Rendel, Palmer & Tritton*

The Paper describes a number of applications of diaphragm walling to quays and retaining walls on works designed by the Author's firm which include examples of self stable structures.

INTRODUCTION

The term diaphragm wall is derived from the first application of the bentonite mud displacement process for the construction of walls to separate different media. The principal use was in cut-off walls under dams. From this the use was extended to basement walls, foundations and cofferdams. More recently the bentonite displacement system has been applied to the construction of tanks, quays and retaining walls. In most concepts for retaining walls the panelling has been regarded as a substitute for sheet piling and, although a number of free-standing cantilever walls have been built, the approach to higher walls is generally on the conventional basis of a plane face and horizontal ground anchors.

2. In many cases this is a satisfactory solution particularly where a secure anchorage can be assured. Instances can arise, however, in which horizontal anchors are not acceptable, for example where the anchors might prejudice future construction or services behind the wall or because of limitations of site. Consideration can then be given to construction in depth from simply stiffening the plane wall with Ts to the formation of cells or curved shapes. For these the technique is a versatile tool and the stability concept ranges from stiffened cantilevers into the realm of gravity structures.

3. Considerations other than those of plane bending then have to be taken into account. Owing to the difficulties inherent in linking the reinforcement across panel joints in the composite bay, the elements must generally be designed to span vertically rather than horizontally. Also, to be able to use the composite moment of inertia in assessing the degree of stability, vertical shear along the joints must be resisted by providing restraint against laminar tilting. This is best achieved by ensuring that the toes of the elements are securely keyed into a ground of suitably high shear value and the tops encastre into an adequate superstructure. An alternative solution can be found in the method used at Redcar, where steel shear connectors were introduced across the joints.

4. The other major consideration is the extent to which frictional or adhesive forces between the soil and the structure can be mobilized. Some theoretical work has been done on this aspect and on the subject of direct pressure, but there is a paucity of such information and much reliance has to be placed on experience and retrospective analysis.

ROYAL SEAFORTH DOCK

5. The first major application of these principles was in the design of the quay walls at Liverpool.[1] The ground conditions were generally favourable, the upper levels being beach deposits of sand and shingle over a considerable thickness of boulder clay in turn overlying Bunter Sandstone.

6. An arch type of closed face wall (Fig. 1) was used for the quays over a length of 2000 m on the eastern side. The front face was constructed with bored cylinders to reduce toe pressure and the arches ensured that most of the forces on the structure were carried in compression. The long back fin provided considerable resistance to overturning, as before the wall could rotate the fin would have to be drawn like a tooth. Initially it had been intended to add rock anchors to the back fins but as a result

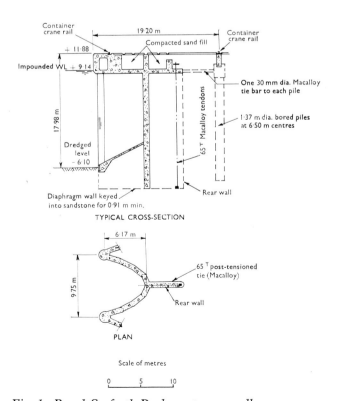

Fig. 1. Royal Seaforth Dock: east quay wall

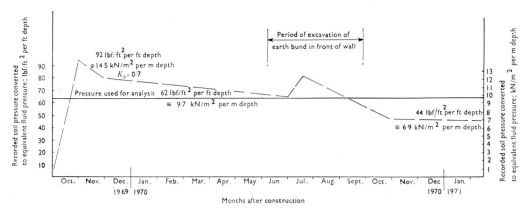

Fig. 2. Test bay 275, east wall Seaforth: average of four load cell readings

of the work on a prototype bay these were considered unnecessary. Resistance to laminar displacement was provided by dowelling the tops of the walls into a very substantial deck structure. The exposed height of the walls was 18 m and the walls were sunk 1–1·5 m into the sandstone.

7. The initial design was begun in 1965. At that time the unknown factors were the magnitude of earth pressure and the amount of wall friction or adhesion that could be called upon after the excavation of the dock basin. It was assumed that the structure would be rigid and unable to yield sufficiently to mobilize the full shear resistance of the soil. For this condition CP 2 recommends designing for at rest pressure.

8. Results of the tests carried out on a prototype bay, constructed in advance of the main contract, showed this assumption to be invalid. This was later confirmed by

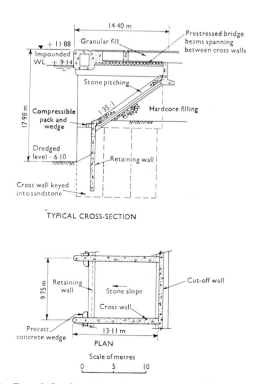

Fig. 3. Royal Seaforth Dock: west quay wall

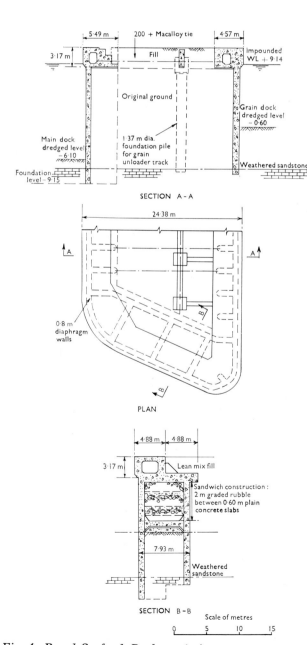

Fig. 4. Royal Seaforth Dock: grain jetty

similar tests on a bay in the permanent wall which, after the removal of the earth from the front face of the structure, recorded a yield of 50 mm on a height of 22·5 m. This would suggest a rotation about the front toe. As the result of these investigations the structure was finally analysed for an equivalent fluid pressure of specific gravity 1·0. Earth pressure readings on the bay are given in Fig. 2, adjusted to show equivalent fluid pressure. The construction instrumentation and findings of the tests on the prototype bay are described in a paper by J. F. Uff.[2]

9. The shape was devised to encourage progressive arching in the ground behind the wall and it is considered that this was effectual in reducing the resultant pressure. The quay was designed for heavy vertical loading which would increase its resistance to overturning. The most severe horizontal loading occurred in the construction stage when the dock had been fully excavated and before the water was admitted, when the unbalanced ground-water head exceeded 15 m.

10. On the western side of the dock the existing beach was 10 m below the final deck level so that a different design approach was needed and an open quay principle was adopted (Fig. 3). From beach level a series of cross walls were constructed by the diaphragm technique and these were extended up to the soffit of the deck using normal reinforced concrete. Between the cross walls, on the dock face, diaphragm walls were constructed to retain the ground between the beach level and the dock bottom, a height of about 8 m. The tops of these walls were integrated into capping beams spanning between the cross walls and retained by wedges with compressible packings inserted into slots in the cross walls. Above the level of the retaining walls earth filling was placed and sloped to meet the underside of the deck at the rear of the structure. The slope was protected by stone pitching.

11. Along the grain dock the foundations of the adjacent silos provided a ready-made secure anchor for the retaining walls. This permitted the use of a lighter form of construction arranged in panels T shaped in plan. The stalk of the T was tied back to the silo foundations. A similar form of construction was used in the grain jetty. Here the T shaped panels on each face of the jetty were cross connected (Fig. 4) providing a very economical section.

12. At the head of the grain dock the T shaped panels were continued but, instead of ties and anchors, the stalks of the Ts were extended and full use was made of ground friction, which by that time could be confidently predicted.

REDCAR

13. At Redcar, the first phase of the British Steel Corporation's major steelworks development called for the provision of a new 10 million tonnes per year ore unloading terminal, to accommodate vessels of up to 200 000 dwt capacity requiring a minimum depth of 20 m of water and an ultimate dredged depth of 28 m below cope. Construction began in late 1970 and was completed in the spring of 1973. The substructure of reinforced concrete diaphragm walls, forming the foundations of the wharf, was completed in 21 months.

14. The main berth is 320 m long by 31 m wide and forms part of the wharf frontage 518 m long. Depth of construction from cope to founding level of the diaphragm walls is about 42 m and this is thought to be probably the deepest solid type of quay yet constructed. The general arrangement is shown in Fig. 5.

15. Before the diaphragm wall construction could be started, the site of the wharf had to be reclaimed from the tidal sands on the south side of the River Tees adjacent to the river entrance. A working platform was formed by means of 20 000 m³ of tipped slag bunds to contain some 19 000 m³ of sand pumped ashore from maintenance dredging operations. The top of the working platform was carpeted with slag to 300 mm thickness at a level of about 1 m above MHWS.

16. The main wharf substructure of 320 m length comprises 21 equal cells of reinforced concrete walls formed by diaphragm walling. Each cell measures about 30 m from front arch wall to rear arch wall and 15 m between centres of transverse walls, as shown in Fig. 6. Front and rear arches are 800 mm thick and the transverse walls are 1150 mm thick on the front ends reducing to 800 mm at the rear. The front arch is stiffened against bending forces, imposed by the contained soil in the cells after dredging in front of the wall, by means of vertical ribs of diaphragm wall extending out from the arch. Front

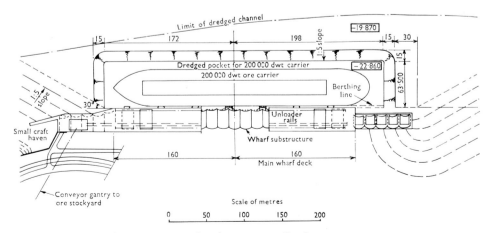

Fig. 5. Redcar ore terminal: general arrangement showing quay wall substructure

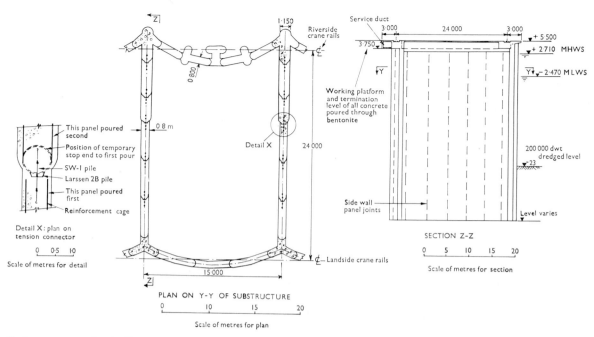

Fig. 6. Redcar ore terminal: diaphragm wall cell details

arch and transverse walls penetrate into Keuper Marl bedrock.

17. Because of the unusual depth of the wall panels and the very heavy surcharge loading to be taken on the back of the berth from the ore stockpiles, it was estimated that the superstructure members would not provide sufficient restraint against laminar tilting. A panel connector was therefore devised and installed in the cross walls to transmit tensile forces along the whole depth of the wall. It is made up from standard trough section and straight-web piling and can be extended as necessary into the panel.

18. It was considered unnecessary to provide a solid deck to the berth. The tops of all diaphragm walls are built into an integrating superstructure framework comprising heavy reinforced concrete beams in the form of a Vierendeel girder, the flanges of which support the unloader tracks and complete the structural system.

19. At the leading corners of the main berth the diaphragm walling is formed into cells, which could be regarded as vertical box girders spanning between the founding bedrock and the concrete deck beams. During construction these cellular shafts were excavated inside to a depth of 24 m without appreciable ingress of water, and were subsequently backfilled with alternating layers of compacted sand and membranes of plain concrete. The concrete membranes acted as internal stiffeners, thus providing strong points for berthing forces.

20. Beyond the northern end of the main wharf a system of open-fronted diaphragm wall cells forms the basis of a future extension for a second berth.

WEST DOCK SCHEME, BRISTOL

21. In 1970 it was required to develop a design for this dock which would reduce costs substantially below those of conventional methods and so allow a basic scheme for a lock and impounded dock to gain authorization.

The time limit was reduced to 3½ years and this was also a considerable factor in indicating the order of the works.

22. The site is on saltings where the soil succession is of soft alluvial clays and silty gravel overlying Keuper Marl. The 28 ha dock basin can be protected from future tidal flooding but for the lock a temporary embankment —as was used at Seaforth and Redcar—had to be con-

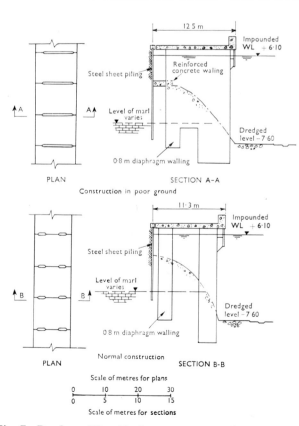

Fig. 7. Portbury West Dock: quay construction

structed across the foreshore. Economy required that the Keuper Marl should be recovered quickly for the temporary embankment of the lock, and the overlying gravel was required for a stockpile and for use in working platforms. It was, therefore, preferable to carry out the excavation of the dock in the dry, but a cut-off wall using diaphragm walling was not economical and could not have been formed in time.

23. It was decided to adopt an open quay form of construction—as on the west wall at Seaforth—but at the rear to put a cut-off of sheet piling, driven through the gravel into marl, so that the dock excavation could proceed in the dry. It was considered that the Keuper Marl would form a satisfactory slope in front of the piling down to dock bottom 13 m below cope. Penetration of conventional piles into Keuper Marl would have meant very hard driving. Circular bored piles would have been too numerous so advantage was taken of the diaphragm grab to form a pier which was strongly resistant to transverse bending. Two such piers formed the legs of a portal capable of

resisting the bending intensity imposed with bents generally 10 m apart. This is shown in Fig. 7.

24. The structural deck, conveniently cast on the ground, did not allow incorporation of a deep section reinforced concrete crane beam, as was the case at Seaforth. The design was, therefore, kept flexible to allow the introduction of crane rails at any reasonable gauge at a later stage. This resulted in a slab which was fairly highly stressed but which was thin enough to allow post-tensioned concrete crane beams to be superimposed later. The slab, filling and any crane load will provide additional loading on the rear or tension leg and will increase the factor of safety.

25. Provision is made for revetting the slope under the quay. This work is being done in the dry, following the success of the groundwater lowering system of continuously pumping deep wells. The quay is designed to stand in the dry state for many months and also with full surcharge loading to withstand rapid drawdown of the dock to low water level in an emergency.

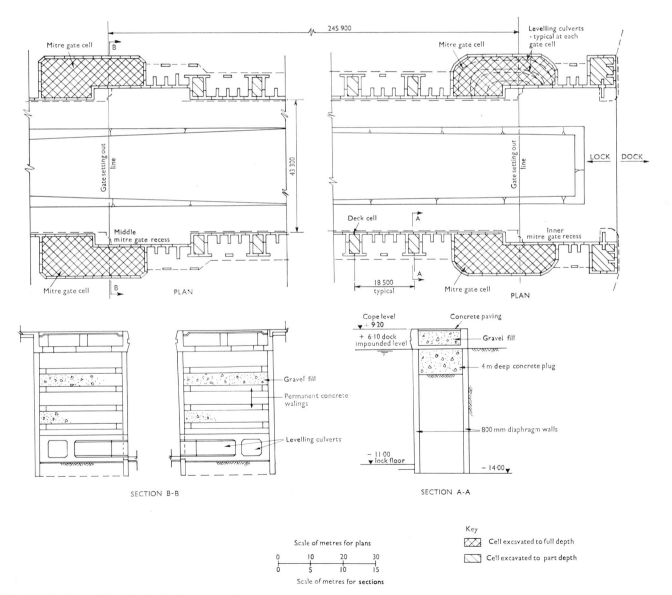

Fig. 8. Portbury West Dock: lock construction

Lock

26. The lock is 420 m long, 43 m wide and 20 m deep from cope to cill. It required a continuous face wall and behind this was developed a series of cellular strong points with wider chambers at the gate cells as shown in Fig. 8. The strong points were spaced as far apart as possible and were envisaged, not so much as monoliths depending on weight, but as narrow portals, with the outer half of the cell giving a considerable resistance to uplift due to the embedment of the whole box some 10 m into Keuper Marl. For reasons of economy the front wall is only 3 m below floor. The intervening panels between strong points span vertically with free-ended conditions at the toe. Therefore they require to be Ts in order to resist bending. The top reaction and structural continuity between strong points and T panels is provided by the trough deck slab. For this it was again convenient to cast at ground level, in this case on a temporary embankment. As the economic embankment level was below high spring tide level the embankment had to be protected by an outer bund.

27. Services are buried or ducted in the gravel filling. This filling assists in resisting overturning.

28. Economy required a structure as narrow as possible and the width of foundation has been maintained at 9 m, the same as the minimum width of decking.

29. The marl varies on this site from grade I, fine mudstone, to grade IV, weathered and predominately clay material. The soil values were, therefore, a matter of judgement. The floor of the lock is really only a scour protection to the marl. The most onerous conditions arise while the floor is being cast and the walls are subjected to full height loading from the ground outside. As in the dock, this situation will last about $1\frac{1}{2}$ years. It is hoped to avoid the complication of strutting across the considerable width of 42 m during this period.

30. The assessed factor of safety is regarded as a function of penetration depth into hard marl together with the achievement of an adequate frictional value. In the dock the horizon of suitably hard material was determined from the working of the grabs, but such evidence was not obvious in the lock. Consequently, after the construction of the wide gate cell, it was decided to introduce casings into the outer wall of the deck cell strong points, so that at a later date, should the need arise, vertical rock anchors could be placed down these casings with little drilling.

31. As the work progressed seawards it was known that the surface of the marl dipped slightly. When only grade III material was present at the normal dig level of -14 m it was decided to increase the depth to -17 m.

32. The gate cell excavation is a good example of the surface of diaphragm wall concrete and of the use of panels as a cofferdam. The overall shape of the cell was decided by grab dimensions after discussion with the successful tenderer. The individual units comprising the T and deck cells were similarly agreed.

HARROW-ON-THE-HILL RESERVOIR

33. This reservoir had to be built into the side of a hill, three sides being in cutting and the fourth side open. The enclosed sides were bounded by private gardens and established trees which had to be avoided. The walls had to be located as close to the boundary fences as was practically possible and no ground anchors were permitted within the adjoining private property. As there was no room for any open cutting and as the noise from conventional pile-driving would have been unacceptable, diaphragm walling appeared to be the appropriate solution. However, the structure had to be capable of retaining 17 m of soil, including surcharge, during the construction stage. Furthermore, although it was the intention to place the floor immediately, in practice this meant several months' duration. The ideal requirement, therefore, was for a free-standing cantilever wall that would allow excavation to proceed unhindered.

34. The soil was well suited to diaphragm wall construction. It consisted of brown clay overlying blue clay

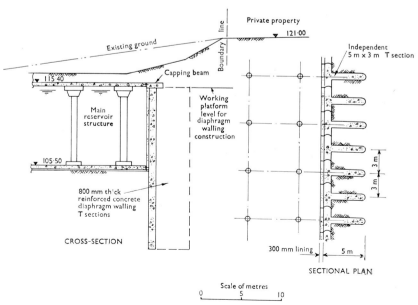

Fig. 9. Harrow-on-the-Hill Reservoir: arrangement of perimeter walls

and although there was a thin water-bearing sandy layer some 11 m below ground level this was not considered to inhibit the diaphragm wall method.

35. The design finally evolved is shown in Fig. 9. It consists of a series of independent T sections of 800 mm wall thickness, 3 m wide and with fins 5 m deep, reinforced to form a counterforted but baseless cantilever wall, founded about 10 m below the reservoir floor level in blue clay, the overall height being on average 21 m.

36. It was estimated that adhesion between the fins and the soil would be sufficient to permit the weight of the soil column between the fins to be taken into account in calculating the forces on the wall. A total of 84 Ts were used to complete the three enclosed sides of the reservoir. The fourth side, being open, is a piled counterfort structure.

37. The assumptions were proved to be justified and the resulting deflexions were of a very small order. The walls were successfully completed in a satisfactory time and at reasonable cost and were incorporated into the structure with an inner lining secured to the face by splice bars plucked from the diaphragm panels.

ACKNOWLEDGEMENTS

38. The Paper is presented by permission of the following authorities and engineers for whom the work was constructed.

Royal Seaforth Dock: The Mersey Docks and Harbour Company, Chief Developments Engineer R. N. Norfolk, MBE, FICE.

Redcar Iron Ore Terminal: Tees and Hartlepool Port Authority, Chief Engineer J. Clark, MICE, AMInstHE.

West Dock, Avonmouth: Port of Bristol Authority, W. J. Sivewright, MA, FICE.

Harrow-on-the-Hill Reservoir: Colne Valley Water Company, Consulting Engineers John Taylor and Sons.

REFERENCES

1. AGAR M. and IRWIN-CHILDS F. Seaforth Dock, Liverpool: planning and design. *Proc. Instn Civ. Engrs*, Part 1, 1973, **54**, May, 255–274.
2. UFF J. F. Insitu measurements of earth pressure for a quay wall at Seaforth, Liverpool. *Insitu investigations in soils and rocks*. British Geotechnical Society, London, 1970, 229–239.

Diaphragm walls as load bearing foundations

H. KIENBERGER, Dipl Ing, *Research and Teaching Assistant, Technische Hochschule, Graz, Austria*

The usefulness of the diaphragm wall as a load bearing element is considered and comparisons are made with cased bored piles. The problem of load transfer is treated and finally tests are reported which give an increased understanding of the complex question of load carrying capacity and allow an estimate to be made of the influence of the construction process with and without the use of bentonite.

APPLICATION OF THE DIAPHRAGM WALL TECHNIQUE

The diaphragm wall technique dates back about 25 years. Its first application was in the construction of cut-off walls, where the main objective was to avoid difficulties arising from the use of steel casing tubes for the drilling of intersecting piles of great depth and in widely varying soil strata.

2. Continuing development and increase of practical experience led to a change of equipment, the chisel using a circulating drilling mud being replaced by a special grab. The circular pile as the basic element was replaced by a prismatic panel.

3. Research work on concrete technology and on bond stress between steel and concrete allowed the diaphragm wall to develop into a highly qualified reinforced concrete foundation element, which is most suitable for the construction of protection walls, quay walls, subway tunnel walls and retaining walls. The main function is to take up horizontal forces from earth and water pressure. The diaphragm wall partially replaces steel sheet piling where its rigidity and its applicability as part of the final construction make it preferable both technically and economically.

4. A new range of application for diaphragm wall elements is their use as load bearing foundations, as an alternative to piles drilled by use of a steel casing. The use of a special grab allows the construction of 'piles' of unconventional shape which offer many possibilities beyond the scope of cylindrical piles drilled by casing (see Fig. 1). As the cost for site installation and diaphragm wall construction is frequently about 5–15% less than for large bored piles, and the difference in design for the superstructure may save another 10–15%, it seems important to investigate the problems of load transfer by diaphragm wall elements.

5. As a general system of deep foundation the diaphragm wall is applicable in principle wherever the use of cast in situ piles is advisable. It offers special advantages when heavy loads call for large monolithic concrete structures; when very deep foundations are required, making the use of steel casing difficult or even impossible; when heavy vertical loads are combined with large horizontal forces and bending moments; and when

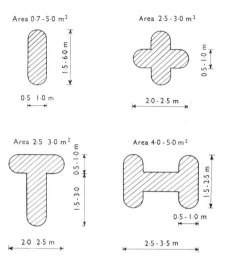

Fig. 1. Cross-sections of typical load bearing elements

special soil conditions are encountered, e.g. saturated silt and sand.

PROBLEM OF LOAD TRANSFER INTO SOIL

Horizontal forces

6. The load transfer for horizontal forces is achieved by lateral earth pressure, which depends on the soil properties (friction, cohesion, density) and the shape and size of the elements, but hardly at all on the method of construction. The computation of horizontal bearing is equal for piles and diaphragm wall elements and is not dealt with in this Paper.

Base resistance

7. Load transfer by base resistance depends on the depth and size of the pile, the soil properties, disturbance of soil at the pile tip by the excavation method, and contact between the concrete and undisturbed soil.

8. Loosening of the soil, as may occur with piles drilled by a casing in saturated soils, is unlikely with the diaphragm wall technique as this always uses an excess hydraulic head which can easily be controlled. Care has to be taken, however, to avoid any inclusion of bentonite mud and sedimented soil particles under the tip when replacing the drilling mud by tremie concrete.

9. Comparing two piles with and without the presence of bentonite mud, Chadeisson[1] found larger initial settlements for the pile drilled with a steel casing. After compaction of the loosened soil by the first loading the behaviour of both piles was the same under the second load cycle. Reese *et al.*[2] examined three test piles drilled with mud and found the base resistance insufficiently

high. This was attributed to entrapment of sedimented loose soil particles which had not been replaced due to a disadvantageous concreting technique.

10. It can therefore be stated that the diaphragm wall method, by achieving minimum soil disturbance even in saturated soils, offers good conditions for base resistance, but it demands care and control in the construction process.

Skin friction

11. Under the working load of structures, which is far below the ultimate load, load transfer from piles into the soil is mainly achieved by skin friction. This depends primarily on the concrete surface (roughness) and on the soil properties (friction, cohesion, density), and secondarily on the contact between soil and concrete.

12. The basic questions concerning load transfer by skin friction are whether the bentonite mud is or can be completely replaced by the concrete, and whether or in which case or to what amount the skin friction is affected by any bentonite filter cake remaining on the contact face between soil and concrete.

13. These questions can be answered only by tests and practical experience. In laboratory tests on skin friction under various confining stresses, Farmer *et al.*[3] found a negligible difference between test bodies produced with and without the presence of bentonite mud, at least for the rather short standing time of 1·0–1·5 h. Reese *et al.*[2] found a noticeable skin friction in the case of three test piles drilled with bentonite mud. Chadeisson[1] observed a better load–settlement behaviour for one pile drilled with mud than for the other pile drilled by a casing. Though he did not separate the results for base resistance and skin friction, the almost identical settlement curves for the second load cycle indicate a rather similar skin friction for both piles.

14. Additional test results obtained from test piles in Vienna are presented in §§15–21.

LOAD TESTS ON PILES

15. The design of UNO City in Vienna is based on Y shaped structures 56–116 m high grouped round a conference pavilion and connected with each other by

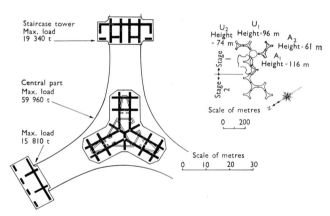

Fig. 2. *UNO City, Vienna: plan showing foundation system with insert showing location and height of buildings*

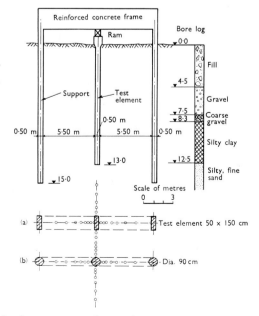

Fig. 3. *Arrangement of experiment*

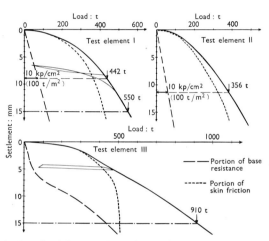

Fig. 4. *Load–settlement behaviour of test elements*

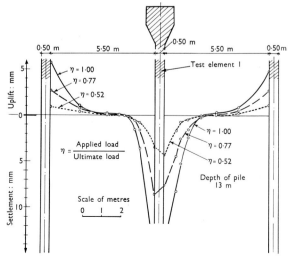

Fig. 5. *Soil deformation observed 2 m below ground surface*

staircase towers. The result is a concentration of loads at the middle and ends of the Y, so that the widely spaced constructions are statically indeterminately supported at four points. The prestressed construction allows a maximum differential settlement of 20 mm. Because of the very changeable, unfavourable soil conditions, deep foundations were chosen. The arrangement of the diaphragm wall panels is shown in Fig. 2.

16. Apart from the usual soil investigation by means of boreholes, and laboratory tests and standard penetration tests, in situ load bearing tests were carried out. Three test elements, 15 m apart from each other, were situated in virtually the same soil conditions. Properties and loads are shown in Table 1. Tension as well as compression loads were used. The reaction piles or panels were arranged at a distance of 6 m between centres. This seemed enough to prevent influencing the test panels. Elementary test procedure and soil strata are shown in Fig. 3.

17. The significant calculation data were pressure, settlement of the pile heads, base pressure on the pile ends, and deformation of the soil at four profiles 2 m below the ground surface. Apart from the soil deformations each measurement was obtained from two sources. The end bearing pressure was ascertained from a hydraulic source —system Glötzl—by visual reading of a manometer; all other sources worked inductively and were recorded on punched tape with a high speed measuring device.

18. In Fig. 4 the simplified load–settlement curves for the compression tests are shown. Test panels I and II, which are of equal depth but different shape and cross-section, do not allow a definite comparison since the base cross-sections are in a relation of $0.9 : 1.0$ and the skin surfaces $0.8 : 1.0$, but still permit an estimate to be made of the influence of the drilling mud on the skin friction.

19. Since in both cases the point resistance built up evenly from the beginning, at a value of 10 kp/cm^2 (100 t/m^2 base pressure) a comparison can be drawn. At this load stage the total load carried by the pile was 356 t, of which 293 t were accounted for by skin friction. This corresponds to an average figure of 7.85 t/m^2. With the diaphragm wall the total load was 442 t and the skin friction amounted to 372 t; the average shear was therefore 8.0 t/m^2. The average shear values differ by only 2% and can therefore be reckoned as practically the same. The relevant settlements were 9 mm for the diaphragm wall and 11·4 mm for the pile. The difference is not considerable and may be explained in accordance with Reese.[2]

Although the base was cleaned about 20 min before concrete was poured, a certain sedimentation of fine sand in water must have followed. In spite of pouring with a tremie pipe this could not be forced out. With the diaphragm wall panel, on the other hand, the bentonite mud prevented a rapid sedimentation, and no appreciable compressible layer could be formed.

20. With the pull-out test (not set out here), both test elements of 13 m depth showed practically the same skin friction with the same top displacement. In a comparison of the diaphragm wall panels of 13 m and 24 m depth, a displacement of 15 mm (i.e. the maximum attained displacement of panel I) gave, for the zone of silty clay and fine sand which reached a depth of over 70 m, an average figure of 4.2 t/m^2. For the upper zone of gravel and fill an average figure is meaningless. It would be in the region of 8.5 t/m^2.

21. The soil deformation near the surface is interesting (Fig. 5). The values for equal settlement of the panels were nearly the same, i.e. in both profiles at right angles to each other. It is evident that not only were the reaction piles far enough apart but also that the zone of influence in the soil nearly disappeared within a relatively short distance. The zone of influence covers about $1\frac{1}{2}$ times that of the pile diameter or a corresponding circle for the diaphragm wall. It must be pointed out that the zones of influence for both diaphragm wall panels of very different depths (13 m and 24 m) were virtually the same. A similar observation was made in 1969 with pull-out tests on diaphragm wall panels of the same cross-section (50 cm × 150 cm) for the foundation of pylons. The panels, constructed in very uniform, loose, fine silty sand at depths of 3·60 m, 4·50 m and 5·50 m, showed the same zone of influence, which was however only 0·5–0·6 m. The distance of influence of larger soil deformations seems accordingly to depend more on the type of subsoil than on the depth of the foundation. However, the fact that at the tests in Vienna this distance was the same for the pile and diaphragm wall indicates also that the skin friction was mobilized in the same way and was not influenced by the bentonite mud.

CONCLUSION

22. Fundamentally it can be stated that, with observance of certain design principles and with careful and rapid construction, diaphragm walls as load bearing foundations are at least as efficient in load bearing capacity as in situ concrete pile foundations, while from other points of view they have distinct advantages.

REFERENCES
1. CHADEISSON R. Influence du mode de perforation sur le comportement des pieux forés et moulés dans le sol. *Proc. 5th Int. Conf. Soil Mech., Paris, 1961.*
2. REESE L. C. *et al.* Bored piles installed by slurry displacement. *Proc. 8th Int. Conf. Soil Mech., Moscow, 1973.*
3. FARMER J. W. *et al.* The effect of bentonite on the skin friction of cast in place piles. *Behaviour of piles.* Instn Civ. Engrs, London, 1971, 115–120.

Table 1. Test elements

No.	Description	Test load, t	Depth, m	Depth of reaction pile or panel, m
I	50 cm × 150 cm diaphragm wall panel	500	13	15
II	90 cm dia. pile	500	13	15
III	50 cm × 150 cm diaphragm wall panel	1000	24	40

In situ diaphragm walls for embankment dams

A. L. LITTLE, BSc(Eng), FICE, FIStructE, FGS, MConsE, *Consultant, Binnie and Partners*

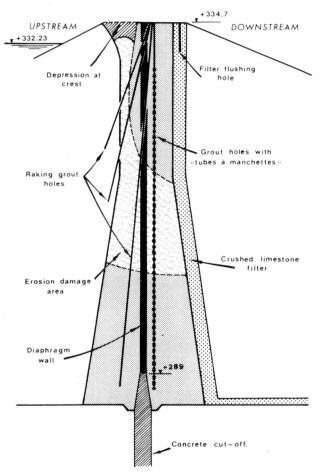

Fig. 1. *Balderhead: cross-section of core*

The Paper describes slurry trench type diaphragms used at Balderhead, Lluest Wen, Withens Clough and Upper Peirce dams. The first three were installed as remedial works, using plastic concrete backfill in the slurry trenches. The last example is of a dam at Singapore where the cut-off was formed using the self-setting slurry or grout wall technique.

INTRODUCTION

Four embankment dams with diaphragm walls are described in the Paper. In the first three cases, tube-à-manchette grouting was used as a first aid measure and to restore lateral and vertical stresses in the cores. As well as grouting, it was desired to have a more positive, erosion-free membrane; plastic concrete or grout membranes cast in slurry trenches were selected.

BALDERHEAD

2. Balderhead Dam, which was completed in 1965, has a maximum height of 48 m and is composed of rolled shale fill with a central rolled boulder clay core which, for the top 18 m, is of a constant width of 6 m. A concrete spearhead connects the core to the foundation (Fig. 1).[1-3]

3. Serious leakage developed in 1966–67 after first filling and swallow holes developed in the fill. The reservoir was drawn down to el. 326 m (9 m below the crest), when leakage was greatly reduced.

4. Examination showed that severe erosion damage had developed in places. After intensive investigation, it was surmised, although not proved directly, that hydraulic fracture of the core had occurred, leading to erosion. It was decided to use tube-à-manchette grouting in the

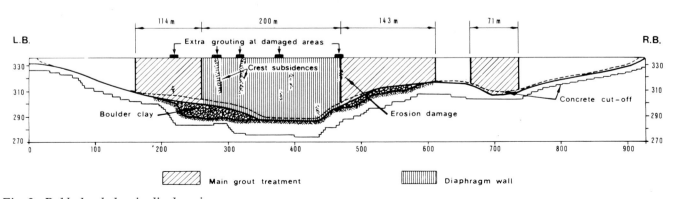

Fig. 2. *Balderhead: longitudinal section*

core over a length of 530 m (rather more than half the crest length of 925 m) and, over a length of 200 m in the more damaged areas, to form a diaphragm wall resting on the top of the concrete spearhead (Fig. 2). The swallow holes were filled with a flyash filler grout. Excavation in the core was under bentonite slurry using kelly bar rigs, supplemented by reverse circulation rigs in the deeper sections to a maximum of 46·4 m. Panels excavated were 0·6 m thick and 6 m long with stop ends formed by 0·6 m dia. steel tubes. The choice of these comparatively short lengths was to minimize the seriousness of a collapse, should it occur, since the reservoir still contained 10·4 × 10^6 m³ of water. Although the difficulty of excavation was greatly increased by the presence of boulders in the core, all the excavated panels (except one which was about 15 cm out of line) landed on the 0·6 m wide flattened top of the spearhead.

5. The excavated panels were backfilled with a plastic concrete mix (see Table 1) tremied into position under the slurry. In mixing, the bentonite was first hydrated, then cement was added and finally the aggregate. Each panel was cast continuously from the bottom and no difficulty was experienced during the concreting. To accommodate expected movements, the plastic concrete was designed to

have an E value less than 147 kN/m² (1·5 × 10^4 kg/cm²).

6. Dam performance after treatment was monitored by measuring leakage and by readings from 69 piezometers spread along the dam on the upstream side of the core. After first refilling these showed piezometric levels within 10 cm of reservoir level but subsequently a depression of some of them in an untreated section of the dam was traced to the effects of further cracking; this was remedied by grouting alone.[4]

LLUEST WEN

7. Lluest Wen Dam was constructed about 1890 with the traditional puddle clay core of the time. Its early years had been troubled by various signs of weakness, and remedial grouting had been undertaken as early as 1912–1916. Apparently this and other measures had been effective and its history forgotten.

8. In the winter of 1969–70, a serious situation developed.[5] Puddle clay from the core of the dam was found to be extruding at the downstream end of a 6 in. drainage pipe which passed through the dam from the valve shaft upstream to the tunnel downstream (Fig. 3). A swallow hole appeared in the crest of the dam. An emergency was

Table 1. Properties of plastic mixes used in the four dams

	Balderhead		Lluest Wen		Withens Clough		Upper Peirce	
Weight per unit volume	2039 kg/m³	100%	1956 kg/m³	100%	1845 kg/m³	100%	1252·3 kg/m²	100%
Lignosulphite	—	—	—	—	—	—	1·3	0·1
Bentonite (B)	44	2·2	24	1·2	25	1·3	32	2·6
Water (W)	400	19·6	405	20·7	409	22·2	841	67·1
Cement (C)	195	9·6	227	11·6	61	3·3	252	20·1
Flyash (Fa)	—	—	—	—	300	16·3	126	10·1
Aggregate (A)	1400	68·6	1300	66·5	1050	56·9	—	—
W/C	2·05		1·78		6·7		3·35	
W/solid	0·24		0·26		0·28		2·06	
W/(C + Fa)	—		—		1·13		2·23	
W/(C + Fa/2)	—		—		1·93		2·67	
B/(C + Fa)	—		—		0·07		0·09	
B/(C + Fa/2)	—		—		0·12		0·10	
Permeability k, *cm/s*	0·6 × 10^{-7} – 2 × 10^{-7}		10^{-3} – 10^{-4}		—		1·26 × 10^{-8} – 2·07 × 10^{-8}	
Specimen diameter	101 mm		101 mm and 57 mm		100 mm		76 mm	
Average axial strain at at peak, ϵ_p	ϵ_p, %	σ_3, kN/m²	ϵ_p, %	σ_3, kN/m²	ϵ_p, %	σ_3, kN/m²	ϵ_p, %	σ_3, kN/m²
14 days	—	—	1·3	206	8·0	210	—	—
14 days	—	—	1·7	412	14·8	420	—	—
28 days	—	—	1·2	206	5·8	210	—	—
28 days	7·9	482–520	1·6	412	9·1	420	—	—
90 days	1·3	241	1·0	206	2·9	210	3·38	197
90 days	4·3	482–520	1·3	412	4·8	420	3·83	296
Average peak deviator stress, q	q, kN/m²	σ_3, kN/m²	q, kN/m²	σ_3, kN/m²	q, kN/m²	σ_3, kN/m²	q, kN/m²	σ_3, kN/m²
14 days	—	—	2930	206	689	210	—	—
14 days	—	—	3585	410	1344	420	—	—
28 days	1379	241	3240	206	896	210	—	—
28 days	2482	482	4205	412	1620	420	—	—
90 days	2826	241	4481	206	1654	210	1808	197
90 days	3309	482	6481	412	2240	420	2286	296
Average tangent modulus at 1%, E_t	E_t, kN/m²				E_t, kN/m²	σ_3, kN/m²		
28 days	53 780	E_t indepen-			24 959	210		
28 days	53 780	dent of σ_3			34 475	420		
90 days	56 539				81 361	210		
90 days	56 539				106 734	420		

declared in the valley below the dam with partial evacuation; a coal mine downstream had to be temporarily closed. The reservoir level was lowered 30 ft, mostly by pumping.

9. As a first aid measure, grouting was done, no less than 18 tons* of clay–cement grout being injected with a single grout pipe from the crest into the core in the neighbourhood of the 6 in. drainage pipe. Subsequent grouting into the rest of the core injected a further 32 tons* of grout.

10. An investigation having disclosed many cracks in the puddle core and zones of erosion damage not unlike those at Balderhead, it was decided to adopt a diaphragm through the core along its whole length after grouting.

11. The trench of width 60 cm (2 ft) was excavated along the dam in 6 m (20 ft) panels with conventional stop-end joints. The trench was backfilled with plastic concrete (see Table 1). In the neighbourhood of the old valve shaft the slurry trench was discontinued and an exploratory 20 ft dia. shaft was excavated 70 ft down to the tunnel to establish the cause of the trouble. It was found that there was a 2 in. gap between the brickwork of the tunnel and the back of the valve shaft masonry. As this was on the upstream side, reservoir water pressure had been extruding puddle clay through the gap and into the 6 in. drain pipe through a crack in the pipe, which unfortunately communicated with the gap at the back of the valve shaft. As the remedial work progressed, the exploratory shaft was backfilled and the plastic concrete diaphragm was continued across it by placing the concrete in shuttering.

* Measured as dry materials.

12. The dam was also strengthened with a downstream rock toe and the spillway and outlet arrangements were improved at the same time.

WITHENS CLOUGH

13. Withens Clough Dam has a maximum height of 36 m and was constructed in 1895. High piezometric pressures had been registered in the downstream shoulder for some time. Exploratory borings through the puddle core showed that in places it was very stiff and riddled with small holes showing clear evidence of the passage of water; it was decided to construct a diaphragm along the whole length of the dam after tube-à-manchette grouting to seal the worst areas.

14. A slurry trench was excavated across the dam in 6 m panels with stop ends and backfilled with plastic concrete (Table 1). Tests had shown the reservoir water to have a pH of 3·8; accordingly, arrangements were made to produce a concrete that would resist such a highly aggressive water. Tests of various mixes and additives showed that concrete containing an appreciable proportion of flyash maintained the required permeability and resisted disintegration (Table 1).

15. Observations of the downstream piezometers showed that the pressures dropped immediately the diaphragm had been constructed in their neighbourhood even before it had been completed across the dam. It was necessary to break through the old outlet culvert and pipe during construction of the slurry trench; this was done by entering the tunnel from the upstream side and providing a clay–cement seal. Fortunately, the reservoir

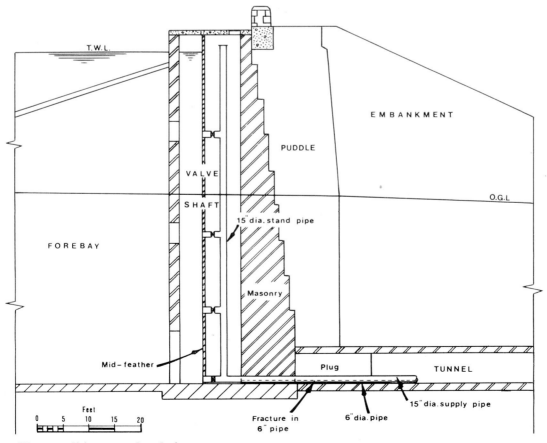

Fig. 3. Lluest Wen: conditions at valve shaft

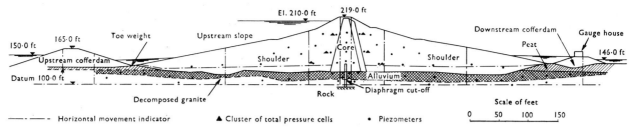

Fig. 4. Upper Peirce Dam: cross-section

had to be emptied to construct new draw-off works at the same time.

16. The reservoir is now two thirds refilled and dam performance is satisfactory.

UPPER PEIRCE DAM

17. Upper Peirce Dam, which is for the water supply of Singapore, is now nearing completion. It has to be constructed within the upper end of the existing Peirce reservoir without interrupting water supply from the existing reservoir (Fig. 4). The final height of the dam will be 90 ft above the downstream toe.

18. Because the dam site is within an existing reservoir, it was not considered practicable to excavate to a depth of more than 45 ft below reservoir level and remove the soft alluvial material from the foundations; the dam is being constructed on this soft material, only the organic surface material first being removed with hydraulic jets. In view of the expected settlements and the difficulty of conventional excavation, it was decided that a plastic non-erodible type of cut-off constructed in slurry trench was the only practicable solution. Such a cut-off would need to penetrate the core fill in the dam and be taken down into the decomposed granite underlying the alluvium.

19. Upstream and downstream cofferdams were constructed by placing decomposed granite with clamshells across the reservoir. Sheet piling was driven through the cofferdams to form temporary cut-offs. An existing tunnel formed the overflow for the upper part of the reservoir so cut off.

20. After the space between the cofferdams had been pumped dry, removal of the surface organic material began, followed by fill placing.

21. When the fill in the rolled clay core reached an elevation of 135 ft a slurry trench was excavated through it down to the level of the decomposed granite. Unlike the other three diaphragm walls described in this Paper, the diaphragm was not formed by displacing the slurry with tremied concrete. Instead, cement had already been added to the slurry in the trench; the slurry set and formed the permanent cut-off. A suitable retarder prevented premature setting of the slurry. Stop ends are not required because the set material is of the same composition as the slurry and a key can be formed simply by excavating into it. This so called grout wall method has been used many times in the formation of walls for underground construction such as car parks and for hydraulic works (at Gambsheim in Alsace, for example) but, as far as is known, this is the first time it has been used for an embankment dam cut-off. The mix used in the cut-off is given in Table 1.

22. An extensive instrumentation system has been installed in the dam to monitor movements and seepage as well as the more usual construction pore pressures and settlements.

CONCLUSIONS

23. The three dams where plastic concrete membranes have been used as remedial measures are behaving satisfactorily. The fourth dam, Upper Peirce, had not been impounded at the time of writing.

24. Although the results of Table 1 show some erratic variations, it can be concluded that the higher water/cement ratios produce more plastic as well as weaker concretes. It is also apparent that plasticity falls and strength rises with age of the concrete. Strength and plasticity do not correlate with flyash content. Neither is there a good correlation with bentonite content.

ACKNOWLEDGEMENTS

25. Soletanche were the contractors for the diaphragm work at all four sites. The work at Balderhead was engineered jointly by Sandeman, Kennard and Partners (now Rofe, Kennard and Lapworth) and Binnie and Partners. The Author is indebted to the Partners of Binnie and Partners, to Mr M. Kennard, Partner, Rofe, Kennard and Lapworth, to the Clerk and Chief Executive Officer, Tees Valley and Cleveland Water Board, to the Engineer and Chief Executive, the Taf Fechan Water Board, to the Engineer and Manager, Wakefield & District Water Board and to the Public Utilities Board, Government of Singapore for permission to publish this Paper. He is also grateful to Mr R. L. Brown who assembled the data in Table 1.

REFERENCES

1. KENNARD M. The construction of Balderhead Reservoir. *Civ. Engng Publ. Wks Rev.*, 1964, ICOLD Suppl., 35.
2. KENNARD M. *et al.* The geotechnical properties and behaviour of carboniferous shale at the Balderhead Dam. *Q. Jl Engng Geol.*, 1967, **I**, Sep., No. 1, 3.
3. VAUGHAN P. *et al.* Cracking and erosion of the rolled clay core of Balderhead Dam and the remedial works adopted for its repair. *10th Congr. Large Dams, Montreal, 1970*, **I**, Question 36, Report 5.
4. LOVENBURY H. T. The detection of leakage through the core of an existing dam. *Field Instrumentation in Geotechnical Engineering*, Part I, Butterworths, London, 1973.
5. GAMBLIN D. G. and LITTLE A. L. Emergency measures at Lluest Wen Reservoir. *Wat. & Wat. Engng*, 1970, **74**, Mar., 93.

Discussion on Papers 1–4

Reported by A. D. M. PENMAN

The session chairman, Mr M. J. Tomlinson, said that the outstanding feature of Papers 1–4 was their illustration of the immense versatility of the diaphragm wall system as a construction technique. Paper 1 mainly covered the application of the system in its orthodox role of a basement retaining wall; Paper 2 described its use for the construction of massive quay walls and locks; Paper 3 described the use of diaphragm walls as load bearing elements, that is, as a form of piling; and Paper 4 dealt with the effectiveness of the technique for forming deep water-tight cut-off walls for earth dams. He thought it difficult to conceive of another below-ground construction system with such a wide range of application in such a wide variety of ground conditions.

2. Turning to the shortcomings of the system, Mr Tomlinson maintained that the method created as many problems as it solved. This was evident from economic considerations, as shown in Paper 22 by Mr Puller. Not the least of the problems was the construction of the guide walls. This part of the job was conveniently avoided by the specialist subcontractor and became the worry of the main contractor. The price for the guide walls was frequently not included in the diaphragm wall contractor's price, and this might give a misleading concept of the true cost of the system.

3. Mr Tomlinson said that in spite of his somewhat critical remarks he believed that the diaphragm wall, with all its applications, was one of the most outstanding innovations in foundation engineering in the postwar years, and he was sure that the Conference papers and discussions would be a useful contribution to its continuing development.

SOFT CLAYS

4. Dr DiBiagio, of the Norwegian Geotechnical Institute, described an investigation into the applicability of the slurry trench method in soft Norwegian clays. Full-scale experiments had been carried out on instrumented trenches to study the effects of decreasing mud densities. The specific gravity of the mud was reduced from 1·24 to 1·1, then to 1·0 (water), and finally the water was replaced by fuel oil with a density of only 0·83. The corresponding closing rates for the trench are given in

Table 1

Specific gravity of slurry	Rate of reduction in trench width, *mm/day*
1·24	0·6
1·0 (water)	3·9
0·83 (oil)	16·9

Table 1. Failure was induced while the trench was full of water by lowering the water level inside the trench by 8 m, and a special study was made of the failure mechanism. Dr DiBiagio said that semi-empirical design rules had been developed by the Norwegian Geotechnical Institute.

PORTBURY WEST DOCK

5. Mr Mansfield expressed interest in the comments on very soft clays made by Messrs Sliwinski and Fleming in Paper 1, particularly after experiences at Bristol on the Portbury West Dock. The scheme is described in Paper 2. Great difficulties were experienced with trench collapses at the seaward end of the entrance lock where the walls were being constructed in ground previously in the tidal range. Collapses occurred so frequently during excavation that the risk of collapse during concreting became unacceptable.

6. The ground consisted of a 4 m thick Keuper Marl fill working platform overlying 8 m of very soft silty clay. Under the clay was 11 m of undisturbed Keuper Marl of hardness varying from grade 4 (weathered and very friable). There was thus a sandwich consisting of good material, Keuper Marl, at the top and bottom with a soft centre of very soft clay. This clay brought about trench collapses in two ways—initially by squeezing into the trench caused by local overstress and subsequently through a deep seated movement suggesting slips. The bentonite slurry alone would not support the trench sides despite many experiments performed in an endeavour to achieve optimum slurry properties. Other expedients, such as reduced panel lengths and partial replacement of the soft clay, were tried but none proved successful.

7. The problem was eventually solved by the provision of lean mix concrete strip piles or barrettes placed behind the guide walls as shown in Fig. 1. These barrettes supported the weight of the marl fill platform and carried that load through the soft clay to the undisturbed Keuper Marl below. Furthermore, the physical properties of the barrettes were such that they contributed considerably to the shear resistance against rotational slip.

8. Mr Mansfield described the technique as being of particular interest in that it extended the construction of diaphragm walling at the lower end of the range of soft grounds. It was thought to be the first time that this particular application had been used. The technique was currently being considered as a method of slope stabilization in deep excavations.

9. In a written reply, Mr Fisher said that the barrettes which he referred to, which were used to support the excavation for the diaphragm wall panels in soft ground at Bristol, were in fact diaphragm walls themselves.

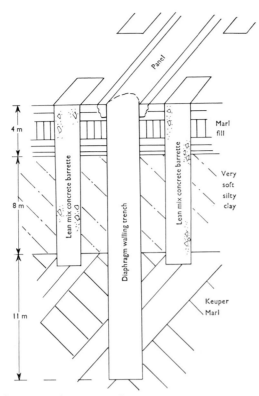

Fig. 1. Lean mix concrete barrettes

GENERAL DISCUSSION ON PROJECTS DESCRIBED IN PAPER 2

10. Mr Irwin-Childs spoke of modern methods of construction developing from earlier design concepts, with new media increasing the potential. He mentioned plane diaphragm wall panels spanning between anchors extending the scope of steel sheet piling to a limited extent. For deeper steel-sheeted walls steel heavy sections had become compounds (Ts). Much stiffer Ts could be made with diaphragm walls so that deeper walls were possible with top anchors only, which could often be set conveniently above groundwater level. Thus the wall for Harrow-on-the-Hill reservoir, described in Paper 2, could be seen to be satisfactory with top ties.

11. But one of the most valuable properties of diaphragm walling was related to the degree of soil adhesion, and this worked horizontally as well as vertically. This factor could be utilized to considerable advantage for structural stability when the elements were set normal to the face, and was an essential component of the examples quoted. The Ts at Harrow not only stiffened the wall but enabled it to stand for some months with no measurable movement without struts or ties. Seaforth Passage was a further example.

12. Extending the principle further by widening the structure, the penetration could be reduced and the concept became more akin to that of a gravity wall, with adhesion serving the function of weight, as in the Seaforth east wall or Bristol lock.

13. Many retaining walls were required to carry surcharge loading behind them. If this was applied to the ground and the wall was tied, the overturning moment was increased. Vertical loading, however, applied to a wider integral structure had the opposite effect, and one berth at Seaforth[1] was designed for a loading of about 100 t/m².

14. Mr Irwin-Childs described the wall at Redcar as a strange hybrid, based on the earlier concept of steel-sheeted diaphragm cells but taken beyond the limit of their capability. With concrete arches front and rear it still needed tension provision in the cross walls: this was the reason for the panel connectors. The deep walls front and rear provided continuous support for the heavy cranes and because of the depth ensured that differential deflexion between the rails would be negligible.

15. Mr Irwin-Childs spoke of the importance of return walls to make the whole wall structure into a self-supporting retaining wall. Because the slurry trench walls had to be built in panels, the shearing forces that the return walls had to resist tended to induce relative sliding at the joints between panels. In reply to a query he said that the steel panel connector joined adjacent panels by the interlocked clutches of two steel sheet piles; it would provide a tension link if the panels tried to move apart, but the existing frictional resistance to sliding at the joint would be increased only by the limited resistance caused by the clutch being embedded in concrete.

16. Economic design related to making elements serve more than one function and it would be apparent from Paper 2 that this had been done in a number of cases. All the structures were built complete from the surface with the attendant advantages of safety and ease of construction.

17. Mr Stacey expressed interest in Mr Fisher's description of the Seaforth and Redcar wharves (Paper 2) where the strength and stability of composite arrangements of diaphragm wall panels had been explored, and said that he had been involved in the design of the Redcar wharf scheme in 1969. At Redcar the walls were arranged to form a series of cells as outlined in Fig. 6 of Paper 2. The use of an arch wall at the front of the cell meant that the transverse walls were put into tension. These walls acted primarily as anchors over the full depth of the wharf. A connector to carry tensile forces between adjacent panels had to be devised. This connexion was made at up to 42 m below dock level with no means of inspection.

18. Mr Stacey said he thought it preferable for adjacent diaphragm wall elements to be arranged to act either independently or in compression with one another; however, the development of a reliable panel connector did widen the application of diaphragm wall structures. He asked if the panel connexions were difficult to make; also if any measurements of the behaviour of the wharf had been made during or after construction and dredging and, if so, whether they suggested that the cell module could be modified to give a more economical arrangement.

19. In a written reply to Mr Stacey's questions Mr Fisher said that the panel connexions were not difficult to make, and out of 140 joints only two gave trouble. In both cases this was believed to be due to failure to properly tighten bolts at the spliced connexion at mid depth, with the result that the lower length of connector wandered off line.

20. Inclinometer measurements of tilting and deflexion of front panels had been kept to check the behaviour of the wharf. Movements during and after dredging had been well within acceptable limits. The cell width from front to back could have been marginally narrower, but

the design dimensions meant that the rear wall could be used as the foundation for the rear unloader rail.

21. Referring to Paper 2, Mr Pawulski said that Figs 4 and 6 of the Paper suggested that curved panels were used both at Seaforth Dock and at Redcar. Mr Fisher replied that in fact the walls consisted of a series of straight panels.

ULLSWATER PUMPING STATION

22. Mr Pawulski described the use of the diaphragm wall technique in constructing Ullswater pumping station in the Lake District. Figure 2 shows the pumping station during construction. The location of the pumping station was determined from general geological data before any detailed site investigation could be carried out. Subsequent boreholes revealed variable strata: silty sand, gravel with pockets of soft silt, boulders and lenses of clay with groundwater near the surface. The configuration of the pumphouse was such that it required about 20 m depth of excavation. After investigating various methods of construction it was decided that the use of a diaphragm wall technique was justified on the grounds of

(*a*) safety during construction, including control of groundwater;
(*b*) economy (incorporating the temporary works in the permanent construction);
(*c*) relatively short period of construction.

23. The main part of the diaphragm wall was circular in plan, 45 m dia., made up from 37 straight panels of depth 23·8–26·8 m. The bottom 3 m of the wall was not reinforced and served as a cut-off. The panels were 0·9 m thick, generally 4 m long and straight in plan. The circular shape was chosen in order to eliminate internal strutting and ground anchors. No ring beams were provided as these would have restricted the wall from acting as cylinder. Figure 3 shows the exposed inner surface of the wall. In addition to the circular wall, short lengths of straight wall were also required. These were generally 12·8–18 m deep and were designed as propped cantilevers. The roof provided the required top support.

24. Most of the excavation for the diaphragm walls was carried out by rope-operated grabs. Although a hydraulically operated grab on a kelly bar was available, this machine proved inadequate for the job. To break through obstructions a 2 t standard chiselling tool was used and the joints were formed using 0·9 m steel tubes.

25. In order to ascertain the deflexion of the diaphragm wall it was decided to cast rectangular hollow steel tubes in some of the panels. Inclinometer readings were taken during and after excavation of soil enclosed by the diaphragm wall.[2] The deflexion of individual panels varied from 8 mm to 25 mm with a maximum occurring towards the top of the wall. These figures suggested among other things that very little bentonite was trapped in the joints between individual panels.

26. Once the diaphragm wall was constructed the contractor had no problems in constructing the rest of the pumping station and dealing with groundwater at the bottom of the excavation.

27. The construction of this diaphragm wall was carried out between December 1968 and March 1969.

Fig. 2. Ullswater pumping station

Fig. 3. Ullswater pumping station

DISPOSAL OF BENTONITE

28. Referring to Paper 1, Mr Pawulski said that an important consideration which was not included in the Paper was the matter of disposal of used bentonite slurry and excavated material contaminated with bentonite. Discharge of bentonite slurry into drains was not normally permitted, and quarry owners might insist that free bentonite is stabilized before accepting excavated material as fill. Mr Pawulski referred to a project where a certain quantity of excavated material was stabilized with approximately 7% lime and then the filled areas almost immediately soiled and put back into agricultural use. However, most of the excavated material was deposited untreated at a quarry and it took almost two years for a

sufficiently strong crust to form so that the area could be soiled and seeded.

CUT-OFF WALLS

29. Mr Kipps spoke of the application of the dia-phragm walling technique to the construction of cut-offs in hydraulic structures using 'plastic' filling material as described in Paper 4. He said that it was not so well known in the UK as in other countries, and deserved wider consideration.

Fig. 4. Pierre Benite scheme

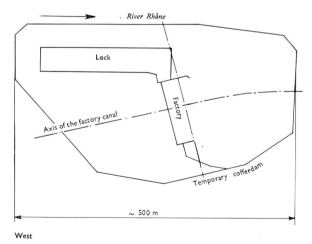

30. As an example of such application, Mr Kipps described the Huxtable pumping station, Arkansas, USA, the cut-off for which was constructed in 1971–72. Situated on the St Francis River at its confluence with the Mississippi, the pumping station served to maintain a low water level in the St Francis River when the Mississippi was in flood. The geology of the site comprised a super-ficial clay layer overlying sand to a depth of 80 ft, below which was impervious clay. The design of the pumping station required a cut-off to prevent under-seepage due to the differential head when the Mississippi was in flood, and a cofferdam was also required during construction to enable the excavation to a depth of 40 ft in the sand to be carried out in the dry.

31. The first plan devised was to use steel sheet piling to act as both temporary cofferdam and permanent cut-off in the form of a box about 800 ft square surrounding the site. In addition, a large number of deep walls would have been required to reduce the groundwater level.

32. This plan was rejected in favour of a 5 ft thick slurry trench. The trench was excavated by first drilling 5 ft dia. holes at 15 ft centres; the intermediate sections were then removed by clamshell. The backfill material consisted of a plastic mix of clay–gravel (3 in. maximum size) and slurry. Fines passing the No. 200 sieve com-prised 10–25% of the mix. After cleaning the bottom of the trench by air-lift, backfilling commenced using a clamshell. The backfill was allowed to rise at its natural angle of repose until the first section was full to ground level, and thereafter backfilling was continued by simply blading the fill material into the trench down this initial slope. To complete the design, a blanket of clay 3 ft thick was laid as a permanent cap across the top of the slurry trench and up to the walls of the pumping station.

33. The slurry trench cut-off itself was more costly than the steel sheet piling would have been, but the almost total elimination of dewatering expenses resulted in an overall saving on the cost of the project estimated at over $1 000 000.

34. Mr Doscher asked about Upper Peirce Dam (Paper 4), where cement was added to the trench-support-ing suspension to form the permanent cut-off.

(a) What was the water content of that diaphragm material after about 30 days?
(b) Was there a relationship between water content and permeability?
(c) Could the high water content of that diaphragm material change in course of time, for instance by a special high flow gradient?

35. Mr Little wrote in reply that the diaphragm at Upper Peirce was buried under embankment fill and was

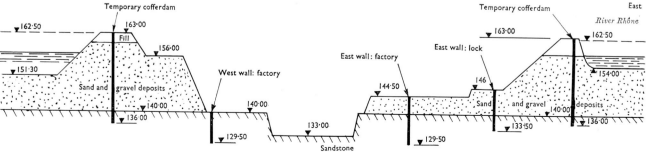

Fig. 5. Pierre Benite scheme: plan, and section along longitudinal axis of factory

not accessible for testing. Consequently, all tests had to be made in the laboratory on test specimens made from the construction mix and other mixes.

36. Accelerated curing was used for the test specimens and the results are considered to relate to 90 days rather than 30 days. The cured water contents, as determined by drying the specimens at 105°C, were generally in the range 105–125%, compared with an initial mixing value of 205%, the difference being mostly due to hydration as precautions were taken to minimize water losses. A volume reduction of about 6% occurred during curing.

37. The permeability was specified to be less than 10^{-8} m/s. This was easily accomplished by specimens tested with a water content of 170% but at a water content of 240% the specimen tested, although initially satisfactory, underwent a marked and unacceptable increase in permeability after about 500 h under a hydraulic gradient of 1000. The permeability tests were done with acidified water having a pH of 5, reflecting the corrosive nature of the reservoir water.

38. The mix design at Upper Peirce dam was based on previous experience at Withens Clough, where the reservoir water was even more corrosive.

39. Mr le Sciellour said that it was rather surprising that the large majority of Papers presented at the Conference dealt with reinforced concrete walls, the accent being placed on the structural behaviour of the diaphragm walls. He said there was no doubt that reinforced concrete diaphragm walls used as retaining walls or bearing structures had given engineers and architects a new construction tool. However, another remarkable possibility was the use of the diaphragm wall as an impervious cut-off. Only Mr Little (Paper 4) mentioned examples where 'plastic diaphragm walls' had been used for embankment dams. Mr le Sciellour said that this process deserved further consideration as it brought economical solutions to various problems.

40. When a diaphragm wall was used as an impervious membrane to protect the sides of an excavation, or as a core in a dam, or as a cut-off underneath an embankment dam, the structural strength of the material used was secondary in importance to its deformability characteristics. If the material placed in the trench was good concrete, the stresses developed either during the dewatering of the excavation or, in the case of a dam, by settlement of the embankment, would result in breaking the wall material. The need was to have a highly deformable and impervious material.

41. Originally (i.e. in the late 1950s) these characteristics were obtained by using clay or bentonite to replace some of the aggregates and part of the cement normally used in concrete. This deformable concrete was then placed in the trench in the same manner as ordinary concrete, i.e. using tremie and stop-end tubes. This technique had been extensively applied, as at Pierre Benite (Figs 4 and 5) to a depth of 36·50 m to protect an excavation of 12 ha, and at Vallabregues (16 ha). These were two sites in the Rhône valley and wall area exceeded 35 000 m² in each case. At Le Havre (Fig. 6), for the construction of a lock capable of receiving 250 000 t tankers, the excavation covered an area of 27 ha and the average depth of the wall was 37·50 m (74 000 m² of wall). Major examples of such plastic concrete walls as permanent structures were the Everlaste dam cut-off in Austria in 1966 and the three dams in the UK mentioned by Mr

Fig. 6. Le Havre: lock excavation

Table 2. Plastic concrete mix

Cement	60–100 kg/m³ of concrete
Microaggregates	0·900–0·950 m³/m³
Bentonite	15–25 kg/m³

Little. A normal evolution of the process took place at Razaza Dyke (Iraq) where, to provide a satisfactory core and cut-off to a dam which was then heightened for the third time in 20 years, the major part of the diaphragm wall was constructed with a concrete where sand and silty clay were the only aggregates. Table 2 shows the content of a 'plastic concrete' mix, which would provide a deformability of 5–10% under a lateral stress of 100 kN/m².

42. The latest development of the process was the self setting slurry or grout wall mentioned by Mr Little, where the fluid used during the excavation of the wall was left in place when the trench had reached its full depth, and set at a predetermined time. This method saved the cost of placing a new material in the trench after excavation. Once set, the material remained soft for a time; the joints between panels, therefore, were easily made by 'biting' into the set material existing in an adjacent panel. This eliminated the risk of faulty joints, a necessary condition when constructing cut-offs.

CONTRIBUTORS

Dr DiBiagio, Norwegian Geotechnical Institute.
Mr H. S. Doscher, Bundesanstalt für Wasserbau, Karlsruhe.
Mr F. Irwin-Childs, Rendel, Palmer and Tritton.
Mr O. Kipps, ICOS (Great Britain) Ltd.

Mr F. le Sciellour, Soletanche Co. Ltd.
Mr A. J. S. Mansfield, Edmund Nuttall Ltd.
Mr J. K. Pawulski, Rofe, Kennard and Lapworth.
Dr A. D. M. Penman, Building Research Establishment.
Mr D. B. Stacey, Binnie and Partners.
Mr M. J. Tomlinson, Wimpey Laboratories Ltd.

REFERENCES

1. AGAR M. and IRWIN-CHILDS F. Seaforth Dock, Liverpool: planning and design. *Proc. Instn Civ. Engrs*, Part 1, 1973, **54**, 255–274.

2. PENMAN A. D. M. and CHARLES J. A. Measuring movements of engineering structures. *Proc. 13th Int. Congr. Surveyors, Wiesbaden, 1971*, Commission 6, Paper 605.4.

The properties of bentonite slurries used in diaphragm walling and their control

M. T. HUTCHINSON, MA, DipChemEng, *Cementation Research Ltd*

G. P. DAW, BSc, PhD, MInstP, AInstPet, *Cementation Research Ltd*

P. G. SHOTTON, BSc, PhD, *Cementation Piling and Foundations Ltd*

A. N. JAMES, BSc, PhD, FRIC, *Cementation Research Ltd*

Bentonite slurries used in diaphragm walling have to carry out a wide variety of functions, some of which place conflicting requirements upon their properties. For instance, a slurry which forms a good filter cake upon the wall of an excavation, and thus enables the hydraulic pressure of the mud to be exerted for stabilization of the face, may well be too resistant to flow to enable it to be cleanly displaced from reinforcing bars by concrete. On the other hand, a slurry which is easily displaced may not build up a suitable filter cake. Also, a slurry which is dense, and exerts a large hydrostatic head for stabilization purposes, may become trapped in the bottom of the trench during displacement because of its density. This Paper describes the fundamental flow and filter cake building properties of muds, and presents results of laboratory and field measurements of these parameters. It also examines the relevance of these properties to the diaphragm walling process, and puts into perspective current drilling mud technology practice. As far as freshly prepared and hydrated bentonite slurry is concerned, it is shown that there is a fairly well defined concentration limit below which the rate of filter cake build-up is very slow, and an upper limit above which the slurry is very difficult to handle. It is also shown that the mechanism by which a membrane is developed, enabling the hydrostatic pressure to be exerted on the soil, varies from surface filtration through deep filtration to rheological blocking of pores as the soil varies in type from fine sand to open gravel. Contamination by detritus, groundwater or cement can completely alter the relevant properties of slurries, and in the Paper the allowable levels of pollution are discussed, with respect to both field and laboratory measurements. The use of chemicals to control the effect of contamination is also considered. Finally, the Authors present the proposed new bentonite slurry specification and test procedure for diaphragm walling, in terms of the plastic viscosity, density, pH, shear strength and filtering performance of the mud. This specification takes into account the research and experience of Cementation and the drilling mud technology which has been built up over the years.

INTRODUCTION

Bentonite slurries, of the type used for supporting excavations, can vary widely in properties, both physical and chemical. They must, however

(a) support the excavation by exerting hydrostatic pressure on its walls

(b) remain in the excavation, and not flow into the soil

(c) suspend detritus to avoid sludgy layers building up at the excavation base.

In addition these slurries must allow for

(d) clean displacement by concrete, with no subsequent interference with the bond between reinforcement and set concrete

(e) screening or hydrocycloning to remove detritus and enable recycling

(f) easy pumping.

2. In general, items (a)–(c) require thick, dense slurries, while items (d)–(f) need very fluid slurries. There is, therefore, a conflict of requirements which must be resolved before an acceptable specification for slurry properties can be drawn up. It is the purpose of this Paper to examine the functions required of the mud in the light of experimentation carried out by Cementation Research Ltd (CRL) (and with reference to the practice of drilling mud specialists where relevant), and to propose a compromise bentonite slurry specification for use in piling and diaphragm walling. The Paper also recommends further work to remedy areas where this specification is suspect.

3. In §§ 6–41 consideration is given to the effect of slurry properties on each function. This enables limits to be put on most of the slurry properties, defining a slurry which is acceptable for each function and for the overall excavation process.

4. The properties of the bentonite slurry considered here are given in Table 1, with definitions and current laboratory test methods.

5. The primary aim of the specification is to ensure that the slurry is capable of fulfilling functions (a)–(d) of § 1, without deleterious effect upon the finished pile or wall. In addition, for reasons of economy, the maximum re-use of the slurry and easy disposal of used materials is required. For these reasons the widest possible specification is needed, in terms of allowable limits for slurry properties, rather than optimum conditions. In most cases it is apparent that, for any given function, there are maximum or minimum allowable values of slurry properties, and limits have been derived from these in order to

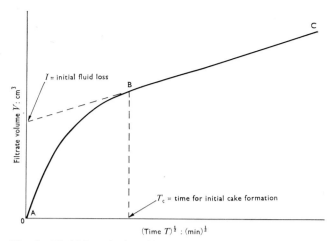

Fig. 1. Fluid loss during filter cake formation

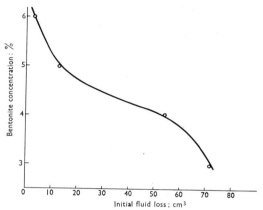

Fig. 2. Initial fluid loss versus bentonite concentration

Table 1. Bentonite slurry properties

Property	Definition	Current test method
Concentration	kg of bentonite per 100 kg water	—
Density	Mass of given volume of slurry	Mud balance (e.g. by baroid)
Plastic viscosity Apparent viscosity Yield stress	For a slurry (behaving as a Bingham body) under shearing conditions: shear stress $= T + V_p S$ where T = yield stress V_p = plastic viscosity S = shear rate; apparent viscosity = shear stress/shear rate and is dependent upon shear rate for a Bingham body	Fann viscometer
Marsh cone viscosity	Time for fixed volume of slurry to drain from a standard cone	Standard Marsh cone, as used by the drilling companies
10 min. gel strength	Shear strength attained by the slurry after quiescient period of 10 min (slurry violently sheared before starting)	Fann viscometer Falling tube shearometer (Note: these two measurements give answers which commonly differ by up to a factor of 2.)
pH	Logarithm of the reciprocal of the hydrogen ion concentration	pH meter, pH papers can give unreliable results
Sand content	Percentage of sand greater than 200 mesh in suspension	API sand content test (basically 200 mesh screen)
Fluid loss	Volume of fluid lost in set time from fixed volume of slurry when filtered at set pressure through standard filter medium	Standard fluid loss apparatus as used by drilling companies (600 cm³ mud, 100 lb/in², 30 min, filter paper)
Filter cake thickness	Thickness of filter cake built up under standard conditions	Measure filter cake built up in fluid loss test

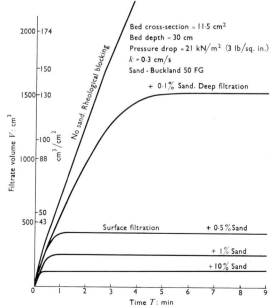

Fig. 3. Effect of added sand on filtration of 5% Berkbent through fine gravel bed

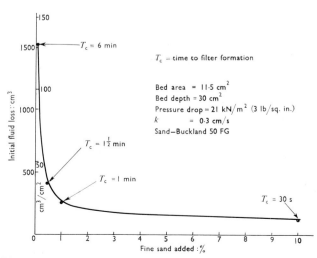

Fig. 4. Effect of sand content on initial fluid loss

develop specification. All measurements and comments refer to Berkbent CE, one of the commonly used bentonites of the converted type.

EFFECT OF SLURRY PROPERTIES ON REQUIRED FUNCTIONS

Excavation support

6. In order to exert a stabilizing pressure on permeable walls of an excavation, the bentonite slurry must form a seal on the surfaces which it contacts. This avoids both loss of slurry into the soil, with consequent reduction in angle of friction, and rise in pore water pressure, and the mud can give maximum stabilizing effect.

7. Three different mechanisms can occur under various conditions during the formation of the seal.

8. Surface filtration occurs when a classical filter cake is initiated by the bridging of hydrated bentonite particles at the entrance to pores in the soil, with negligible penetration of bentonite. During and after formation of the filter cake, water continues to percolate through it from the slurry into the soil. Water lost in this way is referred to as fluid loss. Experimentation at CRL has shown that loss of fluid during surface filtration can be divided into two quite distinct parts: initial fluid loss, which occurs during filter formation, and subsequent fluid loss, which obeys the usual law

$$q = mT^{-1/2}$$

where q = flow rate
T = time after filter formation
m = constant.

This effect is shown in Fig. 1.

9. Deep filtration occurs when slurry penetrates into the soil, slowly clogging the pores and building up a filter cake within them. The seal in this case may extend several centimetres into the soil.

10. In both surface and deep filtration the bentonite concentration in the filter cake is greater than in the slurry (typically 15% for a slurry containing 5% bentonite).

11. Rheological blocking occurs when slurry flows into the ground until restrained by its shear strength. This is the mechanism operating when bentonite is found many feet away from the excavation wall. Penetration distance can be written as

$$L = \frac{Pa}{2G}$$

where L = distance penetrated
P = driving pressure
G = slurry shear strength
a = effective pore radius.

12. Of these three mechanisms, surface filtration is much to be preferred, since the seal is formed most rapidly with no penetration of bentonite into the soil. In §§ 13–16 conditions are considered which result in this behaviour.

13. There is a cut-off concentration of 4–4½% bentonite (see Fig. 2) below which initial fluid loss rises sharply even in ground of quite low permeability $(5 \times 10^{-3}$ cm/s); the Authors therefore propose

Minimum allowable bentonite concentration = 4½%.

14. The presence of small quantities of fine sand in the slurry can change the sealing mechanism in open ground (~ 1 cm/s) from deep to surface filtration (see Fig. 3) with a consequent dramatic reduction in initial fluid loss (see Fig. 4). Although during excavation there will always be suspended material in the slurry, the Authors consider that to obtain good performance the bentonite feeding the excavation should have

Fine sand concentration (<100 mesh) $> 1\%$ w/w.

15. At a given concentration, it is necessary for the slurry to be well mixed and hydrated before it can form a good filter cake. The method of mixing is particularly important in this respect. A very sensitive measure of degree of hydration is the 10 min gel strength of the slurry, and the Authors propose that the minimum acceptable level for this is 50% of the fully hydrated value of a 4½% slurry. Thus

Minimum 10 min gel strength (measured on Fann
viscometer) = 36 dyn/cm² (7·5 lb/100 ft²).

Fig. 5 shows the importance of an efficient mixing system. Bentonite prepared with a high shear mixer has a much faster rate of hydration—and a higher final shear strength —than that prepared with an anchor stirrer. To a certain extent, the presence of fine sand can counteract incomplete hydration in the formation of a filter cake.

16. From the point of view of wall stabilization, the higher the density of the slurry the better. Therefore there is only a lower limit to the density allowable for this function, equivalent to a slurry containing 4½% bentonite and 1% sand.

Minimum allowable density = 1·034 g/cm³.

Excavation sealing

17. The ground is envisaged as a reasonably homogeneous granular medium with no gross voids or fissures; consequently, if a slurry forms a surface filter cake then it will also form an effective seal against losing slurry from the excavation. Thus all limits set in §§ 13–16 apply equally well to this function of the slurry.

18. In addition, even slow loss of water (rather than slurry) from the excavation must be considered, as this can lead to increases in bentonite concentration. Estimation of this is made very difficult by the nature of bentonite filter cakes, which can be so compressible that leakage is

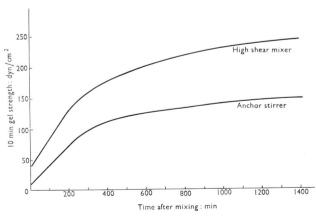

Fig. 5. Effect of mixing on hydration of slurry (5% bentonite)

reduced rather than increased by increasing the driving pressure. Under normal circumstances one finds that the rate of fluid loss is not particularly pressure sensitive[1] and this has been assumed in the calculations.

19. Estimated water losses based on laboratory tests are given in Fig. 6. They show, once again, a sharp change in behaviour at bentonite concentrations of 4–5%. Thus

Minimum bentonite concentration allowable for good

sealing $= 4\frac{1}{2}\%$.

20. It would be an advantage to propose a fluid loss figure also on this basis; however, the Authors feel that the fluid loss test in use at present is not sufficiently relevant to make this possible (see §§ 46 and 47).

Suspension of detritus

21. During the excavation process, it is inevitable that detritus will become mixed with the slurry. Apart from raising the average density of the slurry, this can give rise to slowly settling layers of sand and silt which can cause density gradients in the slurry and a build-up of sludge at the excavation base. This is precisely the situation which will cause displacement difficulties, when the tremied

concrete is unable to push the slurry cleanly from the excavation bottom. Moreover, the heavy sludgy layers at the bottom will tend to have high viscosities also, with deleterious effects on displacement past the reinforcing cage and walls (see §§ 26–32).

22. It is thus advantageous to allow the minimum of settling of detritus in the slurry, even if this means that the slurry is *on average* denser than it would otherwise be.

23. With freshly prepared bentonite slurries a sharp change in behaviour occurs when the bentonite concentration is above 4% (see Fig. 7). Below this concentration much higher percentages of sand settle out, and the Authors consider that slurry should therefore have suspending properties corresponding to at least a fresh 4% bentonite slurry, i.e. 10 min gel strength 25 dyn/cm².

24. Similarly, settlement in hoses, pipes, pumps and storage tanks leads to blockages and handling difficulties which can be lessened by ensuring that the slurry has sufficient gel strength.

25. Therefore, the recommendations are

Minimum bentonite concentration $= 4\%$
Minimum 10 min gel strength $= 25$ dyn/cm²

(6 lb/100 ft²).

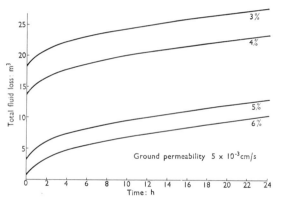

Fig. 6. Variation of total fluid loss with time from a slurry trench panel 15 ft × 100 ft × 2 ft

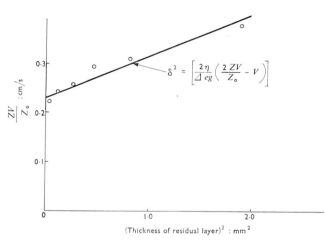

Fig. 8. Thickness of residual bentonite layer

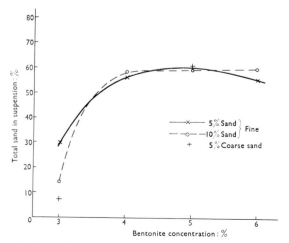

Fig. 7. Variation of sand in suspension with bentonite concentration

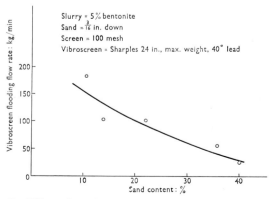

Fig. 9. Effect of sand content on vibroscreen flow rate

Displacement by concrete

26. After the excavation has been made and any reinforcement necessary to the finished structure placed in it, concrete must be tremied into the trench to displace the slurry. The displacement process itself is extremely complicated, and has been studied both theoretically and practically at CRL in recent years, and at CIRIA.[2]

27. At CRL the approach has been to consider the displacement as taking place in two distinct phases: firstly, displacement at the base of the excavation (mainly lateral displacement) and secondly, displacement from the walls of the excavation and from the reinforcement bars (mostly vertical displacement).

28. The following conclusions have been reached from this work.

(a) The density of the slurry must not rise above 1·3 g/cm³. This was determined empirically from experiments in shallow (3 ft deep) simulated pile bases (24 in. dia.), by observing the occurrence of incomplete displacement.

(b) The plastic viscosity of the slurry must not rise above 20 cP (equivalent to a maximum bentonite concentration of ~15%). This was derived from a theoretical analysis of the displacement of slurry from vertical surfaces, supported by some model laboratory experiments.

29. The theory assumed that slurry behaves as a Bingham body with a yield stress very much lower than that of the displacing concrete. On this assumption one can develop the following approximation for the thickness of slurry remaining on a vertical face.

$$\delta = \left[\frac{2\eta V}{\Delta \rho g} \left(\frac{2Z}{Z_0} - 1 \right) \right]^{\frac{1}{2}}$$

where δ = thickness of layer
η = plastic viscosity of slurry
V = velocity of displacing front
$\Delta \rho$ = density difference between concrete and slurry
Z_0 = height of interface above initial position
Z = distance above initial position.

30. Fig. 8 shows a graph of ZV/Z_0 against δ^2, obtained from some laboratory experiments which used a model Bingham body material (oil+PFA) in place of bentonite slurry. Also on the graph is the theoretical line from the equation in § 29. The figure of 20 cP given in § 28 was derived from a further assumption, that a layer of slurry on the surface would be absorbed by the concrete if it were less than 10^{-2} cm thick, and thus not interfere with the bond between concrete and reinforcement.

31. Neither conclusion (a) nor (b) can be considered watertight, and it is worth illustrating our reservations. Firstly, in the simulated pile bases, only the effect of mud density (increased by sand addition) was studied, whereas it is apparent that the yield stress and viscosity—which were also changed—will affect the process. This point would become of vital importance in consideration of support of weak ground by muds weighted with barytes or pyrites to increase the stabilizing hydrostatic pressure. Secondly, in the work done on displacement from vertical surfaces, the conclusions were extrapolated from model experiments, in which a fluid rheologically similar (i.e. Bingham body) to a bentonite slurry was used. This model may be inadequate since concrete contamination

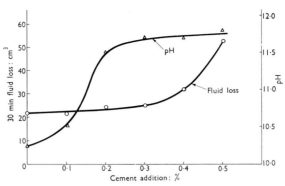

Fig. 10. Effect of cement contamination on fluid loss

can modify mud properties markedly at the interface between slurry and concrete.

32. Nevertheless, the Authors consider both conclusions to be good first approximations. In view of the importance of the displacement to the overall process, however, it is felt that more work should go to establishing limiting mud properties.

Detritus removal

33. In order to allow for recirculation of used bentonite slurries, they must be cleaned either after or during excavation. Experience has shown that it is usually sufficient to remove all particles larger than 100 mesh (although occasionally this can give rise to high levels of silt contamination).

34. Contamination of slurries by local clay particles, which produces unsatisfactory slurries, has not proved amenable to physical separation. The possibility of chemical treatment, although doubtful, should not be ruled out.

35. The screening efficiency of vibroscreens is much affected by solids content (see Fig. 9), and the Authors consider that 25–30% contamination by sand is the maximum that is treatable (i.e. density less than 1·21 g/cm³).

Effects of contamination

36. During the process, slurry can become chemically contaminated either from the soil itself, or from groundwater, or from spillage. The effect that a contaminant of this sort will have on the slurry is, of course, determined by the chemistry of the bentonite itself. By far the most common form of contamination is by cement, which invariably occurs to some degree during tremieing. Cement contamination results in thick, permeable filter cakes, very high fluid losses and high viscosities.

37. Laboratory tests show that, with a standard 5% bentonite slurry, the effects start to become noticeable at about 0·3% cement contamination (see Fig. 10), which is equivalent to a slurry pH of about 11·7. The Authors therefore propose that

Maximum slurry pH value = 11·7.

Pumping and distribution

38. A great deal has been written about slurry pumping.[3] The properties of slurries considered here are only those which affect flow in pipelines, not in the pumps themselves.

39. Under normal conditions, flow is turbulent in nature (flow > 9000 litres/h, viscosity 10 cP, 10 cm pipe

gives Re > 2000) and controlled largely by density. However, because of the thixotropic nature of the slurries, pressures are highest when starting up. It is these initial conditions which are dealt with below. The two effects are both controlled by the shear strength of the slurry.

40. Firstly, to avoid settling in the lines, a minimum shear strength G is required. This is given by

$$G = \frac{\Delta \rho g R}{3}$$

where R = particle radius

$\Delta \rho$ = difference in density between sand and slurry.[4]

For example, 1 mm dia. particles require a value of 25 dyn/cm^2 to avoid settling.

41. Secondly, it must be possible to restart flow in a pipeline after a shutdown. In this case the shear stress at the pipe wall due to pumping pressure must exceed the shear strength of the slurry:

$$G < \frac{PR}{2L}$$

where P = pump pressure allowing for hydrostatic head

L = pipeline length

With $P = 280$ kN/m^2 (40 lb/in^2), $R = 5$ cm, $L = 300$ m, this gives $G < 200$ dyn/cm^2 (40 lb/100 ft^2). Therefore pipelines containing a slurry with a 10 min gel strength of 200 dyn/cm^2 should be emptied if left standing for more than 1 hour.

DISCUSSION ON PROPOSED SPECIFICATION

42. The conclusions reached in §§ 6–41 are summarized in Table 2. It will be seen that critical limits are proposed for six out of the eleven properties originally considered. The five properties not specified are apparent viscosity, Marsh cone viscosity, yield strength, fluid loss and filter cake thickness. It is important to consider here the relevance of these parameters.

43. Apparent viscosity is an ill-defined property whose value will depend on the shear rate of the measuring system. According to the presentation in this Paper it does not seem to be of prime importance to the process, and the Authors therefore feel that it is unnecessary to limit—or measure—it.

44. Yield strength is also a property of the slurry which does not appear to be of prime importance to the process. On the other hand, 10 min gel strength—which is a very closely related property—has importance in three out of the six operations occurring in the process. The Authors believe that 10 min gel strength should be measured and not the yield strength.

45. The Marsh cone test does not measure the viscosity of a mud but a complex property relating to density, viscosity and shear strength. Therefore it cannot be related to the criteria used in this Paper. Nevertheless it is so simple that it could be used as an empirical test if an extensive programme of testing provided useful correlations.

46. The Authors feel that the usual fluid loss test is inapplicable to muds used for stabilizing alluvia. The filter paper used in the test is so different in nature from the granular media in the trench wall that different methods of filter cake formation can easily occur in each. A slurry which performs well against a fine-grained porous material may thus be quite inadequate to form a surface filter cake against ground of permeability to water of, say, 1 cm/s. An alteration in the test is thus required, so that the material against which the filter cake is formed is sufficiently similar to the soil to give realistic behaviour, using pressures which are more in keeping with the process.

47. Limits to fluid loss and the associated filter cake thickness are the outstanding items missing from Table 1, and represent the main parameters still needed to complete a slurry specification.

Table 2. Bentonite limiting propertie for converted calcium montmorillonite (e.g. Berpbent CF)

	Bentonite concentration	Density	Plastic viscosity	Apparent viscosity	Marsh cone viscosity	Yield strength	10 min gel strength (Fann)	pH	Fluid loss	Sand content
Excavation support	> 4½% § 13	> 1·034 g/cm³ § 16					> 36 dyn/cm² § 15			> 1% § 14
Excavation sealing	> 4½% § 19									
Detritus suspension	> 4% § 25						> 25 dyn/cm² § 25			
Displacement by concrete	< 15% § 28	< 1·3 g/cm³ (requires further verification) § 28	< 20 cP (requires further verification) § 28					< 11·7 § 37		< 35%
Physical cleaning		< 1·21 g/cm³ § 35								< 25% § 35
Pumping							> 25 dyn/cm² § 40 < 200 dyn/cm² § 41			
Limits	> 4½% < 15%	> 1·034 g/cm³ < 1·21 g/cm³	< 20 cP				> 36 dyn/cm² < 200 dyn/cm²	< 11·7		> 1% < 25%

Column notes (spanning rows): Apparent viscosity: Not a primary parameter. Marsh cone viscosity: Regarded only as a qualitative test. Yield strength: Regarded as less important than 10 min gel strength. Fluid loss: Results can be deceptive, with present type of test.

CONCLUSIONS AND RECOMMENDATIONS

48. Proposals can be put forward for a slurry specification in all aspects except that of fluid loss and filter cake thickness. The specification is

Bentonite concentration	$> 4\frac{1}{2}\%$
Density	$> 1\cdot034$ g/cm^3
	$< 1\cdot25$ g/cm^3
Plastic viscosity	< 20 cP
10 min gel strength (Fann)	> 50 dyn/cm^2
	< 200 dyn/cm^2
pH	$< 11\cdot7$
Sand content	$> 1\%$
	$< 25\%$

There are several areas where further work might sharpen up these limits, and these are detailed in §§ 50 and 51.

49. Control of the process is gained by adhering to this recommended specification and carrying out a systematic schedule of tests to ensure that this is done. Clearly the appropriate personnel must be available on site to make the measurements on a routine basis.

50. The most important area for further work is in estimating the importance of fluid loss to the process, and in developing a site method of assessing it. At this stage it is envisaged that the usual fluid loss apparatus would be used, but with the filter paper replaced by a simulated sand bed—possibly sintered metal—and much lower pressures (10–25 lb/in^2).

51. Other areas where work could profitably be carried out are in the effects of sand in the slurry (see § 14), where better stabilization of open soils might be accomplished, in the effects of contamination by cement upon the properties, and in the problems of displacement of the slurry by concrete (see §§ 26–32).

REFERENCES

1. ROGERS W. F. *Composition and properties of oil well drilling fluids.* Gulf Publishing Co., Houston, Texas, 1963 (3rd edn), 328.
2. CONSTRUCTION INDUSTRY RESEARCH AND INFORMATION ASSOCIATION. *The effect of bentonite on the bond between steel reinforcement and concrete.* CIRIA, London, 1967, Interim Research Report 9.
3. CHENG D. C. H. A design procedure for pipeline flow of non-Newtonian dispersed systems. *1st Int. Conf. on Hydraulic Transport of Solids in Pipes, Coventry, 1970,* Brit. Hydromech. Res. Assn, Cranfield.
4. ROGERS W. F. *Composition and properties of oil well drilling fluids.* Gulf Publishing Co., Houston, Texas, 1963 (3rd edn), 315.

A design study of diaphragm walls

E. L. JAMES, MSc, MICE, *Soils Engineer, Cementation Ground Engineering Ltd*

B. J. JACK, MICE, MIStructE, *Partner, Jack and Letman Associates*

The Paper presents an empirical solution to multi-tied wall analysis. This method may be used in the design office without the necessity of determining the stress/strain characteristics of the soil. Large-scale tests on flexible and rigid walls (10 ft × 8 ft) are described. Measurements of wall deflexions were carried out for the former and strain measurements were taken for the latter. From this data evidence is provided to support the present empirical design method and suggest possible improvements.

NOTATION

E	Young's modulus
H	horizontal force
I	second moment of area
M	moment
P	load
P_a	resultant of active pressure envelope
P_p	resultant of passive pressure envelope
R	resultant tie force
$T_1, T_2, T_3, \ldots$	tie forces at levels 1, 2, 3, ...

INTRODUCTION

Deep excavations with vertical faces often require lateral support to prevent excessive movement of the surrounding soil. The form of support may be that of a simple cantilever or may necessitate the incorporation of one or more ties. When more than one tie is required the empirical trapezoidal pressure distribution is often assumed. This provides a convenient and straightforward design method for uniform soil strata, but when design involves variable soils the enclosing trapezoidal pressure distribution becomes a matter of interpretation and can often produce variable solutions dependent upon the judgement of the engineer.

2. The Paper outlines a simple empirical solution; it uses the same standard lateral pressure coefficient as does the trapezoidal method to establish a fundamental pressure distribution; then, instead of assessing an envelope to determine maximum forces, it carries through a stage by stage analysis simulating the method of construction carried out in the field. Each stage of excavation is analysed by assuming an equivalent single-tied wall method of analysis using a new centre of rotation for each excavation depth. The results using this approach compare favourably with those obtained by model and full-scale field tests.

3. This design approach has been used successfully over the last six years in the design of numerous multi-tied walls, some of which have been instrumented by James in conjunction with the model tests reported in this Paper. A case history study is given in Paper 15 presented by Littlejohn and Macfarlane.

4. Models provide a powerful tool for examining the influence of certain parameters on the behaviour of multi-tied retaining walls since tests incurring a risk of overall failure or having an undue influence on subsequent wall behaviour cannot, in general, be performed on full-scale walls.

5. The two model tests described here were performed to study specific wall configurations under conditions closely approaching those pertaining during the construction of full-scale anchored walls. The tests were carried out on models having linear dimensions of approximately an eighth full size. An excessively flexible model was used in the first test to provide information on the deformed profile of such a wall and the loads developed in the passive, restraining anchors. The model employed in the other test was designed to simulate the stiffness of a typical diaphragm wall anchored at three levels and subjected to anchor stressing to design loads during excavation.

INTERACTION OF SOIL AND WALL

6. The interpretation of soil parameters determined by tests must be considered carefully in conjunction with the assumptions made in this simplified design approach. This method allows the design of multi-tied walls to be computerized with the advantage that the engineer may quickly compare the effects of the more variable parameters of this problem and so arrive at a meaningful solution.

7. The influence of soil and wall flexibility on the design of single-tied walls has been investigated by Rowe.[1,2] He established flexibility coefficients for single-tied walls which provide a simple and efficient design office method of adjusting the free earth support method of analysis to take the wall stiffness into account. From results obtained from model analyses, flexibility can be seen to have similar effects in the case of multi-tied wall design. Unfortunately, due to the additional variables, coefficients in the form presented by Rowe are not yet available for this case.

8. In order to assess the effects of wall flexibility in conjunction with wall stiffness it is necessary for the engineer to adopt an approach using simplified soil strata.

9. The advent of computer systems available for the use of the designer has led to the development of programs permitting theoretical studies of soil and wall interactions. However, the data required for such solutions is rarely available and such rigorous design using unfamiliar parameters often lulls an engineer into a false sense of security. The analysis described below provides excellent facilities to assist an engineer to understand the actual behaviour of multi-tied walls and can be used to determine

flexibility coefficients for multi-tied walls as used in single tie analysis.

THEORETICAL ANALYSIS

10. The fundamental differential equation of elasticity may be written as

$$P = \frac{\mathrm{d}^4 y}{\mathrm{d}x^4} EI$$

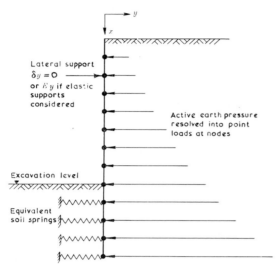

Fig. 1. Simulation of soil–wall interaction

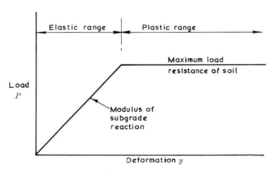

Fig. 2. Simplified stress/strain diagram

where P is the load. When P becomes a function of deformation,

$$P = \frac{\mathrm{d}^4 y}{\mathrm{d}x^4} EI = yr$$

where r is the equivalent spring stiffness of the soil.

11. The equation may be solved by many well known standard solutions. Since numerous computer programs are available for structural frame analysis, the stiffness method probably provides the simplest approach.

12. The wall may be considered as a series of members connected by nodes; at these points horizontal members simulating the soil stiffness or support points may be provided as illustrated in Fig. 1.

13. By incorporating a simple iteration routine into the standard programs and using a simplified stress/strain diagram, the elastic/plastic effects of the soils (Fig. 2) may be simulated. Estimates of P are repeatedly calculated for each deflexion profile until convergence to a condition of equilibrium is achieved.

Equivalent single tie approach

14. The following design method for multi-tied walls has been developed to predict the necessary tie bar forces by considering the effects of the temporary support produced by the passive pressure at intermediate excavation stages.

15. It involves a procedure which calculates the position and magnitude of a resultant tie at any stage of excavation by treating the wall as a single-tied structure.

16. The method is based on the following assumptions:

(a) the mobilizing and resisting soil forces are those determined from Rankine's earth pressure;
(b) at failure there is a unique point of rotation in the plane of the wall;
(c) the wall is only of sufficient length to mobilize a factor of safety of unity against rotation at any stage of excavation.

17. The first assumption is made to simplify the calculations, and is the usual one made when calculating earth pressures in the design office.

18. Assumption (b) enables the use of a simple procedure (§§ 20–24) to calculate the additional tie bar forces

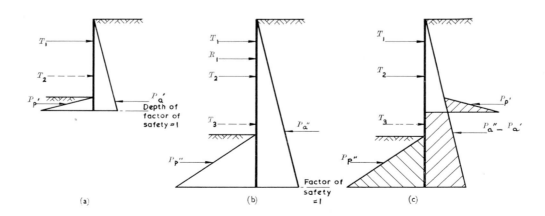

Fig. 3

produced when the passive pressure is transferred to the active side during the next stage of excavation.

19. Assumption (c) allows for maximum resultant tie forces to be resolved by considering free earth support, and thus fixity produced by continuity of the wall is neglected.

20. Consider the equilibrium conditions of the system shown in (a) and (b) of Fig. 3. From (a):

$\Sigma H = 0$ is satisfied when $T_1 = P_a' - P_p'$ (1)

$\Sigma M = 0$ is satisfied when $M_p = M_a$ (about position of T_1).

From (b) of Fig. 3:

$\Sigma H = 0$ is satisfied when $T_1 + T_2 = P''_a - P''_p$

$\Sigma M = 0$ is satisfied when $M''_p = M_a''$ (about the centroid of T_1 and T_2).

21. If we now consider the resultant tie R_1, shown in (b) of Fig. 3, acting at the centroid of T_1 and T_2, then

$$R_1 = T_1 + T_2 = P''_a - P''_p$$

Substituting for T_1 from equation (1):

$$P_a' - P_p' + T_2 = P''_a - P''_p$$

and hence

$$T_2 = P''_a - P''_p - P_a' + P_p'$$

22. From this it can be seen that the additional pressure transmitted to T_2 is temporary support offered during the previous stage. The equivalent beam loading to T_2 is illustrated hatched in (c) of Fig. 3. As T_2 is unknown, so also are both the position and magnitude of R_1. By assuming one, the other may be calculated and the initial assumption checked. Accuracy is essential in calculating the lever arm of R_1 about the position of factor of safety of unity, as this dimension is usually small in comparison

with the magnitude of R_1. The following iteration procedure (§ 23) is recommended to ensure convergence to the correct value.

23. Figure 4 shows the case where the excavation level has been reduced to a position for the insertion of the fourth tie. Considering the equilibrium of the system:

$\Sigma H = 0$ is satisfied when $T_4 = R_n - R$

$\Sigma M = 0$ is satisfied when $Rx - T_4 y = 0$ (2)

where $y = D - x - z$.

Substituting in (2):

$$f(x) = Rx - T_4(D - x - z)$$
$$= Rx - T_4 D + T_4 x + T_4 z$$

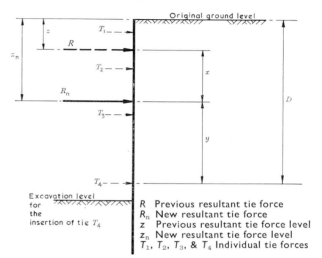

R   Previous resultant tie force
R_n New resultant tie force
z   Previous resultant tie force level
z_n New resultant tie force level
T_1, T_2, T_3, & T_4 Individual tie forces

Fig. 4

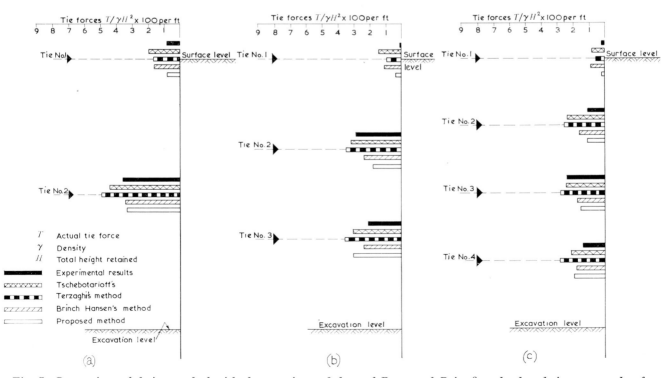

Fig. 5. Comparison of design method with the experimental data of Rowe and Briggs[3] and other design approaches for wall with (a) two ties (b) three ties (c) four ties

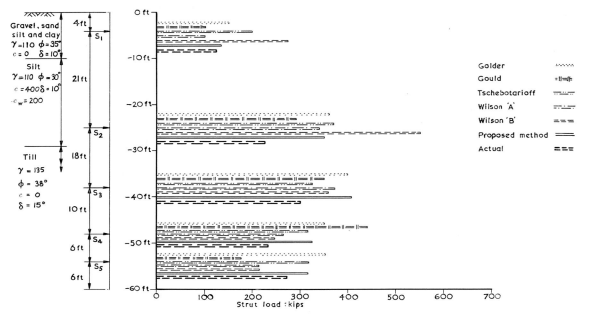

Fig. 6. Comparison of strut loads on retaining wall (Journal of the Soil Mechanics and Foundations Division, American Society of Civil Engineers, May 1970[4])

Substituting in the Newton–Raphson iteration formula:

$$x_{n+1} = x_n - \frac{Rx - T_4 D + T_4 x + T_4 z}{R + T_4}$$

where x_{n+1} is the new estimate of x and x_n is the previous estimate of x.

24. Comparisons between this design procedure and the results of published experimental work suggest that the design provides a good estimate of the horizontal forces involved. Figure 5 presents the tie forces predicted by this design approach and compares them with the results of the model tests performed by Rowe and Briggs.[3] The correlation between this equivalent single tie approach and the corresponding tie forces predicted by various other theories is presented in Fig. 6.[4]

MODEL TESTS

25. Many workers[3,5] have described model tests on retaining walls. In general, recent work has been concentrated on fairly small-scale tests. These are useful because they can be performed relatively quickly and a close control can be exercised on test conditions. Notable exceptions to this scale include the almost full-scale tests being performed by Tcheng.[6]

26. The two model tests described here were performed on walls of approximately an eighth full size and having widely differing flexibilities. Whilst numerous parameters were studied, the main aim of both tests was to measure anchor loads together with wall and soil movements.

27. Neither model was capable of simulating the modes of failure associated with full-scale retaining walls due to the high factor of safety and rigidity of the anchors.

Test facilities

28. The overall plan and elevation of the testing pit is shown in Fig. 7. The extension of an existing test pit by

the construction of a 4 ft high side wall and a 9 ft front wall enabled a test region 10 ft × 10 ft × 9 ft to be enclosed. Walls were constructed of thick timber boards, lapped to avoid leakage, slotted into rigid I section stanchions. A high degree of rigidity was required for all the enclosing walls to reduce any changes in soil pressure brought about by a moving outer boundary. The inner surfaces of the side walls were rendered plane by plastering. Horizontal lines were marked on the sides at 3 in. intervals and the inner boundary was then lined with heavy duty polythene to reduce side wall friction and to keep out moisture.

29. The walls in both tests were 10 ft wide and 8 ft deep. They were composed of two panels 4 ft × 8 ft and one of 2 ft × 8 ft, arranged so that a central large panel matched the off central position of the tank at the rear of the pit.

30. Both walls were suspended vertically in the pit at approximately equal distances from the rear and front (excavated side) boundaries. These distances, of about 5 ft, were controlled by the available resources and proved a limiting factor in the effective simulation of site conditions, particularly in relation to ground movements. The test walls, spanning the full width of the pit, were subjected to almost perfect plane strain conditions by minimizing side wall friction and by the provision of flexible seals at the ends. Test measurements were, however, concentrated on the central wall panel in both tests.

Flexible model wall test

31. *Wall construction and instrumentation.* Steel sheet of thickness 0·048 in. (18 SWG) was used to fabricate the flexible wall. The sheets were suspended from a horizontally supported steel strip by pop riveting their top edge centre points, and the strip was then reset horizontally by means of ten wire supports. This method of construction ensured that the vertical edges of adjoining sheets butted perfectly and thus eliminated distortion of

the sheets. Joints were taped on both sides and further riveting was deployed to complete the wall fabrication, the butting edges being stitched together with small metal strips. Square section hollow steel stiffeners were bolted across the full width of the wall at the anchoring levels to reduce transverse distortion during anchoring. Four layers, each containing five anchors, were employed with this model wall, these being at depths of 0·5 ft, 2·0 ft, 3·5 ft and 5·0 ft below the wall crest.

32. Changes of the wall profile were required at each construction stage for points below, as well as above, dredge level. A system was devised, incorporating Bowden cables, whereby the total movement of a point on the model wall was transmitted through the mass of sand and measured remotely as rotation of a protractor. Twenty-four such devices were mounted on a common shaft in order of elevation of their corresponding measurement points on the model wall. The dimensions of the system were such that a rotation of one degree of a protractor represented a wall movement of 0·017 in. In practice, friction in the system due to dust, small kinks in the wire, and a possible amount of slack and distortion of the shear, reduced the estimated accuracy of the system to about 0·02 in.

33. Anchor loads were monitored for all eight anchors in the central wall panel, by means of in line load cells. These consisted of metal strips upon which were mounted electrical strain gauges. The strips were completely enclosed within the anchor stiffening members together with wires and compensating gauges, and they served to transmit the loads applied by the wall to the anchor rods.

34. All measurements of vertical wall and sand movements were made using dial gauges on a rigid steel structure. Perspex discs were mounted on the stems to detect movements of the sand and surface.

35. *Sand placement.* The degree of compaction attained in a mass of sand depends upon the rate and velocity of deposition for a given sand. The total weight of sand required for the test was between 30 and 40 tons, depending upon the degree of packing attained. Oven dry Leighton Buzzard sand of 16–30 grade was selected.

36. In order to overcome the difficulty of maintaining a constant drop height, a pneumatic method of placing was selected, where the sand particles attained a sufficiently high velocity for the effect of elevation changes to be negligible. This method also enabled the sand to be moved almost completely mechanically. An airlift conveyor consisting of a Venturi throat was buried within the mass of sand in the sealed storage tank.

37. This method had some major disadvantages. The dust created during the filling made supervision of the exact placing impossible. Most of the dust was created by breakdown of intact particles in transit so that a change occurred in the mechanical properties of the sand.

38. No manual levelling of the sand was performed except to smooth out the final upper surface.

39. All the anchors, which consisted of 0·25 in. dia. rods, were laid horizontally in position before filling, but complete freedom of the wall was permitted in a direction normal to its plane. The wall was then released from its supporting girder and a full set of readings was recorded.

40. *Test process.* Excavation was performed by drawing the sand away from the wall and over the front timber retaining wall into a hopper.

41. At each of the levels represented by the wall mark-

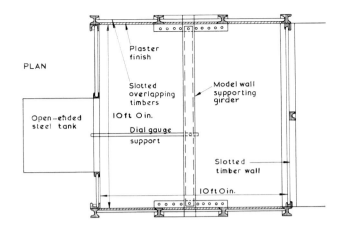

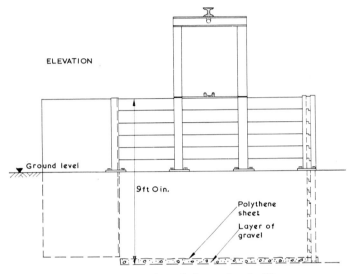

Fig. 7 Plan and elevation of the testing facility

ings the sand surface was drawn smooth and all the readings, including temperature, were recorded after a 15 min lapse. These levels were positioned so that each anchor overdig was 1·5 in. Upon exposing a horizontal stiffener the normal set of readings were recorded and the anchors at this level were then locked off with the nuts slightly more than finger tight.

42. Excavation was continued to a depth of 7 ft 6in. (6 in. above the toe of the wall), below which it was considered unsafe for further dredging.

43. No passive failure at the toe occurred, the wall tending to buckle below the lower anchoring level. A shear plane could be detected running parallel to the wall at a distance of about 9 in. behind the wall.

44. *Results.* The early stages of excavation caused movement of the wall crest into the excavation. Upon application of the restraint of the upper anchors, further excavation gave rise to rotation about the anchor level due to the high wall flexibility. This behaviour cycle was observed to a lesser extent at each subsequent anchoring stage.

45. The deflected profile at the design excavation depth (as evaluated by the design method, applying a factor of safety of unity) is presented in Fig. 8. It can be seen that total movements at anchoring levels were a

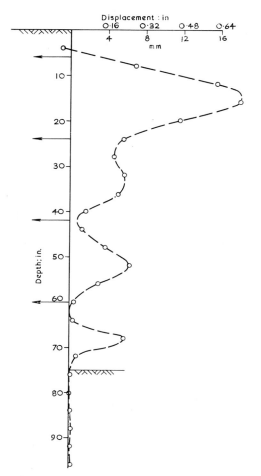

Fig. 8. Deflexion profile of the flexible wall at the design condition of excavation

maximum for the upper layer and decreased with the depth of each layer. This reflects mainly the total wall movement that occurred prior to locking each anchor since movements required to mobilize the measured anchor loads were calculated to be insignificant.

46. For each row of anchors the load increased rapidly with initial further excavation, this coinciding with a period of rapid wall deformation, but became almost constant after locking the next anchor row. A further rapid escalation of anchor loads occurred during excavation beyond the design depth.

47. Vertical movements of the surface of the retained sand were observed even at the rear of the pit at the very early stages of excavation. Simultaneously, larger movements were observed of the region within 20 in. of the wall, culminating in the development of a shear plane 12 in. behind the wall after excavating beyond the design depth. Profiles of the ground surface behind the wall at various excavation depths are presented in Fig. 9.

48. The in situ density of the sand on the excavated side was assessed during the test and was found to increase with depth down to 4 ft excavation and then decrease, indicating that some heaving of the passive zone took place. A mean density of 110 lb/cu. ft, representing a porosity of about 33%, was indicated. Specimens of the sand before and after the test possessed angles of internal

friction of 36° when tested at this initial density and an angle of wall friction of 22° was recorded in a drag friction test.

Rigid model wall test

49. The main aims of the second test were to measure the variation in anchor loads after prestressing, the ground surface movements, and the bending moments developed within a wall of flexibility comparable to that of a typical diaphragm wall during the process of excavation.

50. *Model wall.* The model wall size was again fixed at 10 ft wide and 8 ft deep. For a typical diaphragm wall 64 ft deep, this gave an actual to model size ratio of about 8. Rowe's flexibility number[1] for this typical wall was equated to that of the model to enable the second moment of area of the model to be evaluated. (The flexibility number, evaluated using Rowe's dimensions and based on the overall design retaining depth of the model, was found to be $\log_{10}\rho = -3.08$.)

51. A sandwich type of construction was selected since it produced a lighter, cheaper wall, it enabled strain gauges to be fixed to the inner protected surface and it resulted in an amplification of strains.

52. Two sheets of steel 0.064 in. thick were bolted at a distance of 0.976 in. apart, the space being fixed by tubular spacers at each bolt set in a 3 in. square mesh. This spacing was chosen by optimizing cost against the maximum anticipated relative deformation of the wall face under the soil pressure. Panels were carefully constructed in the vertical position and all stiffener bolts were stressed to a torque of 12 ft lb starting at the centre ones and moving outwards as a wave front.

53. The panels were suspended above the lined test pit using a uniformly distributed counterweight system. The three panels were placed in line and the end seals were fixed to the pit side walls. Upper and lower edge beams were fixed in position giving continuity between panels. Butting edges were taped and some edge stiffener bolts were used to secure short steel strips thus holding the panels firmly together.

54. Three layers of five anchors were employed, these consisting of lengths of threaded rod, one end of each being joined to the model wall by means of specially prepared bolts incorporating tubular stiffeners. The other ends passed through the rear wall of the pit. The three anchor layers were at depths of 1 ft, 3 ft and 5 ft below the wall crest, the anchors being evenly spaced at 2 ft intervals horizontally.

55. In an attempt to maintain fairly even temperature within the test area, the enclosing building was fully lined with expanded polystyrene.

56. *Instrumentation.* Because of the rigidity of this model the deflexion measuring devices used on the first model were not employed. Instrumentation of this model consisted of electrical strain gauges within the wall central panel, improved anchor load monitoring and deflexion measurements of the sand and wall crest.

57. The design bending moment diagram suggested that some regions of the wall would suffer high and rapid variation of bending moments. Reference was made to this diagram in selecting the locations of the 55 gauges installed. In each case gauges were fixed in pairs as near as possible to the centre points of the squares cornered by stiffeners, one of each pair being horizontal and the other

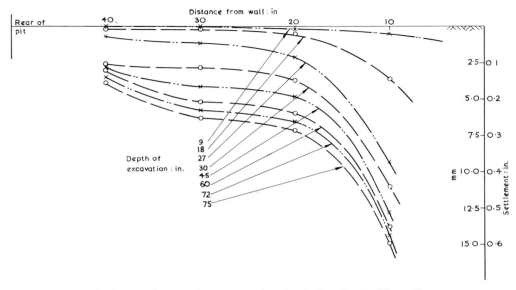

Fig. 9. Surface profiles of retained ground at various excavation depths for the flexible wall

vertical. The gauged panel was calibrated under pure bending and sand pressure condition prior to installation in the pit.

58. Flexible rings were threaded diametrically onto the free ends of the anchor rods beyond the rear of the pit to permit some yield under load. For the monitored anchors in the central panel these rings were strain gauged to operate as small proving rings. This type of ring was also used to set the prestress level of the non-monitored anchors.

59. A layout of dial gauges similar to that employed on the flexible wall test was again used.

60. *Sand placement.* For this test, sand was placed by a rapid discharge method. Dry sand was discharged from the storage silo into a skip. This was lowered by

crane through the roof of the test area and was quickly deposited from a free height of 5 ft. The raising of the sand level was kept even for both sides of the wall. Only slight changes occurred between the strain readings for the free hanging wall and those for the wall buried in sand. Upon completion of the sand placing, the upper surface was levelled off manually. This method proved much quicker and less dusty than the air blown technique. It must, however, have resulted in a looser packing and possibly to a less homogeneous mass.

61. *Procedure.* Due to the excessive edge stress at the base of the model wall, caused by its own weight, the wall was fully counterbalanced throughout the test in a manner permitting vertical movement due to downdrag.

62. The overall test procedure was identical to that

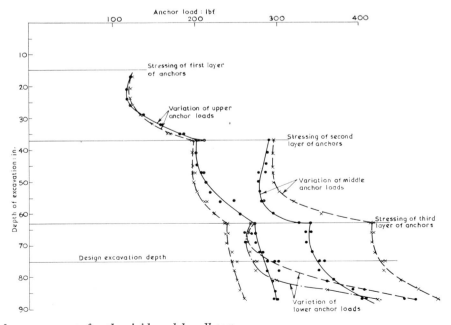

Fig. 10. Anchor load measurements for the rigid model wall test

for the flexible model except in one important respect. Upon reaching the design overdig level for each row of anchors, these anchors were stressed up to design loads instead of being left as passive anchors. At each level of excavation, readings were recorded of all the wall strain gauges, anchor proving rings and the dial gauges. Measurements of temperature at five points in the test region were made before and after each set of readings. Excavation continued to a depth of 7 ft 3 in. which was 1 ft lower than the design depth for a factor of safety of unity.

63. *Results.* The results of measurements of anchor loads are presented in Fig. 10 which shows the variation in loads caused by further excavation after stressing. The curves demonstrate that a gradual increase in anchor loads for a given stressed layer occurred upon subsequent further excavation. The later acceleration of this behaviour was invariably stemmed by the action of stressing the next layer, after which point the cycle was repeated. This interrelated effect of anchor layers was not observed in the flexible wall test. These results were greatly influenced by the moderately high rigidity of the anchors despite the insertion of the flexible rings. Insufficient information was available on actual field behaviour to permit effective model simulation of anchor load deformation behaviour. At the design excavation depth loads in each row of anchors exceeded the design (locked-in) values.

64. After excavating to 7 ft 3 in. the wall was left in position for a considerable time. After a period of six weeks anchor loads were recorded and, with one exception, reductions in each were observed, the mean anchor loads for each layer being 136%, 53% and 74% of the locked-in loads.

65. Anchor stressing was observed to have a direct influence on the settlement of the retained sand surface. The general trend of settlements were very similar to those observed in the flexible wall test. However, the results presented in Fig. 11 demonstrate that the action of stressing each layer of anchors was to reduce the rate of settlement. Despite the retarding influence of anchor stressing on the rate of settlement, it was noted that the total settlement at the design excavation depth was almost identical to that measured in the flexible model wall. No shear plane was detected in this test.

66. The high rigidity of this model and the relatively low bending moments developed gave rise to fairly low strain levels. Due to the considerable temperature fluctuations that unfortunately occurred during the test period, and the long-term drift of the measuring equipment, great difficulty was experienced in discerning between these influences and the actual strains developed. Consequently, whilst bending moments of the same order of magnitude as the design predictions were recorded, no results are presented.

ASSESSMENT OF ANCHOR LOAD MEASUREMENTS

67. The nature of the two tests performed enable comparisons to be drawn between design predictions and actual tie bar forces for the two opposite extremes of wall flexibility.

68. For the very flexible wall the inert anchors were permitted to develop equilibrating loads. A comparison of the results with the design values indicates that the proposed method predicts only 50% of the anchor load actually developed at the upper anchor. This is attributed to the very high wall flexibility which prohibits redistribution of the load to lower anchor levels. This is underlined by the observation that the proposed design method overestimates the loads at each of the remaining anchor levels by an average of 27%.

69. Similar comparisons made for the anchor loads developed in the rigid wall tests indicate that the final loads exceed the design predictions (and prestressing levels), the greatest deviation again occurring at the upper anchors. This is ascribed to the inability of such a rigid wall to relax prior to the stressing of this upper layer and give rise to a reduction of the pressures exerted by the sand. At subsequent anchoring levels the high rigidity of the anchors and their inability to model creep movements becomes the controlling influence.

CONCLUSIONS

70. Model tests by other workers, and measurements on diaphragm walls, in particular those made by James as part of an overall study described in Paper 15 by Littlejohn and Macfarlane, have confirmed the general applicability of the proposed design method.

71. This approach provides the designer with a useful and versatile tool since varying soil strata and random optimizing selection of tie bar levels can be incorporated. Furthermore, it is a repetitive single-tied wall design of a form with which most design engineers are familiar, and it allows the wall penetration to be calculated in a manner which satisfies rotational equilibrium as well as the horizontal equilibrium achieved in the trapezoidal method.

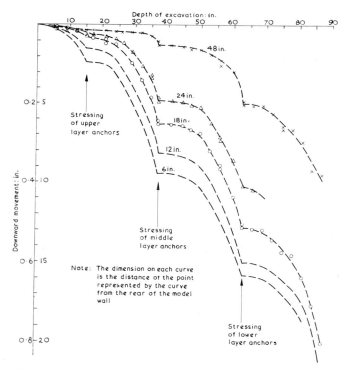

Fig. 11. Settlement of the retained soil surface for the rigid model wall test

72. The model tests described have proved to be useful exploratory studies of the two extremes of flexibility. Comparison between the experimental data and the predictions of design methods, including the one proposed, suggest that conservative factors of safety should be adopted for the design of the upper anchors.

73. Whilst conditions for the more rigid model test matched fairly closely those pertaining in the field, the influence of anchor stressing upon the load variation of anchors already loaded, observed in the test, has not been detected in field measurements of anchor loads. This is attributed to the high stiffness of the model anchors and suggests that more information should be sought on the overall load/deflexion characteristics of ground anchors.

74. Although the two tests are obviously not directly comparable, the observation that total settlements of the surface of the retained mass were almost identical at the design excavation depth may be worthy of further attention.

ACKNOWLEDGEMENTS

75. The model tests described here were performed by James whilst working for Cementation Research Ltd, and the Authors are grateful to the various parties involved for permission to publish this information.

REFERENCES

1. ROWE P. W. A theoretical and experimental analysis of sheet pile walls. *Proc. Inst. Civ. Engrs*, 1955, **4**, Part 1, No. 1, Jan., 36–69.
2. ROWE P. W. The flexibility characteristics of sheet pile walls. *Struct. Engr*, 1955, May, 150–158.
3. ROWE P. W. and BRIGGS A. Measurements on model strutted sheet pile excavations. *Proc. 5th Int. Conf. Soil Mech.*, *1961*, **II**, 473–478.
4. GOLDER H. Q. *et al.* Predicted performance of braced excavations. *Proc. Amer. Soc. Civ. Engrs*, Soil Mech. Fdns Div., 1970, **96**, May, No. SM3.
5. VERDEYEN J. and ROISIN V. Stresses in flexible sheet pile bulkheads due to concentrated loads applied at the surface of the retained mass. *Proc. 6th Int. Conf. Soil Mech., Montreal*, *1965*, **II**, 422–426.
6. TCHENG Y. and ISEUX J. Full scale passive pressure tests and stresses induced on a vertical wall by a rectangular surcharge. *Proc. 5th European Conf. Soil Mech.*, 1972, Sociedad Espanola de Mec. del Suelo y Cimentaciones, Madrid, Session II, Paper 13, 207–214.

Design and construction of a diaphragm wall at Victoria Street, London

F. T. HODGSON, BArch, DIC, FIStructE, ARIBA, *Partner, Clarke Nicholls and Marcel, Consulting Engineers*

The Paper describes the design and construction of a 700 m long diaphragm wall. Reasons for the choice of a diaphragm wall are given, and site restrictions and surcharge loading are described. A discussion of the design criteria covers the factors of safety for the active and passive pressures in the gravel and clay, the allowable stresses for the concrete and reinforcement of the wall during both the temporary and permanent conditions of its construction, and various possibilities considered for the temporary support of the wall. Specification notes cover site laboratory requirements for testing the bentonite properties. The Paper describes the methods adopted for supporting the wall in its temporary condition for each section of the job: these methods were ground anchors, preloaded structural steel struts and a reinforced concrete waling slab. Ground anchor loads and factors of safety are given. Movements of the wall during excavation are recorded. The Paper describes the method of installing the diaphragm wall and excavating the third basement as a mining operation. Diaphragm wall and ground anchor costs are given. Conclusions are drawn regarding the advantages of using this type of construction for this particular job.

GENERAL DESCRIPTION

The site, shown in Fig. 1, is on the south side of Victoria Street and is bounded by Victoria Street, Carlisle Place to the west, Francis Street to the east and to the rear by Howick Place and Ashley Place. The development comprises two blocks, B and C, connected at first basement level. Block B is towards Victoria Station at the west end of the site and block C is at the east end of the site.

2. A total of 700 linear metres of diaphragm wall was constructed from the existing basement level, 4 metres below street level. The wall is generally 500 mm thick, except in the case of the third basement area of block C where a 600 mm thick wall was used. The depth varied between 8 m and 10 m, increasing in the third basement to 15 m. The average panel length was 4·6 m. The overall area of wall constructed was 7078 m².

3. The selection of a diaphragm wall, tied back with ground anchors, was based on the following factors: owing to the site being immediately adjacent to residential property at the rear, the use of any techniques involving heavy vibration or excessive noise was excluded; the support system on the quarter mile frontage of Victoria Street was one metre away from the existing kerb and during excavation movements of the temporary supports could not be tolerated; the use of ground anchors to restrain the diaphragm wall provided an open excavation and this facilitated construction of the complex structure within the double basement; any significant ground movement in the adjacent streets would affect services such as the hydraulic main in Howick Place and Ashley Place working at a pressure of 5·5 MN/m² and the old brick sewer in Victoria Street; and the surcharge on the retaining wall system from the HA and HB loading on the surrounding highways had to be accommodated.

4. The site of block C is shown in Fig. 2 with construction of the diaphragm wall in progress, and Fig. 3

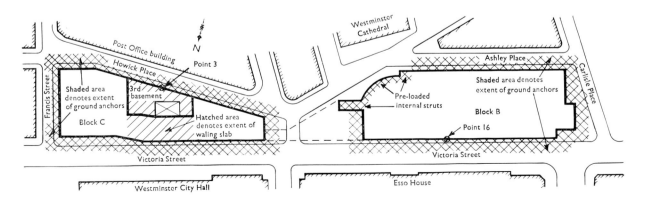

Fig. 1. Plan showing location of site and extent of diaphragm wall

Fig. 2. Block C showing diaphragm wall in progress

Fig. 3. Block C showing diaphragm wall and ground anchors and area of third basement

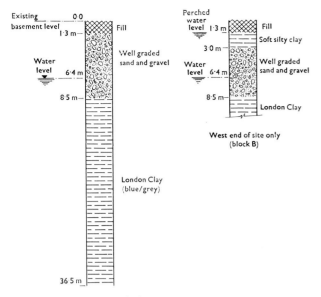

Fig. 4. Typical borehole logs

Table 1. Factors of safety

Support system	Factor of safety on passive pressure	Pressure on earth face
Wall supported by one level of ground anchors	2·0	1·0 × active pressure
Wall supported by more than one level of anchors or struts	1·6	1·3 × active pressure

Table 2. Bentonite requirements

Marsh funnel measurement for viscosity	32–55 seconds
Apparent viscosity	7–15 centipoises
Plastic viscosity	12–30 centipoises
Bingham yield	24–144 dynes/cm²
10 minute gel strength	48–192 dynes/cm²
Density	1·02–1·09 g/cm³
Sand content	Less than 5%
Fluid loss	11–35 cm³
pH	10–12

shows the diaphragm wall and ground anchors and area of third basement. Details of soil conditions are given in Fig. 4.

5. The main contractor was George Wimpey & Co. Ltd. The subcontractor for the diaphragm wall and ground anchors was Cementation Ground Engineering Ltd.

DESIGN CRITERIA

6. The factors of safety for the design of the wall are shown in Table 1. (For wall supports at more than one level, the factor of 1·3 in the pressure on the earth face accounts for the fact that wall deformations are restricted by the support system and so theoretical active pressures cannot develop.)

7. In cohesive soils, earth pressures were calculated assuming that the in situ undrained shear strength was equal to one half the measured strength, obtained from 100 mm dia. specimens taken from U4 sampling tubes, and no allowance was made for wall friction or adhesion on active pressures; for passive pressures, wall friction was taken as 10° in granular soils and wall adhesion as one quarter of undrained strength in cohesive soils.

8. For the condition when the wall was supported at more than one level, the total pressure to be carried on the earth face was based on the recommendation of Terzaghi and Peck[1] for deep excavations with more than one support.

9. The allowable stress in the reinforcement during the temporary condition was 210 N/mm². The bending moments in the wall are greatest during this period of the works. In the permanent condition, however, to take account of long term considerations, the stresses were limited to 140 N/mm².

10. The temporary support of the vaults, by cantilever steel H piles above the existing basement level, produced horizontal forces of a considerable magnitude on the tope of the diaphragm wall. As a result of these loads the position of ground anchors was critical and slight adjustments in their level produced substantially differing bending moments on the wall.

11. In the third basement, the depth of penetration of the wall below the lowest excavation level was determined by considerations of the required passive resistance at the toe.

12. For a multi-tied condition, one method of analysis is to convert the multi-tie forces into one equivalent force, but this produces a very conservative solution for the depth of toe-in required. In this example, the solution was based on strut forces in cuts as outlined by Terzaghi and Peck.[1]

SPECIFICATION NOTES

13. The method of construction of diaphragm walls is by now familiar. However, on this contract a special effort was made to maintain effective control of the bentonite properties. The in situ bentonite tests laid down in the specification for cast in place concrete diaphragm walls issued by the Confederation of Piling Specialists are considered to be inadequate, and for the Victoria Street contract the specification laid down eight separate tests on the mud. These tests, set out in Table 2, were carried out daily in a fully equipped site laboratory, thus ensuring continuous quality control. They provided an effective low-cost method for the control of the bentonite and permitted the re-use of desanded bentonite.

14. The specified tolerances for the wall were 1 in 80 for verticality and a maximum of 75 mm for overbreaks.

15. As the guide trenches invariably form part of the temporary works required to be carried out by the main contractor, it is preferable to keep the design and construction of them under his control.

16. All excavated material was placed in watertight lorries for removal from site, immediately following digging. Stockpiles on site of excavated material were not permitted.

17. The box work for recesses in the inner face of the wall was made up from plywood sheets, wrapped in polythene and secured to the cage of reinforcement by additional bars welded to the main cage. From experience, this was found to be the most practical solution.

TEMPORARY SUPPORT OF WALL

18. At various locations around the site three methods were used for temporary support of the wall. Ground anchors into the gravel and clay strata were used generally, except in one section of block B, where Macalloy bars were employed to facilitate their destressing and removal. Preloaded structural steel struts were used in the area adjacent to Westminster Cathedral, where ground anchors were not practical. A reinforced concrete waling slab was used in the third basement of block C.

19. In blocks B and C, 288 anchors were placed. The calculated maximum working load for the anchors on the Victoria Street elevation was 56 tonnes, using two anchors per panel, but elsewhere along the perimeter the average load per anchor was 45 tonnes. In both cases a free length of 5 m was adopted, with the fixed or dead end being 4·3 m and 3·7 m respectively. The majority of the anchors are at right angles to the wall in plan and have a declination to the horizontal of 25–30°. The dead end of all anchors was outside the soil shear plane for the active pressure wedge.

20. In the third basement of block C, 28 clay anchors were placed. Three anchors, each with a working load of 55 tonnes, were used in all panels.

21. The factors of safety adopted were 3 for clay anchors and 2 for gravel anchors.

22. In the gravel five test anchors were placed and stressed, at different locations around the site. In the wall of the third basement, two anchors in the clay were tested to failure. The failing loads of the test anchors in the clay are shown in Table 3.

Table 3. Failing loads of test anchors in clay

Length of drilled hole, *m*	Free length, *m*	Failing load, *tonnes*
24·5	20	145
18	14	93

23. Because the heavy surcharge from the highways acted intermittently on the wall, the force on the anchors was much greater than that required for the active pressure alone. Movement of the wall due to mobilization of the active pressure was, therefore, largely prevented.

24. Preloaded struts were used to support the wall adjacent to Westminster Cathedral because it was considered inadvisable to use ground anchors in this area. The struts were preloaded using freyssi jacks and the load was maintained until the wall was supported by the permanent structure.

25. The clay anchors, installed in the third basement of block C, were re-checked after a period of three months to ensure that their load was maintained. No relaxation of the anchorages due to creep of the clay had occurred.

26. During all the construction stages the diaphragm wall was monitored for movement. Immediately following construction of the diaphragm wall, and prior to any excavation, a survey station and target were established at each end of the wall, well outside the perimeter of the site. At selected points along this line permanent packs were set into the top of the wall and daily checks were made for possible movement. Table 4 shows a few selected readings. In the case of the third basement, the checks were extended down the face of the wall by means of a plumb bob.

27. During the piling operation the use of a Mueller vibrator was terminated when work immediately adjacent to one section of the wall induced a movement of 25 mm of the wall into the site. The gravel anchors were checked for load and all except one had retained their working load. After further examination, it was concluded that the frequency of the vibrator had coincided with the natural frequency of the gravel, and that the wall and the retained material had moved en masse.

THIRD BASEMENT CONSTRUCTION

28. The basement location and cross-section are shown in Figs 1 and 5. The basement excavation extends 7 m below the level of the second basement slab, giving a total depth from street level of 14 m. Owing to the proximity of the Post Office building, it was decided to support the diaphragm wall with a concrete waling slab at second basement level.

29. Before excavation of the third basement in situ concrete piles were augered from the second basement level. These piles have a cut-off level at the underside of the third basement raft slab. A steel casing was left in up to ground level and enabled a man to get down to position and grout in the temporary steel columns which supported the waling slab at second basement level. A waling slab was constructed on the ground for the full width of the building so that thrusts from the diaphragm walls on each side of the site counter-balanced each other.

30. Excavation of the third basement was then carried out as a mining operation through an opening in the waling slab. After approximately 4 m had been excavated the clay anchors were installed below Howick Place. These anchors were 24 m long with loads of up to 55 tonnes. The excavation and ground anchor installation is shown in Fig. 6. Excavation was then carried down to the formation level of the raft. Figure 7 shows the third basement after completion of the raft slab. The clay anchors were then destressed and construction was carried on above. During the time that this work was being carried out, close observation of the wall was undertaken. Deflexions of the top of the wall are shown in Table 5. During the constructional period of the third basement area, daily plumb checks taken on the wall at point 3 showed zero movement.

COSTS

31. A summary of costs for the diaphragm wall and ground anchors is given in Table 6.

GENERAL CONCLUSIONS

32. The construction period of the wall including the establishment of the plant was seven weeks for each block. There was a considerable time saving on the contract by having only one operation for the installation of the wall. This wall functions not only in the temporary condition as a support in up-holding the roads and the excavation sides, but also in the permanent condition as a retaining wall.

33. The use of ground anchors gave an unobstructed site for the excavation and facilitated the subsequent piling and foundation construction. The only exception

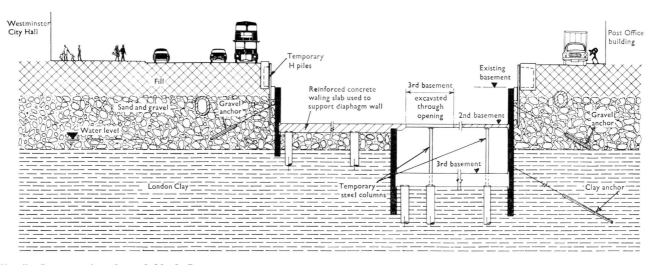

Fig. 5. Cross-section through block C

Table 4. *Readings taken at point 16 during second basement construction adjacent to Victoria Street*

Construction stage	Date	Deflexion at top of wall, *mm*
Diaphragm wall construction	Dec. '71	0
1st stage excavation		+1·5
Ground anchors (gravel)	Feb. '72	0
2nd stage excavation		+3·0
2nd basement slab constructed	Oct. '72	+3·0

+ denotes movement into site.

Table 5. *Readings taken at point 3 during construction of third basement adjacent to Howick Place*

Construction stage	Date	Deflexion at top of wall, *mm*
1st stage excavation to ground anchor level	Feb. '72	0
2nd stage excavation to second basement level	Apr. '72	+3·0
2nd basement slab constructed	Nov.–Dec. '72	0
1st stage excavation to clay ground anchor level 3rd basement	Jan. '73	0
2nd stage excavation to formation level of raft 3rd basement	Feb. '73	0

+ denotes movement into site.

Table 6. *Costs*

Diaphragm wall	Total area of wall	7078 m²
	Total cost of wall	£193 000
	Cost per unit area	£27·26/m²
	Average weight of reinforcement per unit area	55 kg/m²
	Total cost of bentonite testing	£1750
	Cost of bentonite testing per unit area of wall	25p/m²
Ground anchors	Cost of gravel anchors per unit length of wall	£36/m
	Cost of clay anchors per unit length of wall	£150/m

Fig. 6. Third basement excavation and ground anchor installation

Fig. 7. Third basement after completion of foundation raft

to this was in the area adjacent to Westminster Cathedral, where preloaded steel struts were used.

34. In both blocks the wall was constructed without expansion joints, and although it was exposed to the weather for twelve months no cracking was detected. The only instance of any leakage of water was through one or two of the ground anchor holes which were subsequently resealed.

35. Although it was impossible to prevent entirely the ground movement of the soil surrounding the site, the horizontal wall movements were kept to an absolute minimum and damage to the services in the surrounding roads was prevented. It was concluded that the horizontal movement of the top of a wall could be virtually prevented by an adequate amount of force in the ground anchors.

36. On site bentonite testing was rapid and was well worth the small cost involved for maintaining the quality of the bentonite.

REFERENCES

1. TERZAGHI K. and PECK R. B. *Soil mechanics in engineering practice*. Wiley, New York, 1967, 2nd edn.

A load bearing diaphragm wall at Kensington and Chelsea Town Hall, London

B. O. CORBETT, BSc, FICE, MIStructE, *Geotechnics Division, Ove Arup & Partners*

R. V. DAVIES, BSc, *Geotechnics Divison, Ove Arup & Partners*

A. D. LANGFORD, MICE, MIStructE, *Structures Division 3, Ove Arup & Partners*

The basement at Kensington and Chelsea Town Hall is approximately 140 m × 65 m in plan by 13 m deep. The Paper discusses the design, construction and performance of the diaphragm wall, with particular reference to the results of a load test on an individual panel. It is shown that, when compared with the usual method of design for large diameter bored piles, the performance of the test panel was not adversely affected by the use of bentonite.

INTRODUCTION

A new £7 million town hall for the Royal Borough of Kensington and Chelsea is under construction at a site adjacent to Hornton Street, Kensington, London. Above ground level, the scheme essentially consists of three buildings, the largest of which is a five-storey rectangular structure built around a central courtyard (see Fig. 1). This building, to be used for the main civic offices, is approximately 90 m × 70 m in plan, covering about two thirds of the site area. The remaining buildings house an assembly hall and the council chamber, the latter being raised above the general ground level on four concrete columns.

2. Beneath pavement level, there is a three-level basement and underground car park which, with the exception of the central courtyard, covers the entire area of the site (see Figs 2 and 3). The basement is on average 13 m deep and approximately 140 m × 65 m in plan.

3. The internal columns to support both the basement floor levels and superstructure are founded on a 0·9 m thick reinforced concrete raft, locally thickened to cater for the higher column loads arising from the council chamber and main building. However, fairly substantial loads from both the assembly hall and the main building are required to be carried around the periphery of the basement, and it was decided to adopt a diaphragm wall for the basement walls that extend to beneath the lower basement floor level to act as a foundation for these loads. This solution takes maximum advantage of the diaphragm walling technique, since the walls are used in both the permanent and temporary condition as earth-retaining structures, and the extension of the wall to beneath excavation level also assists stability during the temporary works. In addition, the method of construction is quiet and free from vibration, a not unimportant aspect in this area of London.

GROUND CONDITIONS

4. The site is underlain by approximately 50 m of London Clay, the surface of which has been eroded over the southern part of the site and a dense gravel terrace deposited (see Fig. 4). The gravel extends over approximately half of the site and represents the northern edge of the Flood Plain Terrace. Overlying the gravel and, in part, the London Clay, is a deposit of brickearth (a stiff, sandy clay), which thins out towards the northern side of the site. Piezometers installed in the London Clay showed the phreatic surface to be approximately 5 m below ground level, generally increasing almost hydrostatically to just above the base of the clay and therefore appearing to be only marginally affected by underdrainage, which is known to occur in this area of London.[1]

DESIGN

5. The vertical loads on the diaphragm wall were carried both in side friction and end bearing and the load–settlement characteristics of the wall were designed on this basis. Since the method of construction is such that it would be difficult to guarantee good bearing contact at the base of the panels, the design was cross-checked on skin friction alone. Conventional methods of estimating side friction are based almost totally on experience with

Fig. 1. Architect's model viewed from the south

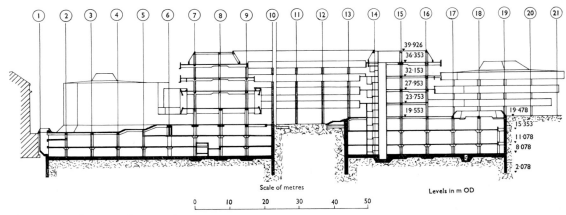

Fig. 2. Section through structure from south to north

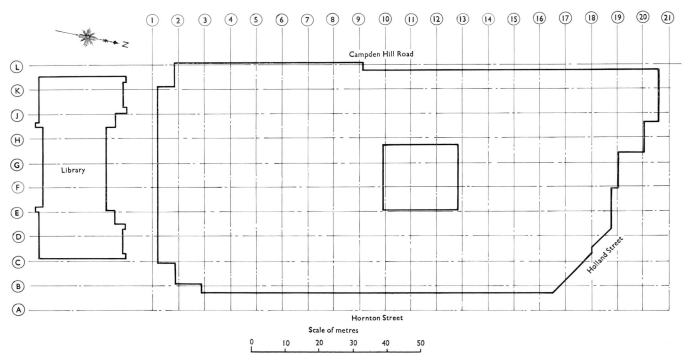

Fig. 3. Plan of basement

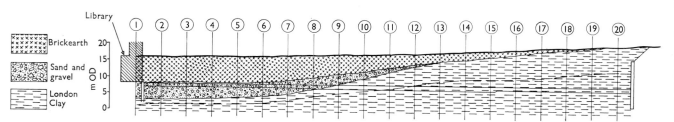

Fig. 4. North–south geological section

piles, and in clays generally relate the ultimate shaft friction or adhesion to the undrained shear strength of the clay. These methods are largely empirical and, as pointed out by Burland,[2] there are dangers in extrapolating them to comparatively new situations.

6. For a diaphragm wall also acting as a foundation, there are two basic factors that should be considered before estimating shaft friction by empirical methods. Firstly, the wall is installed prior to excavation. At this stage, the ultimate shaft friction available may in principle be similar to that for a pile. However, during construction of the basement important stress changes will occur both beneath and around the excavation, and in time an entirely new stress equilibrium will be reached, compatible with the final loading, geometry and groundwater regime. An analysis based on the undrained shear strength of the clay does not recognize these stress changes, and to examine how they may affect the load carrying capacity of the diaphragm wall it is necessary to analyse the problem in terms of effective stress.

7. The use of effective stress methods in estimating shaft friction for clays has been discussed by Chandler[3,4] and more recently by Burland,[2] whilst the stress changes that occur due to an excavation have been considered by Henkel.[5] However, the problem is complex and at the present time this method does not lead to design parameters. Nevertheless this approach does aid in understanding the effects of excavation on shaft friction so that engineering judgement can be used in assessing the applicability of empirical design methods. For the particular conditions of the basement at Kensington and Chelsea Town Hall, an analysis of this type showed that the possible shaft friction available by the time the load was finally applied to the wall was unlikely to be less than that available prior to excavation, although it should be emphasized that this is by no means a general conclusion.

8. The second factor concerns the effect of the bentonite slurry used in constructing diaphragm walls. Many engineers are of the opinion that bentonite could substantially reduce shaft friction. However, this opinion appears to be based primarily on intuition and is not borne out by test results. For example, Burland[6] found from a load test on two diaphragm wall panels, formed with and without bentonite, that the load/deflexion characteristic and ultimate load were virtually identical, whilst Farmer *et al.*[7] found that shaft friction developed on bored cast in situ piles concreted under bentonite was higher than would be anticipated for bored piles formed in the dry. Similar observations have been reported by Fernandez-Renau[8] and Chadeisson.[9] Nevertheless, test data are limited and it was decided to carry out a load test on a diaphragm wall panel.

TEST PANEL

9. The test panel, $1 \cdot 2 \text{ m} \times 0 \cdot 5 \text{ m}$ in plan and $14 \cdot 4 \text{ m}$ deep, was excavated under bentonite using a 500 mm cable operated grab. Excavation took two days and on completion the panel was left open overnight. The bentonite (Fulbent 570) was renewed immediately prior to concreting, and during concreting an average overbreak of $8 \cdot 5\%$ was recorded. The soil profile logged during excavation is shown in Fig. 5, together with test data from a nearby borehole. The panel was allowed to cure for five weeks

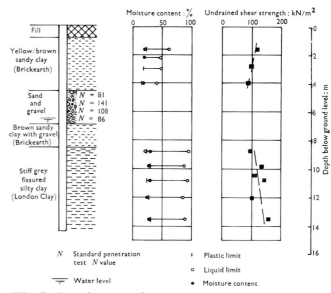

Fig. 5. Log of test panel

prior to load testing, during which time ground anchors were installed to provide a reaction for the test.

10. A preliminary estimate of the ultimate load of the panel was made on the basis of normal semi-empirical design methods appropriate to bored piles formed in the dry. This assumes that the ultimate load carried by the panel (Q_{ult}) is the sum of the ultimate load carried by the shaft (Q_s) and the ultimate load carried by the base (Q_b) and that

$$Q_s = A_1\alpha_1 C_{u1} + A_2\sigma'_v K \tan \delta + A_3\alpha_3 C_{u3} \qquad (1)$$

where A = area of panel in contact with the soil

C_u = undrained shear strength of the clay

α = a factor relating C_u to the ultimate shaft friction

σ'_v = average effective overburden pressure in the gravel

K = ratio of horizontal to vertical effective stress

δ = ultimate angle of friction developed between panel and gravel

and suffices 1, 2 and 3 refer to brickearth, gravel and London Clay respectively. Also,

$$Q_b = A_b C_u N_c \qquad (2)$$

where N_c is the bearing capacity factor. Assuming that $\alpha_1 = 0 \cdot 35$, $\alpha_3 = 0 \cdot 5$, $K \tan\delta = 0 \cdot 7$ and $N_c = 9$ gives, neglecting overbreak,

$$Q_s = 570 + 480 + 1140 = 2190 \text{ kN}$$

$$Q_b = 740 \text{ kN}$$

and thus $Q_{ult} = 2190 + 740 = 2930 \text{ kN}$.

11. The panel was then subjected to six cycles of loading, the first five cycles being incremental to maxima of 1000 kN, 1250 kN, 1500 kN, 1750 kN and 2000 kN, with the maximum load of each cycle being maintained until the rate of settlement was less than 0·05 mm in half an hour. On the sixth cycle the panel was loaded at a constant rate of penetration, for which a maximum safe reaction of 4000 kN was provided.

12. To examine how the load was transmitted to the ground at the various stages of loading, Soletanche Enterprise took the opportunity to instrument the panel with

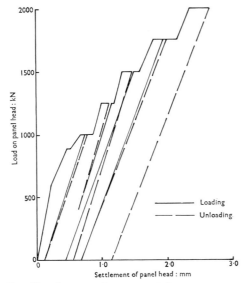

Fig. 6. Load/settlement of test panel (incremental loading)

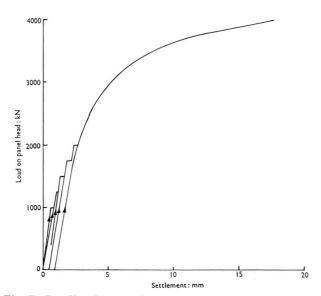

Fig. 7. Load/settlement of test panel (constant rate of penetration test)

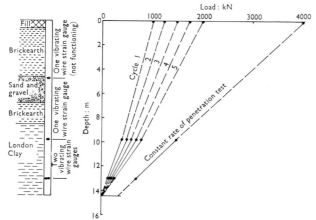

Fig. 8. Load against depth from vibrating wire strain gauges

mechanical displacement gauges and vibrating wire strain gauges. Complete details of the instrumentation and results are given in a report by Soletanche Entreprise[10] and only a brief description and principal results are given here.

13. The instrumentation consisted of Telemac F3 vibrating wire strain gauges, together with mechanical deformation gauges installed at various depths within the panel. Single strain gauges were installed at depths of 0·70 m, 4·70 m and 9·80 m and two at a depth of 13·00 m. For protection during concreting, the gauges were encased in small concrete blocks and installed by attachment to the reinforcement cage. The mechanical displacement gauges consisted of 13 mm dia. steel rods sliding within plastic tubes, fixed at the base at depths corresponding to the strain gauges and extending to ground level. These were intended for direct measurement of axial compression of the panel, but gave inconsistent results and are not discussed further.

14. The load/settlement behaviour of the panel for the incremental loading cycles and constant rate of penetration test is shown in Figs 6 and 7. At various stages during loading and unloading, the strain in the vibrating wire gauges was recorded. Unfortunately, the gauge at a depth of 4·70 m failed to function during the test, probably as a result of damage during concreting. However, the gauges at 9·70 m and 13·00 m gave reasonably consistent results, and the load at these depths calculated from the strain measurements is plotted against depth in Fig. 8 for the maximum load of each cycle. The value of Young's modulus (E_c) used in computing the load was derived from the gauge installed at a depth of 0·70 m.

15. Figure 7 clearly shows that the ultimate load of the panel was considerably more than that calculated, and under constant rate of penetration loading it is probable that the panel would have carried a load slightly in excess of 4000 kN. Nevertheless, the displacement was adequate to mobilize fully the shaft friction and any additional load would have been carried by the base. This is reflected in Fig. 8 where the strain gauge measurements indicate that at a load of 4000 kN the base was only carrying approximately 350 kN, which is just under half the ultimate load anticipated from equation (2).

16. The ultimate load carried by the shaft was therefore approximately 3650 kN, i.e. 67% greater than that calculated with the assumption given previously, and from this test it appears that the use of bentonite increased rather than decreased shaft friction. It could perhaps be argued that when bentonite is used to support the sides of the excavation horizontal stress release is reduced and therefore shaft adhesion/friction is increased. However, variable factors such as the uneven surface between the shaft and the soil due to overbreak, especially in the sand and gravel stratum, could well account for this increase, and without further evidence no general conclusions can be drawn. The test does, however, add to the slowly accumulating evidence that the use of bentonite does not reduce shaft friction.

EXCAVATION

17. During construction of the basement, support to the diaphragm wall was provided by a traditional internal strutting system. Consideration was given to the use of ground anchors, but there were difficulties in obtaining

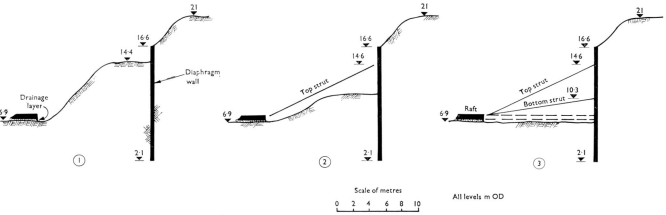

Fig. 9. Excavation sequence for north wall

Fig. 10. Diaphragm wall showing Rendhex struts

the necessary wayleaves. It was also felt that the use of ground anchors for an excavation in London Clay might increase the deformations around the excavation, and that as the excavation would be open for a comparatively long period of time an internal strutting system was preferable.

18. The sequence of construction is shown in Fig. 9 and generally involved the excavation of the centre portion of the site to allow part of the raft to be constructed. During this period a berm was left to support the diaphragm wall; the berm was then reduced to allow the top strut to be installed from the raft. Excavation proceeded in short lengths to allow the lower level of struts to be installed and the raft was then extended up to the diaphragm wall (see Fig. 10). Struts were installed at the joints between wall panels and additional reinforcement

was incorporated in the diaphragm wall to avoid the necessity of a waling beam. Horizontal continuity was also supplied by a capping beam that formed part of the permanent works.

19. Stability of the sides of an excavation in London Clay, together with the ground deformation around the excavation, is largely time dependent.[11] The size of the basement required the excavation to be open for a considerable period of time and it was felt prudent to monitor the movements of the diaphragm wall. This was done to provide a check that the excavation method was not resulting in excessive ground movements and, in case the movements should start to accelerate, to provide an early warning system to potentially unstable areas. In the event, this proved extremely useful since during a critical period of the excavation the site was closed during the national

61

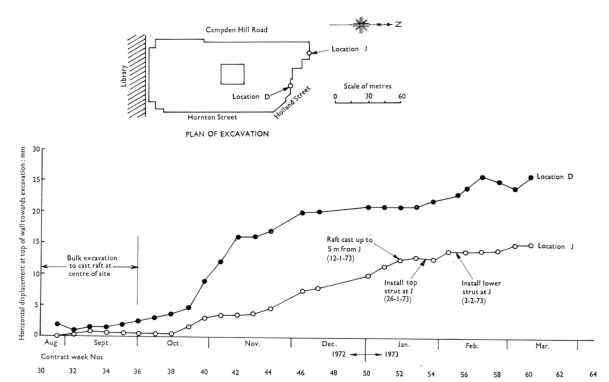

Fig. 11. Horizontal movement of north wall

building workers' strike, and this delay led to concern over the stability of the works.

MONITORING

20. The monitoring system for the movement of the wall was devised and carried out by the main contractor. A total of 14 reference stations, consisting of punched brass plates set into the capping beam, was installed along the north, east and west walls. The horizontal movements of these stations were measured by theodolite from reference points at the north-east and north-west corners, the fixity of these points being checked against three adjacent buildings. The main concentration of effort was in measuring the movement of the north wall, since this was the deepest part of the excavation and almost totally in weathered London Clay. The horizontal movements at the top of the wall are shown in Fig. 11 for two positions, D and J. These two points represent the maximum and the minimum movements recorded along the north wall. As would be expected, the maximum movement (point D) occurred midway along the wall whilst the minimum movement (point J) occurred at the corner.

CONCLUSIONS

21. It is shown that, when compared with the usual method of design for large diameter bored piles, the performance of the test panel was not adversely affected by the use of bentonite.

22. As expected, the maximum movement of the diaphragm wall occurred midway along the wall, whilst the minimum movement occurred at the corner.

ACKNOWLEDGEMENTS

23. The Authors would like to thank Ove Arup & Partners for permission to publish this Paper. The client

is the Royal Borough of Kensington and Chelsea, whose architects are Sir Basil Spence, Bonnington and Collins. The main contractor is Taylor Woodrow Construction Ltd and the nominated subcontractor for the diaphragm wall was Tarmac-Soletanche. The Authors are indebted to Soletanche Entreprise of Paris for permission to publish the load measurements in the test panel.

REFERENCES

1. WILSON G. and GRACE H. Settlement of foundation due to underdrainage of the London Clay. *J. Instn Civ. Engrs*, 1942, **19**, Dec., No. 2, 100–127.
2. BURLAND J. B. Shaft friction of piles in clay—a simple fundamental approach. *Ground Engng*, 1973, **6**, 3.
3. CHANDLER R. J. Discussion. *Large bored piles*. Institution of Civil Engineers, London, 1966, 95.
4. CHANDLER R. J. The shaft friction of piles in cohesive soils in terms of effective stress. *Civ. Engng Publ. Wks Rev.*, 1968, **63**, 48.
5. HENKEL D. J. Geotechnical considerations of lateral stress. *1970 Specialty Conf. on Lateral Stresses in the Ground and Design of Earth-retaining Structures, Cornell*. ASCE, New York, 1–49.
6. BURLAND J. B. Discussion (Session 4—Drilling muds). *Grouts and drilling muds in engineering practice*. Butterworths, London, 1963, 223.
7. FARMER *et al.* The effect of bentonite on the skin friction of cast in place piles. *Behaviour of piles*. Instn Civ. Engrs, London, 1970, 115–120.
8. FERNANDEZ-RENAU L. Discussion (Session 6, Div. 4—Deep foundations). *Proc. 6th Int. Conf. Soil Mech., Montreal*, 1965, **3**, 495.
9. CHADEISSON R. Influence of the boring methods on the behaviour of cast-in-place bored piles. *Proc. 5th Int. Conf. Soil Mech., Paris*, 1961, **2**, 27.
10. SOLETANCHE ENTREPRISE. *Essai de barrette à 400 tons.* 1972. Unpublished.
11. COLE K. W. and BURLAND J. B. Observations of retaining wall movements associated with a large excavation. *Proc. 5th European Conf. Soil Mech. Foundn ¹Engng*. Sociedad Espanola de Mec. del Suelo y Cimentaciones, Madrid, 1972, **1**, 445.

Model performance of an unreinforced diaphragm wall

W. J. RIGDEN, MSc, DIC, MICE, *Lecturer in Engineering, Simon Engineering Laboratories, University of Manchester*

P. W. ROWE, DSc, FICE, *Professor of Soil Mechanics, Simon Engineering Laboratories, University of Manchester*

A circular diaphragm wall, designed as a thin shell in unreinforced concrete, was studied by building 1:80 scale models of the proposed wall and ground conditions and carrying out a series of tests in the Simon Engineering Laboratories' centrifuge. The tests showed that tensile stresses can be largely eliminated by reducing the depth of penetration below formation and the degree of fixity. Any portion of a wall below formation which may be required as a cut-off could be formed with bentonite clay slurry.

INTRODUCTION

The design proposed by Nederhorst Grondtechniek for a multi-storey underground car park in Amsterdam incorporated a circular diaphragm wall 0·8 m thick. This wall was designed as a thin shell and was to be constructed in unreinforced concrete, using conventional diaphragm walling techniques (Fig. 1). Scale 1:80 models of the wall were subjected to the field stress distribution in the 500 g tonne centrifuge at the Simon Engineering Laboratories. Observations were made of the strain distribution in the wall, the overall factor of safety against collapse and the mode of failure. Particular tests were designed to investigate the influence of non-circular construction, fixity below the dredge level and non-uniform loading applied to the ground immediately behind the wall.

MIX DESIGN AND CASTING OF PANELS

2. A maximum aggregate size of 5 mm, an aggregate/cement ratio of $2\frac{1}{2}:1$ and a water/cement ratio of 0·63 were selected to give a workable mix resulting in a dense microconcrete of the specified tensile and compressive strengths, namely 6 N/mm² and 30 N/mm².

3. The steel formwork utilized two radii of 338·1 mm and 328·1 mm separated by silver steel rods, 10 mm dia., which acted as stop ends. The concrete was compacted in the formwork by rodding and vibration. Panel widths of 56 mm were used. The panel thickness varied from 9·95 to 10·15 mm, average 10·00 mm.

ASSEMBLY AND INSTRUMENTATION OF MODEL

4. Five models were constructed, one retaining dry sand, one retaining undisturbed clay, two retaining the same clay remoulded, and one retaining water. A steel former was used to place the soil. The panels were then positioned vertical and tight against the soil.

5. When all the panels were in position they were clamped and grouted. The clamps were then removed and the drainage blankets, perforated tubes and Mersey River sand were placed in position (Figs 2 and 3).

6. Electrical strain gauges were used to measure horizontal and vertical bending and circumferential strain. Two of the precast panels were selected for each type of strain measurement, and gauges were attached to form a full bridge at levels 130, 170, 210, 250 and 290 mm from the top of each panel. Details of the methods adopted to install the gauges and protect them against the soil and water pressure are given by Rigden and Allen.[1]

7. Two calibrations were carried out for each bridge. A theoretical calibration was made using the manufacturer's gauge factor, gauge resistance and the measured elastic modulus of the concrete of 28 kN/mm². An experimental calibration was made by putting the panel under load and measuring the output from the bridge. This agreed with the theoretical strain predicted for that load to within 1%.

8. Displacements of the wall were monitored using linear voltage displacement transducers. At the time of the tests a waterproofed version was not available and hence only displacements at the top of the wall could be measured.

9. Pore water pressures were monitored by placing a porous tip in the soil and connecting this via a small diameter tube to a pressure transducer.

10. The formation of cracks in the concrete panel was detected by painting a thin line on the concrete using Dag (dispersion 915 silver in MIBK) electrically conducting paint. This was connected into an electrical circuit and the output monitored.

THE CENTRIFUGE TESTS

Model 1 test 1

11. Model 1 retained dry sand and was designed to test the experimental procedures. The model stood intact without visible deformation to a scale factor of 135 g, namely an overload scale factor of 1·69, when the maximum direct compressive stress was 3 N/mm² and the vertical bending stress was 0·4 N/mm².

Model 2 test 1

12. Model 2 (Fig. 2) retained the undisturbed sample of organic silty clay from a site typical of that at the future underground car park. The properties of the clay were

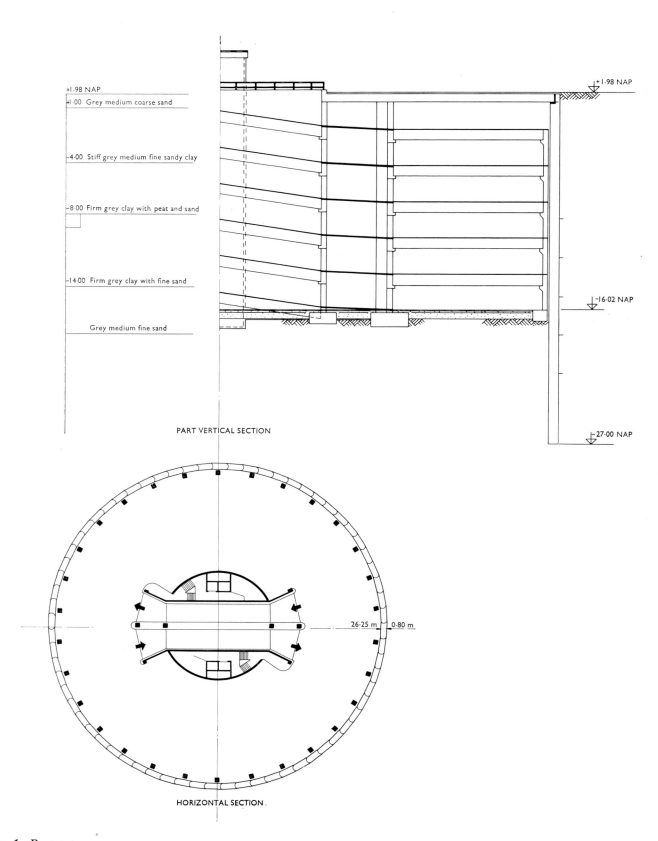

+1·98 NAP

+1·00 Grey medium coarse sand

−4·00 Stiff grey medium fine sandy clay

−8·00 Firm grey clay with peat and sand

−14·00 Firm grey clay with fine sand

Grey medium fine sand

+1·98 NAP

−16·02 NAP

−27·00 NAP

PART VERTICAL SECTION

26·25 m   0·80 m

HORIZONTAL SECTION.

Fig. 1. Prototype garage

$c' = 0$, $\phi' = 31°$, $\rho_s = 1.8$ t/m³. The sand was placed in a dense state $\phi' = 40°$, $\rho_s = 2.06$ t/m³. The coefficient of consolidation under 120 kN/m² was 430 m²/year for vertical flow measured in a 250 mm dia. consolidation cell. The model was accelerated to 80 g over a period of 100 min, and maintained at 80 g for 2 h, when the output from the bridges had become constant. Typical stresses are shown in Fig. 4. A crack was found in the wall at or just above the dredge level. The concrete above the crack had moved forward by some 2 mm relative to the concrete below, and was tilted back into the clay.

Model 2 results

13. The significant stress was that due to vertical bending (Fig. 4). While the model was maintained at 80 g the maximum stress increased to just above 6 N/mm², the limiting value in tension, at a level which agreed with that of the tension crack.

14. A criticism of this test is that an excess pore water pressure occurred at the centre of the clay layer because the differential pressure on the wall was applied during increasing total stress. In contrast, the construction sequence of the prototype maintains the vertical stress constant whilst the lateral stress decreases with excavation. The maximum excess water head recorded was equivalent to 70% of the effective weight of the soil above the tip. Without this excess head the evidence is that the wall would have remained undamaged at a factor of safety against cracking of about 1·35. With outward yield of the wall on crack formation the excess pore pressure decreased rapidly to zero.

Model 2 test 2

15. The cracks in the wall were repaired with fibreglass filler and the model was accelerated until collapse occurred at 134 g, namely with an overload factor of 1·67.

Model 3 test 1

16. Model 3 was designed to minimize the excess water pressure recorded in model 2 and also to investigate the effect of non-circular construction. One diameter of the model was 5 mm larger than the design figure of 676 mm, whilst a second diameter, perpendicular to the first, was 5 mm smaller. This corresponds to an error of ± 400 mm in the prototype.

17. The model was assembled using the same clay as model 2. The clay had been remoulded at twice the liquid limit and consolidated under a final pressure of 140 kN/m² to the field undrained shear strength in a number of 650 mm square cells. The clay was placed in three layers with two intermediate 10 mm thick sand blankets. The sand blankets and lower sand stratum were interconnected with a number of porous plastic drains (Fig. 3). The effective strength parameters remained at $c' = 0$, $\phi' = 31°$, with a saturated density of 1·8 t/m³, so that the long term earth pressures should be similar to those for model 2.

18. On starting the centrifuge water flowed into the centre of the model and only the reduced excess pore water pressure and the lateral earth pressure exerted any net effect on the wall. The model was then accelerated to 80 g and, when the excess water pressure had dissipated, the water in the middle of the model was drained to the dredge level. At this stage the wall was withstanding an

external pressure due to the soil and static groundwater levels, simulating the field construction sequence.

19. After a further 10 min, the centrifuge was accelerated up to 102 g and then the machine was stopped to allow an examination of the wall. No cracks or signs of distress could be seen.

Model 3 test 2

20. The centrifuge was accelerated continuously to 80 g and, after a pause of 8 min, on up to 147 g without flooding the centre of the model. Some swelling of the clay occurred during the pause, indicated by the excess pore pressure recorded in this test.

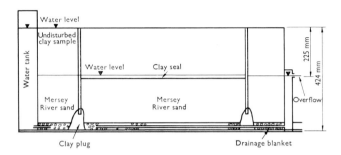

Fig. 2. Model 2: vertical section

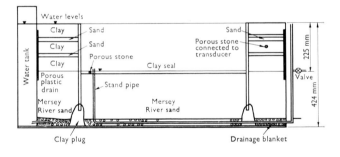

Fig. 3. Model 3: vertical section

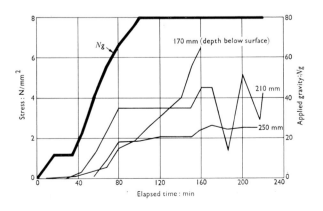

Fig. 4. Model 2 test 1: vertical bending stress/time

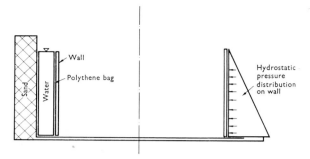

Fig. 5. Model 4: vertical section

Model 3 results

21. The vertical bending stresses measured in model 3 test 1 at the field scale of 80 g near the dredge level varied between 4 and 5 N/mm². At an overload factor of 1·3 (102 g) and with an excess pore water pressure of some 7·5 kN/m² the maximum value rose to just over 5 N/mm².

22. Model 3 in test 2 was undamaged at a scale factor of 147 (an overload of 1·84) when the maximum bending stress on any panel was 5·9 N/mm² and the excess pore water pressure varied between 34 and 49 kN/m², representing an excess head of some 4 m in the field. The bending stress was not as high as would have been expected on the basis of the values at 80 g. An overall settlement of some 10–15 mm occurred during the two tests.

Model 4 test 1

23. The undamaged wall from model 3 was excavated and retested under a full water head of 362 mm on the outside and with no support on the inside (Fig. 5).

24. The model was mounted with a polythene bag inflated by a small excess air pressure. The centrifuge was accelerated up to 30 g and valves were opened to allow water to flow into the polythene bag and the compressed air to flow out. At this stage the centrifuge factor N was large compared to earth's gravity g and the concrete was placed in compression. The model was then accelerated up to 135 g when the test was terminated. Examination of the wall showed cracks at the crown and at the invert running from top to bottom of the wall. These occurred as the machine came to a stop due to the non-uniformity of the water pressure under the transverse earth gravity.

Model 4 test 1 results

25. The highest bending stress measured at 135 g was 0·8 N/mm² at the 250 mm level. Thus the high vertical bending stresses were due to the embedment of the wall in the dense sand stratum.

Model 5 test 1

26. Model 5 (Figs 6 and 7) was similar to the model 3, but with the addition of two surface water tank loads diametrically opposite one another and each acting over one eighth of the model perimeter.

27. The test procedure was identical to that adopted for the third model until 80 g was reached, and then 2 t/m² was applied first on surface A and then B, followed by 4 t/m² on A and B respectively. The model was finally accelerated to 135 g.

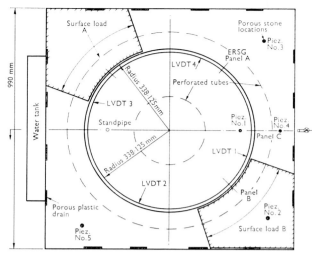

LVDT Linear voltage displacement transducer
ERSG Electrical resistance strain gauge

Fig. 6. Model 5 plan

Fig. 7. Model 5 ready for testing

Model 5 test 1 results

28. The maximum vertical stress recorded at 80 g before any surface load was applied was between 3·8 and 4·5 N/mm² at a depth of 210 mm (Fig. 8). This indicates a minimum factor of safety for this model of 1·33 on the nominal tensile strength of 6 N/mm².

29. The addition of the surface load of 2 t/m² had little effect on the stresses measured at or below the dredge level but caused a marked increase in the stresses recorded at 130 mm and 170 mm. However, at 80 g with two surface loads of 2 t/m² the maximum recorded stress was 4·75 N/mm² at 210 mm. Thus the reduction in factor of safety was from 1·33 to 1·26.

30. At 80 g with 4 t/m² on the surface the maximum recorded stresses were 5·1–5·5 N/mm².

31. When the model was accelerated above 80 g, with both surfaces loaded, certain strain gauge bridges failed when recording a stress between 5 and 6 N/mm² at an acceleration between 85 and 110 g. This is equivalent to local surface pressures of 4·25–5·5 t/m² and an overload on g of 1·08–1·38. The model stood at 134 g.

32. The model was extensively cracked (Fig. 9) although structurally sound.

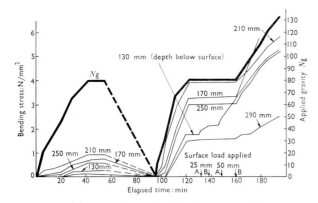

Fig. 8. Model 5 test 1: vertical bending stress/time

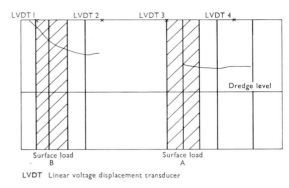

Fig. 9. Model 5 test 1: crack distribution after test

CONCLUSIONS

33. The tests indicated that an unreinforced wall of thickness 0·8 m and free height 18 m, penetrating 11 m into a dense sand, would have stood with a factor of safety of 1·35 against tensile bending failure. Under a local surface stress of 2 t/m² acting over one eighth of the perimeter the factor would have been 1·26. The factor of safety against complete collapse exceeded 1·5. This would be satisfactory for the temporary construction condition.

34. If such a wall were formed with a flexible bentonite clay fill through the sand and with the concrete commencing just below excavation level, the fixity restraint would be removed and the factor of safety against cracking under vertical bending would be raised considerably.

35. For amenity reasons the prototype wall was not built and no field comparison is possible in this particular case. However the use of representative model tests can lead to economy in the use of steel reinforcement and can guide subsequent field control measurements.

ACKNOWLEDGEMENTS

36. The concept of an unreinforced diaphragm wall was due to ir K. F. Brons and the Authors are indebted to him and to Nederhorst Grondtechniek for permission to publish work conducted on their behalf.

REFERENCE

1. RIGDEN W. J. and ALLEN J. H. Some problems encountered in testing models in a large centrifuge. *Strain*, 1973, **9**, No. 3, 119–121.

Discussion on Papers 5–9

Reported by R. W. COOKE

The session Chairman, Mr P. J. C. Buckley, said that Papers 5–9 gave detailed consideration to various design criteria and bentonite specifications currently in use. Two of the Papers dealt with the practical applications of these proposals. Mr Buckley complimented the Authors on some excellent research work and on the frankness of their conclusions and recommendations for site practice. However, he expressed caution that contractors' assertions should be carefully tempered by good practical experience and field research, and both must be complementary in achieving an economic and successful end product.

2. Mr Buckley suggested that current descriptions of the bentonite techniques, which tended to encourage an image of uncontrollable and dirty sites, were misleading. Instead of the phrase 'bentonite mud', he proposed the use of 'bentonite suspensions'.

3. In the two case histories presented, attention was drawn to the environmental acceptability of diaphragm walling techniques, the noise created being minimal in comparison with other methods of construction. Perhaps one of the most interesting points made was that skin friction in diaphragm walling is not reduced to any great extent by the use of bentonite suspensions. This reinforced considered opinion and the results of research carried out within the piling industry.

AUTHORS' INTRODUCTIONS

4. Introducing Paper 5, Mr M. T. Hutchinson said that the objective of the work outlined was to draw up a broad specification for the properties of slurries as used in the diaphragm wall trench. It had become obvious at an early stage in the work that there were marked differences between drilled pile and excavation processes which affected the optimum slurry properties. Of these the most relevant to the excavation process was the formation of the filter cake. Mr Hutchinson described the various mechanisms of filter cake formation: surface filtration in the case of porous rocks and fine sands, deep filtration for more open sands with permeabilities of $10^{-2} – 10^{-1}$ cm/s, and rheological blocking for sands and gravels with permeabilities greater than 10^{-1} cm/s. He said that even in fine soils it was best to keep the concentration above $4\frac{1}{2}\%$, since at smaller concentrations the time for initial filter cake formation rose rapidly.

5. In laboratory tests with fine gravel of permeability 0·3 cm/s, Mr Hutchinson had not found it possible to form a filter cake with a 5% slurry. However, the addition of about 1% of fine sand effected a considerable improvement. The normal filter press test used in drilling mud practice did not demonstrate this improvement and Mr Hutchinson said that an alternative to the filter paper used in the test should be sought.

6. The proposed specification referred to bentonite actually in the trench.

7. Mr B. J. Jack explained that Paper 6 gave an empirical design approach for multi-tied walls, using procedures similar to those for single-tied walls, and described model tests to measure the flexibility characteristics of these walls. He said that normally insufficient soil data was available for more sophisticated designs than the free earth support method for single ties and the trapezoidal method when more than one support was required. Problems arose when estimation of the envelope of pressures became a matter of judgement because of the variability of the soil strata. Mr Jack described the processes and computations for determining the tie forces from the free earth pressure diagram at each stage of excavation. An example of the bending moment envelope obtained from these forces and pressures was given in Fig. 17 of Paper 15 where comparison was made with field observations. Comparisons had also been made with various well-known design approaches and Mr Jack concluded that the new method might provide a simple and consistent design tool.

8. Mr Hodgson, the Author of Paper 7, emphasized that the testing of the bentonite mud was carried out cheaply and efficiently on the site and a high standard of finished wall was achieved.

9. With regard to costs, Mr Hodgson said that diaphragm walls were rarely cheaper than sheet piling combined with in situ reinforced concrete retaining walls on pricing quantities only. However, the installation time for diaphragm walls was much shorter even for comparatively shallow walls, provided the job was extensive enough, and this could save a great deal of money. The time saving could be up to 50%.

10. Mr Hodgson then outlined the first phase of the construction of a 20 m deep basement having walls 1 m thick, in Victoria Street adjacent to the scheme described in the Paper. Ground anchors could not be used and the walls were propped with a waling slab and by passive pressures at the toe. Because of services in the surrounding ground it was particularly important to prevent movement of the top of the wall. The nett loading was to be low and a raft foundation was provided rather than piling. To support the waling slab during excavation, 1·5 m dia. steel casings were augered into the ground and an under-reamed footing formed at the base, as shown in Fig. 1. Concrete was poured to the level of the underside of the raft, and steel stanchions were grouted to the top of the foundation. The waling slab was cast on the ground surface with the top of the steel column cast in to provide support during the excavation processes as illustrated in Fig. 1. Figure 2 shows the completed excavation with the raft concreted to incorporate the steel stanchions forming part of the permanent structure.

11. Construction of the upper floors proceeded as the

excavation was formed and a total time saving of approximately three months was effected by this method of working.

12. Paper 8 on a load bearing diaphragm wall at Kensington and Chelsea Town Hall was introduced by Mr B. O. Corbett who stressed the significance of the comparison, shown in the Paper, between the calculated bearing capacity of a diaphragm wall panel and the result of a constant rate of penetration loading test on it. Mr

Corbett pointed out that assessment of the bearing capacity in terms of undrained strength was unrepresentative of the long term behaviour of the wall after excavation because of the effective stress changes in the soil behind the wall.

13. Mr Corbett then referred to the horizontal displacements measured at the top of the wall at the Kensington and Chelsea Town Hall site and compared these (Table 1) with similar observations at Britannic House

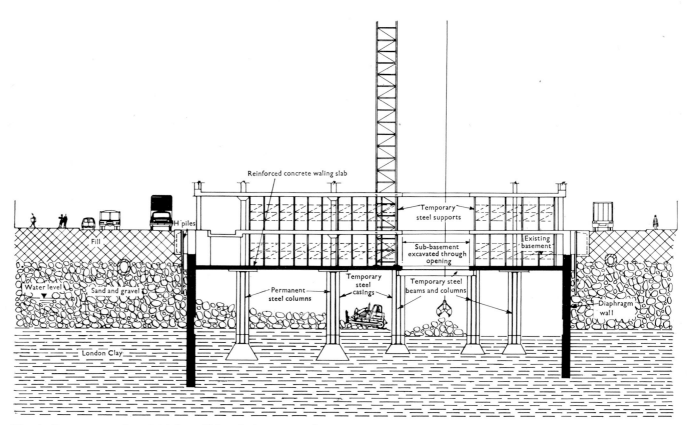

Fig. 1. Basement section: initial condition during excavation

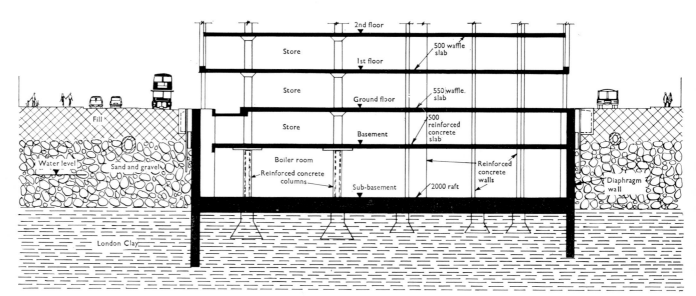

Fig. 2. Basement section: final condition

reported by Cole and Burland.[1] He showed that

$$\text{ratio of squares of depths} = \left(\frac{20}{13}\right)^2 = 2\cdot37$$

$$\text{and ratio of movements} = \frac{61}{26} = 2\cdot35$$

and suggested that there might be an empirical relationship between horizontal movement and the square of the excavation depth.

14. Mr Rigden, whose Paper with Dr P. W. Rowe described centrifuge tests on 1 : 80 scale models of an underground car park having an unreinforced concrete, circular diaphragm wall, said that their task had been to investigate the wall stresses and to suggest methods of reducing them. Any circular drum-shaped structure subjected to external loading without discontinuities would be expected to deform in compression only, so that theoretically no reinforcement would be necessary. However the diaphragm wall for the car park was designed to extend well below the excavation level to provide a cut-off in deep clay layers. The development of internal passive pressures in this case would cause discontinuities of loading at the level of the excavation floor, resulting in bending stresses about horizontal axes. The cracking due to this loading (model 2 test 1) is shown in Fig. 3. To investigate the influence of the depth of the diaphragm below excavation floor level, one of the models was tested with the wall resting on a thin soft clay layer, there being no soil inside the wall. When water pressure to the full head was applied to the outside of the wall no damage occurred and only low bending stresses, due to some fixity provided by the clay, were measured. From these test results Mr Rigden suggested that reinforcement could probably be omitted from circular concrete walls of this type if the walls were terminated close to excavation level.

GENERAL DISCUSSION

Test apparatus

15. The discussion was opened by Professor Veder who took up the point made by Mr Hutchinson and his colleagues on the necessity of using a fluid loss apparatus with either a sand bed or a sintered metal base, and lower pressures. He suggested apparatus (Fig. 4) which simulated the site conditions as closely as possible. He also suggested the use of apparatus which simulated the displacement of bentonite cake when the concrete was poured, as shown in Fig. 5.

Structure of bentonite muds

16. Dr A. N. James said that it was necessary to understand the chemical and physical structure of bentonite muds. He explained that bentonite is not a unique material, but one with a range of properties. The name comes from a particular clay mineral found in the Fort Benton area of Wyoming. Bentonite is a natural sodium montmorillonite, but the term is now used to describe a material which contains a high proportion of montmorillonite artificially converted to the sodium form—as in the UK.

17. Dr James said that montmorillonite crystals were like minute playing cards, composed mainly of the atoms aluminium, silicon and oxygen. The atoms were assembled in such a way that the surfaces of the flat crystals were negatively charged, and the edges positively charged.

Table 1. Comparison of displacements

	Britannic House[1]	Kensington and Chelsea Town Hall
Depth, m	20	13
Horizontal movement, mm	61	26

Fig. 3. Cracks in model

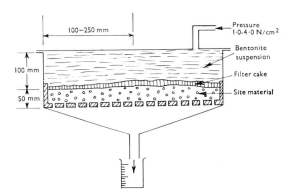

Fig. 4. Proposed low pressure fluid loss apparatus

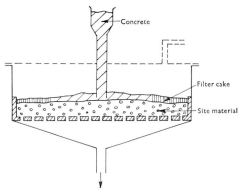

Fig. 5. Apparatus to simulate displacement of filter cake by concrete

Water was absorbed in thick layers on the flat surfaces only if the negative charges were allied to singly charged positive counter-ions, like sodium. Calcium, a doubly charged ion, would not do. He emphasized that the shape of the crystals, the electrical charge density and the counter-ions varied to some extent with the geological source of the mineral and that the thixotropic and water absorption properties of the 'bentonite' depended to a large extent on these factors. By joining their positive edges to their negative surfaces as shown in Fig. 6, the flat crystals sheathed in molecular layers of water were able to form a gel structure capable of behaving in a thixotropic manner. Dr James went on to show how the bonds between the edges and surfaces could be easily broken, resulting in the familiar thinning of muds on stirring. The separate crystals were always subject to some

agitation (Brownian motion) and in time they shuffled back into position so that the edges and surfaces re-joined.

18. Dr James anticipated that the properties required for the support of excavations would be found in the long chains of synthetic polymers already finding favour as drilling fluids.

Lateral movements predicted and observed

19. Dr J. B. Burland outlined the construction of the underground car park at the Houses of Parliament. After the diaphragm retaining walls, the bored piled foundations and the columns had been installed, the ground floor slab was cast and excavation for the basements was carried out as a mining operation. Further slabs were cast as excavation proceeded. The Building Research Station advised on the design and observed earth pressures and movements during excavation. These measurements were compared with a step by step finite element analysis simulating the construction process. The soil properties used in this analysis were based on the results obtained by back-analysing the movements of the retaining wall of another large basement in the City of London.[1]

20. Predicted and observed wall displacements are shown in Figs 7 and 8. They indicate the importance of inward movement of the wall at depth, due as much to release of vertical stress within the excavation as to release of horizontal stress. Dr Burland said that the design method proposed by James and Jack in Paper 6 could not take account of the deep seated movements likely to occur when excavating on deep clay sites.

21. Mr Jack replied that the Authors themselves would adopt a finite element analysis if sufficient data were available. The equivalent spring analysis is not intended as a substitute for finite element analysis when sufficient soil parameters are known, but as an effective method of determining flexibility coefficients as used by Rowe in single-tie wall analysis. It is envisaged that the engineer would determine flexibility coefficients for a single soil stratum of known properties and a particular wall geo-

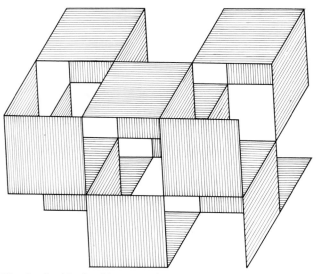

Fig. 6. An ideal cubic network of montmorillonite platelets

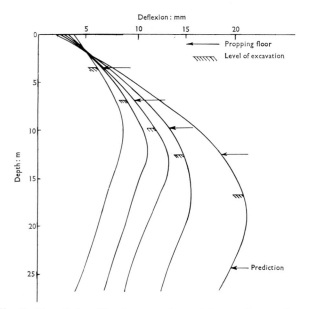

Fig. 7. New Palace Yard car park retaining walls: predicted movements at various stages of excavation

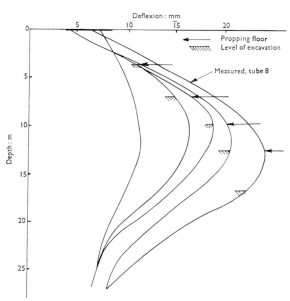

Fig. 8. New Palace Yard car park retaining walls: observed movements at various stages of excavation

metry, from which a reasonable assessment could be made as to the performance of an actual wall under consideration.

22. Lateral movement of a wall can be reduced by using a conservative penetration as results from the empirical method proposed in Paper 6. In the general multitie analysis the penetrations have been shown by site measurements (Paper 15) to produce tolerable settlements behind the wall, and these have been shown to be directly proportional to lateral movement.

Earth pressure

23. Some problems in the design of complex retaining walls without ground anchors, to be constructed for a roadwork scheme close to existing property on London Clay, were described by Mr M. V. Woolley. The most difficult section required counterforts, a prestressed toe and ground beam props (Fig. 9) to limit the movement at the top of the walls which, for obvious reasons, could not be strutted.

24. Perspex model walls in china clay were tested to demonstrate the effectiveness of the various aspects of the design and a finite element analysis was carried out to estimate the wall movement. The most critical design parameter was the value of K_0, and in situ tests were being made with the Cambridge pressure meter to determine the variation of K_0 with time.

25. In a written contribution, Mr J. L. Hislam stated that the main point to emerge from the model tests described in Paper 6 was the form of the lateral pressure profile behind a flexible wall. He said that a trapezoidal design pressure distribution was to be preferred because it tended to equalize strut loads and wall bending moments. Because of the relative rigidities of the two types of wall, Mr Hislam suggested that a trapezoidal design pressure distribution be used for steel sheet piling and a triangular distribution for diaphragm walls. However, for very deep diaphragm walls, trapezoidal pressure distributions might be applicable.

26. In the model tests the action of prestressing the struts reduced the wall movements but led, as expected, to higher strut forces in the early stages of excavation. As excavation progressed, the total strut loads became similar to those observed in equivalent tests without prestressing. Thus prestressing in the early stages was shown to be undesirable since it created a zone of passive pressure in the clay which, for limited heights, was capable of standing unsupported. Mr Hislam's contribution suggested that in a scheme having, say, six strut (or anchor) positions over the excavation depth, the two top supports should not be prestressed.

Scanning electron microscope

27. A written contribution from Mr W. M. Braun was concerned with the use of the scanning electron microscope for recognizing flaws and weaknesses in construction materials and soil and rock samples. Mr Braun's note suggested that by using instant freezing techniques on wet samples it was possible to investigate the influence of water on systems of micron size particles. He had used a Polaroid land camera with the Stereoscan 180 microscope manufactured by the Cambridge Instrument Company. This instrument had shown how a bentonite slurry added to cement and a concrete retarder in a trench

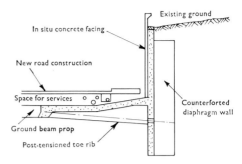

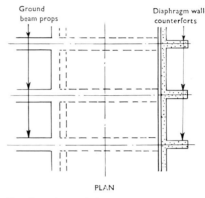

Fig. 9. Roadwork retaining wall scheme

slowly transformed into an impermeable plastic membrane forming a cut-off wall of adequate strength.

Bending moment diagrams

28. Mr I. B. Mackintosh sent a written contribution which raised doubts on the validity of the methods outlined in Papers 6 and 15 for deriving safe bending moment diagrams for strutted diaphragm walls. He pointed out that the variation between the strut loads shown in Fig. 6 of Paper 6 was fairly large. While the prediction method described gave reasonable agreement with the average observed loads in that figure, the structural elements had to be designed for the largest loads and the largest bending moments which could occur. Variations from the average therefore had a significant effect. Mr Mackintosh also questioned the accuracy with which bending moment diagrams could be derived from inclinometer readings and the extent to which concrete creep during the loading period could be taken account of. Finally Mr Mackintosh sought more data from practising engineers on methods of measuring strains in reinforcing bars and on actual strut and anchor loads in relation to the presumed pressure distributions used in design.

Observed deflexions and settlements

29. In Tables 4 and 5 of Paper 7, horizontal movements of the diaphragm wall at Victoria Street were shown. Corresponding observations of the settlement at street level close to the wall were presented by Mr M. J. Tomlinson in a written contribution. In Table 2 these are presented together with Mr Hodgson's data.

30. In contrast to most published data, the settlements were substantially larger than the inward movements of the wall. Mr Tomlinson suggested that this might be due to the fact that the wall movements were measured

2·9 m below street level. The old basement walls and the temporary H piles (shown in Fig. 5 of Paper 7) had deflected during construction of the guide trenches. Due to cantilever action the H piles could have magnified at street level the inward movement of the diaphragm wall. In Mr Tomlinson's opinion, larger wall movements might have been recorded if the observations had been continued a further six months or so. He also suggested that drying shrinkage of the concrete in the basement floors and compression under load would reduce the tendency for the floors to arrest completely the wall movements. Tables 4 and 5 of Paper 7 showed some recovery of wall movement following stressing of the anchors. The settlement measurements reflected this but not to the extent of complete recovery.

31. Mr Hodgson's comments were that the horizontal movements at street level, even if projected from 2·9 m below, would not have approached anywhere near the order of magnitude of the vertical movements recorded by Mr Tomlinson. Neither loading of the waling slab nor shrinkage caused any significant shortening. Additional spot check measurements were taken during February 1974 but showed no change.

SUMMARY

32. While workable design and construction approaches exist, Papers 5–9 and the consequent discussion demonstrate the need for further development and research. Evidence is accumulating, for instance, to suggest that the use of bentonite during excavation has no detrimental effect on the adhesion between diaphragm walls and the adjacent soil. However, this subject has not yet been systematically studied and a series of carefully designed full-scale or large model experiments, such as those made on large bored piles in the early 1960s, could settle the issue fairly quickly. The length of the panel tested by Corbett, Davies and Langford (Paper 8) was rather small in relation to its thickness for too much confidence to be placed on this single result, but it is interesting to note that the settlement observed at a typical working load in the test was approximately 0·3% of the thickness of the panel. In loading tests on friction piles the settlement has frequently been found to be this proportion of the diameter at the working load.

33. Even greater uncertainty is apparent in the distribution of horizontal wall pressures and the corresponding loading in struts and anchors. Very much more work will be necessary both in the laboratory and by way of instrumented construction on site before the simplifications necessary for design can be accepted widely. The power of finite element techniques, and particularly step by step approaches simulating the excavation process, has already been demonstrated for predicting strut forces and wall displacements. The centrifuge is another invaluable tool for body force problems. The use of both of these for studies of specific cases is likely to increase. It is unfortunate that there was no discussion of Mr Rigden's suggestion (Paper 9) that reinforcement might not be vital in simply loaded circular walls, as the potential cost saving could be considerable.

34. On the subject of construction costs, it is becoming increasingly accepted that the opportunity presented by

Table 2. Deflexions and settlements

Construction stage	Date	Deflexion at top of wall, *mm*	Settlement at street level, *mm*
Point 16, Victoria Street			
Diaphragm wall construction	Dec.'71	0	12
1st stage excavation to 5·5 m	Feb. '72	+1·5	21 (after completion of excavation)
Installation of ground anchors at 4·7 m	Feb.– Mar. '72	0	20 (after completion of anchors)
2nd stage excavation to 7·7 m	Aug. '72	+3·0	17 (after completion of excavation)
2nd basement slab completed	Oct. '72	+3·0	15
Last reading	May '73	—	17
Point 3, Howick Place			
1st stage excavation to ground anchor level	Feb. '72	0	15
2nd stage excavation to 5·5 m at second basement level	Apr. '72	+3·0	28
2nd basement slab constructed at 7·9 m	Nov.– Dec. '72	0	17
Excavation to clay anchor level at 11·3 m	Jan. '73	0	17
Excavation to 3rd basement at 13·9 m	Feb. '73	0	18
Last reading	Nov. '73	—	20

the diaphragm wall technique of building the superstructure as the excavation and foundations are formed can substantially speed up the total construction time. However, if full advantage is to be taken of this facility, complete confidence in the materials and workmanship is essential. Methods of continuously monitoring the quality and properties of the slurry in the trench need to be available and there appears to be scope for considerable development in this direction.

CONTRIBUTORS

Mr W. M. Braun, International Construction.
Mr P. J. C. Buckley, Cementation Piling and Foundations Ltd.
Dr J. B. Burland, Building Research Establishment.
Mr R. W. Cooke, Building Research Establishment.
Mr J. L. Hislam, Cementation Ground Engineering Ltd.
Dr A. N. James, Cementation Research Ltd.
Mr I. B. Mackintosh, Balfour Beatty Construction Ltd.
Mr M. J. Tomlinson, Wimpey Laboratories Ltd.
Mr M. V. Woolley, E. W. H. Gifford and Partners.
Professor C. Veder, Technische Hochschule, Graz, Austria.

REFERENCE

1. COLE K. W. and BURLAND J. B. Observations of retaining wall movements associated with a large excavation. *Proc. 5th European Conf. Soil Mechanics and Foundation Engineering.* Sociedad Espanola de Mec. del Suelo y Cimentaciones, Madrid, 1972, **1**, 445.

Some practical aspects of diaphragm wall construction

M. FUCHSBERGER, DiplIng, MSc, *General Manager, ICOS (Great Britain) Ltd*

Now that fluid stabilization of the soil during the trenching process in diaphragm wall construction is being more closely investigated and understood, some practical aspects of the construction technique which contribute to the success of this method of construction need to be emphasized. This Paper deals with the basic requirements of the trenching tools and tolerances, and it refers to the experience gained in using bentonite in certain subsoils including soils with tidal and saline groundwater. It sets out basic requirements of the tremie concrete and refers to the method of jointing individual wall sections or bays, pointing out difficulties which may arise in certain stratified subsoils. Reference is also made to the practical arrangement of steel reinforcement and boxwork inserts, and finally specific areas of future investigation are suggested which could lead to further improvements in the execution and understanding of this relatively new construction technique.

PRINCIPLES OF THE TECHNIQUE

The description diaphragm wall is understood to mean an artificial membrane of finite thickness and depth constructed in the ground by means of a process of trenching with the aid of a fluid support.

2. The construction is carried out from the ground surface by means of mechanical devices of various alternative kinds which permit the progressive excavation of a relatively narrow trench in the ground in such a way that the stabilizing fluid is introduced simultaneously as the trenching operation proceeds. On completion the trench is filled with concrete or other backfill material, thereby displacing the trench-supporting fluid from the bottom upwards. Alternatively, the supporting fluid may be stabilized in situ.

3. The supporting fluid most generally used is a suspension of bentonite, a thixotropic clay, either in its natural form or after processing to enhance its special characteristics. A bentonite suspension has the ability to form a membrane of low permeability at the soil/liquid interface. The membrane allows development of the full hydrostatic pressure of the fluid against the sides of the trench and this, together with limited penetration of the bentonite into the soil, is generally accepted as the predominant factor contributing to the stabilization of the trench.[1-5] The degree of penetration into the soil and thickness of the membrane depend on the excess hydrostatic head, the viscosity and gelling properties of the thixotropic clay suspension and the time allowed for it to build up.[3,5]

4. One side of the trench may have a different degree of stability from the other if external forces and boundary conditions are different for each side.

5. Practical considerations, principally the continuity of supply of concrete or other backfill material, determine the maximum length of trench that may be excavated at any one time, and hence the diaphragm wall eventually constructed is necessarily formed of individual sections or bays, commonly described as panels.

TRENCHING METHOD, TOOLS AND TOLERANCES

6. In theory, any technique for achieving a vertical cut in the soil can be used for the trenching operation. The basic requirement, however, is minimum disturbance of the soil at the cutting face, combined with a cutting or trenching rate slow enough to permit the build-up and maintenance of the membrane or filter cake at the soil/bentonite interface.

7. The type of subsoil, the groundwater, the properties of the bentonite suspension, and the characteristics of the trenching equipment are all important factors that must be taken into consideration in determination of the trenching method.

8. Three basic types of trenching tool are commonly used, either individually or in combination, in average soils. These are percussive, rotary and excavating tools. Only percussive and rotary tools are effective in rock. It is important to select a moderate and controllable cutting speed, with an appropriate rate for the removal of cuttings, in order to avoid violent disturbances at the cutting face and excessive turbulence in the bentonite suspension, both of which can cause localized collapses of the sides of the trench, known as cavitation.

9. Percussive and rotary tools loosen and break the soil into relatively small particles and mix the cuttings with the bentonite suspension at the cutting face. The suspension, laden with soil cuttings, is then transported to ground level by either direct or reverse circulation of the drilling mud. These tools advance in situ and remain at the cutting face until excavation is complete.

10. Excavating tools such as the auger, bucket, shovel, clamshell grab etc. cut the soil in bulk and have to be brought up above ground level for discharge.

11. The grabbing tool operated on either a rope or a kelly bar seems to fulfil the basic requirements outlined above and is the most successful and most commonly used tool for the trenching operation. The success of the grabbing tool may also be attributed to its great efficiency as a tool for bulk excavation in average soil, and in particular to its effective and easily controllable shearing operation when cutting the soil. Combined with a suitable lifting and lowering rate, this avoids undue disturbance at the soil cutting face and in the supporting bentonite.

12. Three principal trenching tolerances have to be considered: deviations from the true vertical and true horizontal alignments, and local deviations from the average trench face. The incidence of these deviations depends on the characteristics of the trenching tool, its shape, size and weight, its support and control mechanism, and also on the type of subsoil encountered.

13. As only vertical trenches can be cut effectively it is logical that the cutting tool should be primarily controlled for verticality by gravity. In this context a heavy tool performs better than a light one and the system of suspension of the tool from the winching rig should preferably utilize the continuous influence of gravity during the trenching process. A repeated lifting and lowering of the tool under gravity has in fact a rectifying influence on deviations from verticality. A check on verticality with increasing depth may be required more frequently for a rigid suspension than for a digging tool suspended freely.

14. Here again the grabbing tool, with its repetitive raising and lowering from its point of suspension, seems to have an advantage over tools advancing in situ, which may require the assistance of a feed-back device, as their verticality of advance may be significantly influenced by varying soil strata of differing penetrability, especially if not horizontal.

15. Horizontal deviations are mainly caused by rotation of the trenching tool about its axis of suspension, and in this respect a big tool with rigid suspension has an advantage over a small one unless special guides or restraints are provided to minimize horizontal rotation of the tool during the trenching operation. If such restraints are not provided a wavy alignment of the trench may result, this tendency increasing with depth.

16. Local deviations of the trench face are closely related to the general stability of the trench; they are the least predictable of all deviations as the factors which influence them are many and generally occur in combination. These factors include

(*a*) the subsoil: whether or not cohesive and its adverse effects on bentonite; its granularity and whether containing boulders or artificial obstructions

(*b*) the bentonite suspension: the excess head over groundwater; its density and other properties

(*c*) the groundwater: its head, whether flowing or static; its liability to contaminate the bentonite suspension

(*d*) the trenching process: the tool, the trenching speed and the method of soil removal

(*e*) subjective influence by the operators of the excavating equipment.

17. Although some of these factors, e.g. subsoil, groundwater and bentonite, can be assessed in advance and kept reasonably under control, a broad practical experience is required to understand and foresee what is likely to happen during the trenching operation, and so allow for remedial measures to be taken when necessary.

18. Some of the factors are discussed below.

BENTONITE

19. The most important property of the stabilizing fluid is its ability to establish almost instantaneously a quasi-impermeable membrane at the soil/liquid interface. In engineering practice a bentonite suspension, with or without additives, is universally used as the stabilizing fluid, and its consistent quality and ability to form such a membrane through the whole trenching operation are the essential features.

20. As has been shown in laboratory tests and now embodied in standard specifications such as that published by the Federation of Piling Specialists in the UK, density, viscosity, gel strength, fluid loss and pH of the bentonite suspension are useful measures of suitability for the process.

21. These properties vary with different sources of bentonite and the type of mixing and time allowed for hydration when producing the bentonite suspension.

22. In addition the properties also vary during the trenching operation as a result of contact with the soil and groundwater. This can cause both mechanical and chemical contamination. A regular check should therefore be made during the trenching operation, especially under variable soil and operating conditions.

23. It should be noted however that the acceptable limits to these variations are relatively wide, and hence too frequent and too stringent controls are seldom justified.

SUBSOIL

24. Practically any type of subsoil normally encountered in the trenching process can be stabilized by a suitable bentonite suspension.

25. Cohesionless soil of high permeability, e.g. poorly graded gravel of large grain size, requires a bentonite suspension of higher viscosity and higher gel strength than do well graded sands and gravels or clays, because of the need to stabilize the individual grains and to limit the penetration of the bentonite into the sides of the trench. In very open ground, inert fillers such as lightweight aggregate, plastic, sawdust etc. may be added to the suspension to avoid excessive loss of bentonite into the soil, which could jeopardize the stability of the trench. A similar expedient may be applicable to highly fissured cohesive soils or rock such as some clays, chalk etc.

26. The operation of excavating soil under a liquid differs substantially from excavating in the open where a stress relief, however localized, can take place. The effect of the compressive force, equivalent to the hydrostatic head of the bentonite suspension on the soil being excavated, is particularly noticeable at depths greater than about 25–30 m, but varies of course with the type of soil. Uniform fine sands, for example, can appear to be rock-hard at such depths.

27. Large boulders embedded in a firm soil at the sides of the trench and inclined soil strata of varying density or hardness may both cause seemingly inexplicable deviations from the vertical due to their power to deviate the digging tool. The operator must keep a continuous close watch for such deviations and apply corrective measures immediately.

GROUNDWATER

28. One of the main advantages of the diaphragm wall technique is the possibility of constructing walls in the presence of groundwater. However, to allow the formation of a filter cake on the sides of the trench, together with a limited penetration of the bentonite into the soil, the bentonite suspension must exert a head in excess of that of the groundwater. This is absolutely essential. The

minimum acceptable excess head will depend on the soil and groundwater conditions. In practice, however, it should not be less than 1·25 m, and allowance must also be made for some variation of the bentonite level during the trenching process.

29. Artesian or sub-artesian groundwater must be suppressed by pumping, or alternatively the ground level from which the work is to proceed may be raised appropriately. Under tidal working conditions the tidal effect on the groundwater level at the trenching line has to be considered, together with the induced (tidal) flow of the groundwater, which may influence the stability of each side of the trench to a different degree.

30. Moving groundwater, in general, alters the effective differential hydrostatic head of the bentonite in the trench by the equivalent head of the seepage pressure, and this must be allowed for. If the line of the trench for a cut-off or diaphragm wall intercepts the normal flow of groundwater for a certain length the trench will act like an underground dam and thus change the flow lines and alter the original groundwater level. This effect can frequently be observed at the closure position on the perimeter of an area which is totally enclosed by a diaphragm wall.

31. During the trenching operation free pore water from the excavated soil mixes with the bentonite suspension. This has to be taken into account when working in soils containing contaminated groundwater or sea water.

32. The effects of sea water dilution on a bentonite suspension have been studied by various proprietary bentonite suppliers who have shown that viscosity and fluid loss are significantly influenced by this dilution.[6] Sea water acts as a mild flocculating agent and initially increases the viscosity of the bentonite in the trench. Excess addition of sea water would lead to complete flocculation, high fluid loss and water separation with consequent loss of thixotropy (see Figs 1 and 2).

33. In practice only some 5–10% sea water contamination of the bentonite slurry need be expected when excavating through ground reclaimed with sea-dredged sand.

Experience on such coastal sites shows that the basic requirement, namely that the bentonite slurry should gel in the soil and form a low permeability membrane or filter cake at the soil/liquid interface, is normally attainable and the adverse effects of salinity are not sufficient to be detrimental to the stability of the trench.

34. Experience in soils with groundwater contaminated with organic material (e.g. foul sewage or undecayed organic matter) on the other hand has indicated that this does have an adverse effect on the stability of bentonite slurry trenches, and the construction of a diaphragm wall under such conditions could, in the extreme case, be rendered impracticable.

CONCRETING

35. For structural diaphragm walls the bentonite-stabilized trench is normally backfilled with concrete of uniform consistency and high quality. Concrete placing is one of the most important and perhaps most difficult stages of the technique. The stabilizing fluid has to be entirely replaced by the concrete and many high risk factors, sometimes conflicting, have an important influence on the success of this operation.

36. A tremie pipe is used to feed the concrete by gravity to the bottom of the trench and is raised in stages as the level of the concrete rises. The difference in density between the bentonite suspension and the concrete is normally sufficient to prevent any mixing of the two except for a layer of about 300–600 mm in the interface zone. The concrete must have a high workability (using whenever possible a well graded rounded aggregate) and a high slump and it behaves therefore like a heavy, viscous fluid. Normal vibration cannot be used as this would cause segregation. The compaction achieved by gravity is adequate.

37. The distance of lateral flow required to fill a panel or section of the wall from any single tremie position should be limited to about 2·50–3·00 m to ensure a uniform flow of concrete. For longer panels two or even

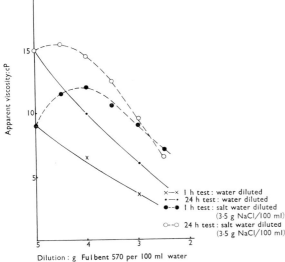

Fig. 1. *Dilution of a 5% Fulbent 570 gel: changes in apparent viscosity*

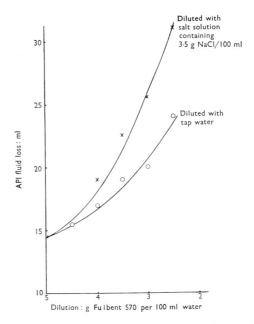

Fig. 2. *Dilution of a 5% Fulbent 570 gel: changes in fluid loss*

more tremies may be used simultaneously, but this introduces further difficulties regarding concrete supply and a potential risk of inclusions in the overlapping zones.

38. It is the upward movement of the fluid concrete with its sweeping action over the soil face and the reinforcement bars and any inserts which removes the bentonite slurry from these surfaces and achieves the essential intimate contact and bond. The success of the diaphragm wall as a structural element depends on this intimate contact. Most deficiencies in diaphragm walls are attributable to insufficient care taken and a lack of understanding of the problem inherent in this important operation. Fortunately most such deficiencies are localized and do not normally impair the overall performance of the wall.

39. In the main, the deficiencies experienced are cold joints, zones of segregated or contaminated concrete, and trappings or inclusions of bentonite mud. The first two are usually the result of interruptions during the concreting operation, which should always be continuous, or of withdrawing the tremie pipe either partially or completely from below the concrete/bentonite interface.

40. Mud trappings usually result from an impediment to the free flow of the concrete by reinforcement bars at too close centres, or by badly designed or badly placed boxing-out, or through the concrete having insufficient workability at some stage during the operation. The occurrence of such trappings may be explained as due to the energy available from the concrete for displacement of the slurry being less than the energy required to move the dense bentonite slurry upward.

41. Since the concrete, behaving like a thick, viscous fluid, flows upwards and laterally from the outlet of the tremie pipe, the energy available from its movement is least at the greatest distance away from the tremie, and as the concrete rises the energy is also falling by an amount equivalent to the reducing differential head of concrete from the full tremie hopper. Therefore trappings of bentonite mud are most usually found at the panel joints and in the upper portion (generally the upper third) of a concreted panel. It follows that shorter panels are less liable to mud inclusions than are longer panels.

42. The nature of the mud trappings is variable. They consist usually of a mixture of soil-laden bentonite and concrete laitance in varying proportions. Although the mud inclusions will be compressed under the weight of concrete above they do not, of course, have significant strength. However, they do exhibit a fairly high degree of impermeability, and as the trappings are generally quite localized this feature should be considered before their removal is undertaken if groundwater pressure is present.

43. There is a natural temptation to ensure the complete displacement of the bentonite by increasing the concrete fluidity until the concrete itself virtually becomes a slurry. Even if an acceptable mix from the point of view of strength can be designed with such excessive workability, the tendency to high shrinkage of such wet mixes must be considered.

CONSTRUCTION JOINTS

44. Structural diaphragm walls can be effectively constructed only in sections or bays. There are various techniques of jointing adjoining bays to provide an effective key and reasonably watertight joint.

45. The most common and perhaps simplest method is the formation of a semi-circular joint by means of a steel tube inserted vertically at the end of a bay as a stop end to the concrete panel. The half-round key on the concrete panel can be used to guide the correspondingly shaped digging tool for the entire depth when excavating the adjacent bay. Any deviation from the vertical of the completed panel is therefore repeated in the adjoining section, but this does help to ensure interlock between adjoining wall panels.

46. If cavitation or overwidth digging has occurred at the end of a bay, concrete will flow around the stop-end tube and cause the formation of a rough concrete ring, which then has to be removed when digging the adjacent bay (Fig. 3). This may become a formidable task and may require the use of a shaped chisel if the concrete is left to set to full strength. Should such concrete rings not be noticed under the bentonite during excavation, complete out of lock misalignment of adjacent wall panels may result. Ingenious devices have been suggested to overcome this problem but their practicability has still to be proven.

REINFORCEMENT AND INSERTS

47. By the very nature of the diaphragm wall technique, reinforcing steel can be placed only in the form of prefabricated cages, which are suspended initially in the slurry-filled trench and held in the correct position horizontally by spacers off the sides of the trench and vertically by support at ground level.

48. In deep diaphragm walls such cages can reach formidable proportions, and may present serious handling problems as generally the only practicable position for prefabrication is the horizontal, from where they must be

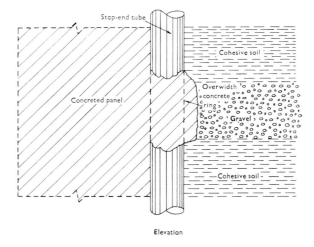

Elevation

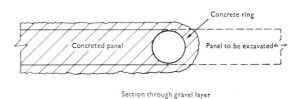

Section through gravel layer

Fig. 3. Penetration of concrete past stop end

hoisted by crane to the vertical position for placing in the trench. There can be very considerable distortion of such a cage under its own dead weight during this hoisting process, and a decision has to be taken, when preparing the detail reinforcement design, whether to make the cage very rigid by means of additional bracing and the welding of bar intersections, or to accept a wide degree of flexibility. The latter solution is generally preferred on the grounds of economy and because in practice a flexible cage is less liable than a rigid one to suffer permanent distortion in handling.

49. For achieving a free upward flow of concrete during placing by tremie, careful attention to detailing can help a great deal. It is advisable to place the main vertical reinforcement outermost, nearer to the sides of the trench, with horizontal distribution steel or links being placed inside. Shear bars and bend-out bars for future structural connexions which cross the wall from front to back should be as far apart as possible or be avoided altogether, particularly at the extremities of a panel where the power of the rising concrete to displace the bentonite is least. The aim in detailing should always be to achieve the maximum void area in plan and end-on views of the reinforcement in order to give least possible obstruction to the free flow of concrete laterally and upwards from the mouth of the tremie.

50. The formation of recesses can be achieved only by means of boxwork inserts attached to the reinforcement cage. As any such inserts restrict the free flow of concrete, their use should be kept to a minimum both in number and size. It is advisable to place inserts only on the outside of the reinforcement cage if possible. Boxwork which penetrates the thickness of the wall to a significant degree can lead to the inclusion of bentonite mud at the edges. The problem of accurately placing the reinforcement cage dictates the need for a fairly generous tolerance in dimension of inserts. The FPS Specification gives 150 mm in both directions, which is considered adequate.

CONCLUSION AND FUTURE OUTLOOK

51. The technique of constructing diaphragm walls as an expedient in the field of civil engineering has been developed in a relatively short span of time, about 25 years, and to a great extent still relies on the experience of a few established firms for its successful execution. Some of the inherent practical problems which must be considered when applying the technique to any particular construction project or when specifying the same have been discussed above.

52. Notwithstanding the most careful attention to detail in design and specification and the exercise of the greatest care in execution, the risk of faults occurring in work entrusted to the inexperienced is high and even the experienced will, if prudent, allow for a measure of remedial work to any exposed walls incorporated in permanent structures. Much scope remains, therefore for further research and for the development of improvements to the technique to eliminate these risks. Currently, however, such efforts are principally directed towards increasing the speed and economy of execution, which has resulted in the development of various diaphragm walling systems. The objectives of speed and economy, whilst desirable in themselves, do not necessarily offer the most fruitful avenue for research.

53. In the Author's opinion, such matters as the improvement of the stabilizing qualities of bentonite suspensions under varying ground conditions, the recovery and recycling of used bentonite and the disposal of spent bentonite, better alignment of the excavation and improvement of verticality, investigation of the concrete cast under bentonite in narrow deep trenches, and the development of practical methods to improve watertightness and to give diaphragm wall structures continuity across panel joints are examples of a field of further enquiry that might yield a handsome bonus in the form of better, cheaper and more universally adaptable diaphragm walls in the next 25 years.

REFERENCES

1. VEDER C. Excavations of trenches in the presence of bentonite suspensions for the construction of impermeable and load bearing diaphragms. *Proc. symp. grouts and drilling muds in engineering practice*, Butterworth, London, 1963, 181.
2. MORGENSTERN N. R. and AMIR-TAHMASSEB I. The stability of slurry trenches in cohesionless soils. *Géotechnique*, 1965, **15**, No. 4, 387–395.
3. WEISS F. *Die Standsicherheit flüssigkeitsgestützter Erdwände*. von Wilhelm, Berlin-München, 1967, Bauingenieur-Praxis, Heft 70.
4. ELSON W. K. Experimental investigation of the stability of slurry trenches. *Géotechnique*, 1968, **18**, No. 1.
5. MÜLLER-KIRCHENBAUER H. Stability of slurry trenches. *Proc. 5th European Conf. Soil Mechanics and Foundation Engineering*, Sociedad Espanola de Mec. del Suelo y Cimentaciones, Madrid, 1972, **1**.
6. LAPORTE INDUSTRIES LTD. *The effect of sea water contamination on the rheological properties of Fulbent 570 slurries*. 1968. Unpublished.

Prefasif prefabricated diaphragm walls

E. COLAS DES FRANCS, Ingénieur de l'Ecole Nationale Supérieure de Géologie de Nancy, *Directeur des Relations Extérieures de l'Entreprise SIF Bachy*

The Prefasif prefabricated diaphragm wall is a new development for construction of a watertight retaining wall. Like the diaphragm wall, the Prefasif method is derived from curtains of interlocking concrete piles, as used in the 1950s at the same time as sheet pile curtains. The established methods of diaphragm wall excavation using bentonite mud are employed. The work is carried out in successive or alternate panels, the size of the panels being determined by the nature of the ground and the dimensions of the Prefasif sections. The guide walls must be carefully constructed since they must align the Prefasif sections and support them in position after they have been lowered. To be sure of a good quality seal, the bentonite mud used for the excavation is replaced by a new grout which is introduced just before the installation of the prefabricated element. This method allows graduation of the characteristics of the grout according to its utilization, even at different levels of the same panel: for instance, a strong mortar can be used at the base of the panel with a different grout further up to guarantee watertightness. The design of the Prefasif elements can be varied widely according to the contract specification. The usual maximum weight of one section is around 30 t. For identical elements the typical design is a rectangular slab with a trapezoidal slot in each end running the full height.

A coupling device is fitted in the bottom of these slots, male one side, female the other. A new section is lowered by crane into the sealing grout, guided by the slot in the previous element, until the coupling device engages. The element remains suspended from the guide walls until the grout has set. The Paper describes the Prefasif method in general and gives three examples of its application.

THE SYSTEM

Cast in place diaphragm wall technology has developed rapidly since 1950, especially in Europe. For example, more than 300 000 m² of diaphragm walling was built in France in 1972, and the method can now be considered as standard practice in civil engineering construction.

2. The prefabricated diaphragm system, which appeared in 1970, is a major new development. It is an extension of the prefabricated interlocking pile which had already been competing with cast in place concrete piles and sheet piling. The Prefasif system is a special type of prefabricated diaphragm walling developed by the Entreprise Bachy.

3. Prefabricated diaphragms have several distinct advantages over other types.

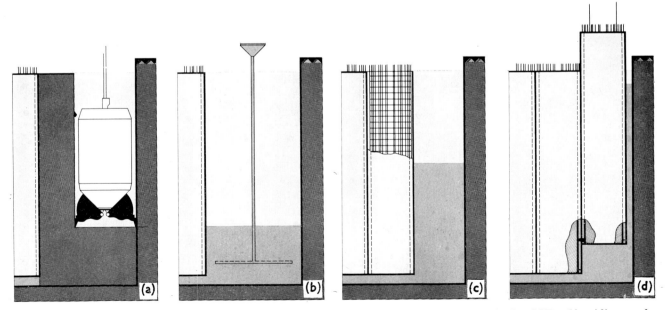

Fig. 1. Stages in the work: (a) excavating for a wall section using bentonite mud and the SIF self-guiding grab; (b) introducing the grout through a spreader; (c) first panel in the sealing grout; (d) lowering the next panel (section showing the guides)

(*a*) There are obvious advantages as regards general appearance. No cutting back is required and the finished surface is even and agreeably clean.

(*b*) The shape of the diaphragm can be tailored to form an integral part of the final structure, satisfying technical and economic considerations.

(*c*) Improved concrete quality and accuracy in placing reinforcement give considerable savings on materials, prefabricated diaphragms being generally 30% thinner than conventional designs.

(*d*) The prefabricated diaphragm can be built and installed in the ground to finer tolerances, and openings can be more accurately positioned.

(*e*) Watertightness at the joints and in the wall itself is better than with conventional diaphragms.

4. Prefabricated panel weight is limited at the moment to 30 tons because of the lifting gear available on building sites, but new developments in lightweight concretes, pretensioning and lifting plant capacity will soon make it possible to use longer panels. The current maximum is 15 m for a width of 2 m.

5. The Prefasif system (Figs 1–4) uses prefabricated reinforced concrete panels built in the casting yard. These are lowered into a slurry trench and bonded to the surrounding ground by means of a grout, which replaces the supporting slurry just before the panels are put into position. The essential features of the system are the displacement by grout of the slurry which supports the trench while it is being dug, a special locking device at the lower end of the prefabricated panels which lines them up with the neighbouring panels (Fig. 5), and a double female joint which is left between adjacent panels and is subsequently regrouted with insertion of a waterstop if required to back up the impermeability of the main grout (Fig. 6).

6. As the final sealing grout is substituted for the trench-supporting slurry, its properties can be chosen to suit each specific case, although of course it must always be at least as strong as the surrounding ground and be capable of properly filling the joint, and its setting characteristics must be suitable for the system.

7. This process has three main advantages as compared with use of the final grout for supporting the trench during excavation. The working programme becomes much more flexible, because the grout is pumped in just before the panels are lowered into the trench; the strength of the grout can be controlled over a wide range, 0·1–10 MPa, which is especially useful if the diaphragm is to carry loads; and there is no danger of the grout being contaminated by the surrounding ground.

8. To reap the full benefit from the displacement technique, one or two precautions must be taken. The grout is generally made up from the supporting slurry by adding cement and extra bentonite if required, and the actual proportions in the slurry on site must be checked before commencement of this work, which proceeds on the basis of reliable data found in the laboratory for the required grout specification. For example, at the Loire Dehaynin site in Paris, the basic data from the laboratory were that the slurry had a density of 1·025 and contained 40 kg/m³ of C_2 bentonite (supplied by the Société Française de Bentonite) and by-products, and that the grout as prepared in the laboratory had a 90 day strength of 2 MPa. The mix for the grout (per cubic metre) was therefore 870 litres water, 62 kg bentonite and 310 kg CLK cement. The properties obtained are shown in Table 1.

9. The triangular diagram (Fig. 7) shows that, with a set cement/water ratio, grouts of different viscosities (i.e. different bentonite contents) possess fairly comparable crushing strengths. As the supporting slurry is used to

(a)

(b)

Fig. 2. Panel prefabrication

Fig. 3. Section hanging by two strap bolts on the guide wall

make up the grout on site, it is essential to control the cement/water ratio more than any other parameter.

10. After the trench has been dug, the slurry is recycled until the sand content is 8% or less, with 1·10 density and 40 s Marsh viscosity.

11. The grout properties are checked first at the batching plant. It is imperative for the grout density to be slightly greater than that of the recycled slurry because it is pumped into the trench through a spreader lowered into the bottom of the trench and displaces the slurry upward. Careful control of the cement/water ratio gives uniform strength in all batches. Further control tests are performed after pumping the grout into the trench to determine the grout density and 7 day and 15 day strengths, with special attention being given to the grout/slurry interface. Once the panels are in place, the last control tests are made at the surface on selected panels.

12. As an example of controls, the following is an extract from the site report during construction of the Loire Dehaynin development in Paris.

Date 3/1/1973   Panels 74, 75, 76
Recycled mud returned to trench before pumping grout
    Density 1·08
    Sand content 2%
    Marsh viscosity 3·6 s

Grout control tests at plant
    870 l/m³ mud
    350 kg/m³ cement
Mud used for grout
    Density 1·10
    Sand content 3·4%
    Marsh viscosity 39 s
Grout characteristics from the mixing plant
    Density 1·29
    Viscosity in 8 mm cone 20 s
    7 day strength 0·32 MPa
    28 day strength 2·15 MPa approx.
    90 day strength 3·24 MPa
Grout characteristics when pumped into trench
    Density 1·29 at bottom, 1·27 at 2 m, 1·25 at 4 m
    7 day strength 0·27 MPa at bottom
    28 day strength 1·91 MPa at bottom
    90 day strength 2·93 MPa at bottom
Grout characteristics with panels in place
    Density 1·25 at surface, 1·25 at 4 m
    7 day strength 0·22 MPa at surface, 0·24 MPa at 4 m
    28 day strength 1·5 MPa at surface, 1·73 MPa at 4 m
    90 day strength, 2·5 MPa at surface, 2·70 MPa at 4 m

Table 1

Cement/water ratio	0·36
Viscosity in 8 mm nozzle cone	16 s
Density	1·25
Crushing strength at 7 days	0·4 MPa
Crushing strength at 28 days	1·3 MPa
Crushing strength at 90 days	2·5 MPa

Fig. 4. Handling a 10 m high panel

Fig. 6 (right). Use of the slots to guarantee a watertight joint: (a) waterstop joint; (b) reinforced concrete key; (c) sealing grout alone

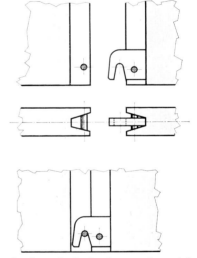

Fig. 5. Securing the foot: the locking hook, and the hook engaging with the locking bar

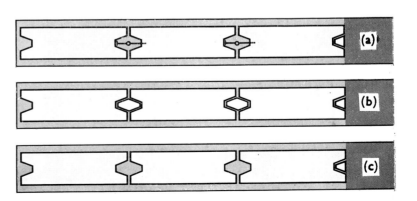

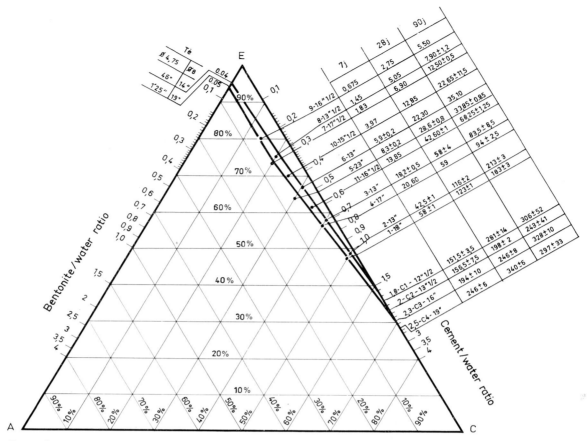

Fig. 7. Bentonite–cement grout triangular diagram. (High turbulence mixing; C2 quality bentonite; blast furnace cement)

Fig. 8. Loire Dehaynin development: view showing row of ground anchors

Fig. 9. Boucry tower: in the foreground is the circular access ramp to the car park, and beyond that the 8 m high prefabricated supporting diaphragm wall

EXAMPLES OF RECENT APPLICATIONS

Loire Dehaynin development

13. This scheme comprises two 12 storey blocks of flats with three basement levels, on the corner of the Rue Dehaynin and the Quai de la Loire in Paris, near the Canal Saint Martin. The substructure retaining wall is a Prefasif prefabricated diaphragm, stabilized during the works with one or two rows of ground anchors (Fig. 8).

14. Ground level lies at 51 m NGF* and the lowest point of the excavation was at 41 m NGF. The diaphragm was built from a platform topped out at approximately 50 m NGF. The geological section through the site shows made ground from 0 to 3·60 m depth, colluvium from 3·60 to 10 m, followed by marl containing certain amounts of gypsum. The diaphragm was anchored 1 m into the marl. Before work started the water table was at 46 m NGF approximately.

15. The patented SIF TMS prestressed ground anchors, each carrying 500 kN load, were anchored in the colluvium over a length of 9 m due to the poor engineering properties of the ground. Unit anchor length was 18 m.

16. Trench area was 2300 m² and diaphragm wall area 2000 m², requiring 77 panels of average size, 12 × 2·50 × 0·35 m, each weighing 23 tons. Seventy ground anchors were installed, making an aggregate length of 1250 m.

17. Space was available near the site for the casting

yard, which comprised six units producing three panels per day.

18. Guide walls were built much stronger than for more conventional diaphragms because they have to take the weight of the panels until the grout hardens. Three panels were installed each day using one excavating machine and one crane. Slurry was provided by a slurry preparation plant with storage and recycling facilities, and grout by a batching plant.

19. Works lasted from November 1972 to February 1973, including the 19 000 m³ of excavation inside the diaphragm.

Boucry tower block

20. The 29 storey tower block and 4 storey block of flats, now under construction on the Rue de Boucry in Paris, near the Porte de la Chapelle, is to have four basement levels with an access ramp to the circular underground car park. Ground level is at 48 m NGF, and the site comprises made ground for 6 m, marl with traces of gypsum from 6 to 11·50 m, Monceau greensand from 11·50 to 12·00 m, and Saint Ouen limestone below this, with its top surface at 36 m NGF. Maximum water table is taken as 38 m NGF.

21. It was decided to build a Prefasif prefabricated diaphragm (Figs 9–11) down to the Saint Ouen limestone at 34·80 m NGF. The bottom of the excavation lies at 36·30 m NGF. The site was first excavated down to 45 m NGF, except along the wall of the neighbouring building where the diaphragm extends up to ground level at 48 m

---

* NGF Nivellement Général de la France.

NGF. Here the wall is 13·50 m high: this is an exceptional size, and the panels weighed 27 tons.

22. The diaphragm was supported during construction by means of a single row of SIF TMS prestressed ground anchors of 600 kN unit capacity anchored in the limestone. Two rows of anchors were used along the party wall, the upper row anchored in the marl and the lower row in the limestone.

Fig. 10. Boucry tower: supporting wall along an existing building (panel height 13·50 m; two rows of TMS ground anchors)

23. The excavated area was 3500 m². The total diaphragm area was 3200 m², requiring 128 panels of width 2·10–2·50 m, thickness 0·35 m and height 10·50–13·50 m, 132 TMS prestressed ground anchors each 15 m long (making an aggregate of 1800 m) and corner braces at the angles.

24. An area was found for the casting yard outside the excavation area with room for 12 casting units producing six panels per day. The site plant included one excavating machine and a crane for handling the panels, although a second excavator was used for a period of three weeks. Construction of the diaphragm required three months from January to April 1973, and the mass excavation and installation of the anchors was completed by 15 July.

25. Weekly control tests were made on diaphragm movements during the mass excavation and it was found that the maximum displacement at the top of the wall was not more than 6 mm.

SCAC office block, Puteaux

26. The SCAC head offices are currently housed in a 15 storey building on footing foundation on the Quai National in Puteaux, but a new 15 storey block with three basement levels is to be built on an adjacent site. A Prefasif prefabricated diaphragm wall was chosen for the excavations.

27. The ground consists of old alluvium from ground level at 30 m NGF to a depth of 13 m (17 m NGF), followed by clay from 13 to 16·50 m (17 to 13·50 m NGF) and Auteuil sand of indeterminate depth below. The water table is about 6 m below ground level.

28. The top 3 m of ground were stripped off and the diaphragm built from a platform at 27 m NGF. Final excavation depth is 20·50 m NGF. The trench was taken

Fig. 11. Boucry tower: Prefasif prefabricated wall. (Surface 3300 m²; four basement levels; excavation depth 11·80 m; ground anchors SIF TMS (60 tons))

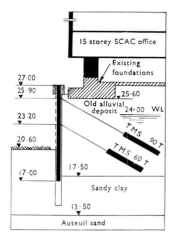

Fig. 12. SCAC office block: section along wall adjacent to 15 storey building

down 2·50 m into the clay, and the space between the bottom of the panels at 17 m NGF and the bottom of the trench at 14·50 m NGF forms a cut-off around the excavation to control groundwater (Fig. 12). The wall is supported during the works by one row of SIF TMS ground anchors of 600 kN unit capacity grouted into the old alluvium. Along the party wall with the present SCAC building an extra row of 900 kN anchors of the same type was installed for permanent support, as the substructure design does not allow the horizontal loads to be taken up by the floor slabs. Also, the diaphragm was thickened to 0·40 m opposite the existing footings, whereas the standard width elsewhere was 0·35 m.

29. The trench was 3050 m² by 0·60 m and the wall 2450 m² by 0·35–0·40 m, made up of 106 panels 10 m long and 2·30–2·60 m wide. The heaviest element weighed 25 tons. There were 96 TMS 600 kN anchors for temporary support and 10 TMS 900 kN anchors for the permanent structure.

30. There was no suitable site nearby for prefabricating the panels and they had to be brought to the site by articulated lorry. Site work required one excavator and a crane for handling the panels. Installation of the panels was completed in two months and after excavating and installing the anchors the site was made over on 15 October 1973.

31. Less than 15 m³/h of leakage has to be pumped out of the excavation.

CONCLUSION

32. Prefabricated diaphragm walls have inherent advantages giving them a great potential for further development. The prefabricated panels can be confined to the open part of the excavation, with the bottom section of the trench used as support or cut-off. It is merely a matter of designing a suitable grout mix.

32. Future developments in handling plant and prestressing techniques will make prefabricated diaphragms suitable for foundations of increasingly greater depth.

Precast diaphragm walls used for the A13 motorway, Paris

M. S. M. LEONARD, MSc(Eng), DIC, AMICE, *Design Engineer, Soletanche Entreprise, France*

The Paper describes the large-scale use of precast wall sections for the construction of the covered motorway linking the A13 (Paris–Rouen) with the Boulevard Périphérique. The precast wall panels were placed on cast in situ concrete in a trench up to 20 m deep founded on sound chalk. The total surface of diaphragm wall was 42 000 m² of which 21 000 m² were precast. The methods are those which have resulted from the experience of Soletanche Entreprise with eight previous sites using precast wall sections. The technique of the preparation and placing of the panels is described and the relative advantages of precast over cast in situ walls are discussed. The Author concludes that precast walls give a better result, both technically and economically, than cast in situ walls, particularly in the case of long term behaviour.

INTRODUCTION

A contract was given for a joint venture of three companies to construct 750 m of underground motorway and 135 m of approach. It was required that 42 000 m² of diaphragm wall and strip piles should be constructed so that earth moving could begin within one year from the date the equipment was brought on site. This Paper describes large-scale use of precast diaphragm walls, and the advantages inherent in the method.

THE PROJECT

2. The motorway was constructed at the western edge of Paris to link the inner ring road (Boulevard Périphérique) with the motorway A13 which runs from Paris to Rouen. The motorway link had to pass through one of the most preserved parts of Paris, notably the Bois de Boulogne and the district near the main Hospital Ambroise Paré (see Figs 1 and 2). The motorway crosses the Seine on a prestressed concrete bridge at St Cloud. It was imperative that the excavation and construction should be accomplished under the following conditions.

(*a*) Nuisance from the site had to be negligible, above all near the hospital. All the large-scale excavations had to be carried out with the roof slab in place.

(*b*) The contract time for work on the wall sections and roof slab was one year.

(*c*) The finished work had to be impeccable and watertight. Any uneven surface of the wall would considerably disturb the air flow in the ventilation ducts formed partly by the precast walls.

The use of precast diaphragm walls was chosen by the client as the best solution to these very demanding conditions.

3. From natural ground level around +34·70 m NGF (Nivellement Général de la France), the excavation equipment encountered the following soils:

   0–1·5 m fill
   1·5–4 m gravels
   4–25 m chalk in varying quality.

Table 1 shows the characteristics adopted for the design. A thin layer of limestone was found on the eastern side of

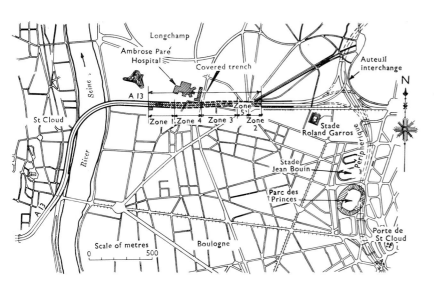

Table 1. Soil characteristics

	ϕ, degrees	c', kN/m^2	Thickness, m
Gravels	30	0	2·5
Weathered chalk	30	10	6
Fractured chalk	30	100–200	9
Sound chalk	30	700	5–10
Chalk below 27 m	30	2500	50

Fig. 1. Map showing the proposed route of the motorway

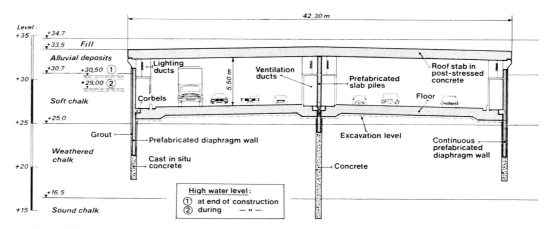

Fig. 2. Cross-section of the covered motorway

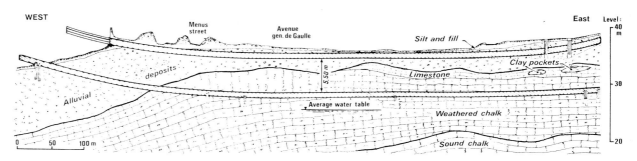

Fig. 3. Geological section over the length of the site

the site (see Fig. 3). The natural groundwater level remained near foundation level, but in time of floods may rise from 25·50 m NGF to 30·50 m NGF. For this reason the design values for groundwater level were 30·50 m NGF under long term conditions and 29·00 m NGF during construction. Twelve piezometers monitored the groundwater levels. An extensive soil reconnaissance included 28 large diameter inspection borings, 4 cored drillings and 12 boreholes for pressure-meter tests. The main problem was to determine the bottom foundation level on the chalk to support the high vertical loads on the walls.

4. The double four lane motorway was constructed by a cover and cut technique (Figs 4 and 5). Two parallel precast diaphragm walls outline the two tunnels which are separated by a central line of prefabricated slab piles. Once the walls and centre slabs were in place, a post-stressed concrete roof was placed before the ground beneath between the walls was removed. The floor of the motorway was then placed together with ventilation ducts on each side of the centre piles.

DESIGN OF THE WALLS AND PILES

5. The centre walls carry part of the vertical load on the roof, and were not designed to support any lateral

earth pressures. The centre walls are basically precast strip piles founded on cast in situ concrete which is carried down to sound chalk 20 m below ground level. The average dimensions of the precast centre piles are 9 m × 1·80 m × 0·40 m. The design vertical loading from the roof to the centre piles varied from 980 to 1600 kN per metre of roof slab. Each pile has a capacity of 400 t and therefore their spacing is not constant. After the excavation of the ground between the walls and piles the gaps between the piles were filled to isolate the ventilation ducts.

6. A cast in situ concrete instead of grout was necessary to form the foundation of the panels for two reasons: firstly, to carry the vertical loading down to sound chalk; secondly, because the vertical stress on the contact between the in situ concrete and precast slab was 6 MN/m², and such a stress could not be supported by the grout used for the excavation of the walls. A core from a test boring in the wall is shown in Fig. 6. It can be seen that the contact between the two types of concrete is excellent.

7. The roof slab provides the lateral support to the outside walls and carries the weight of the backfill placed above together with any traffic loads from roads passing overhead.

8. The outside walls resist the lateral earth and water loads and ensure that the tunnels remain watertight.

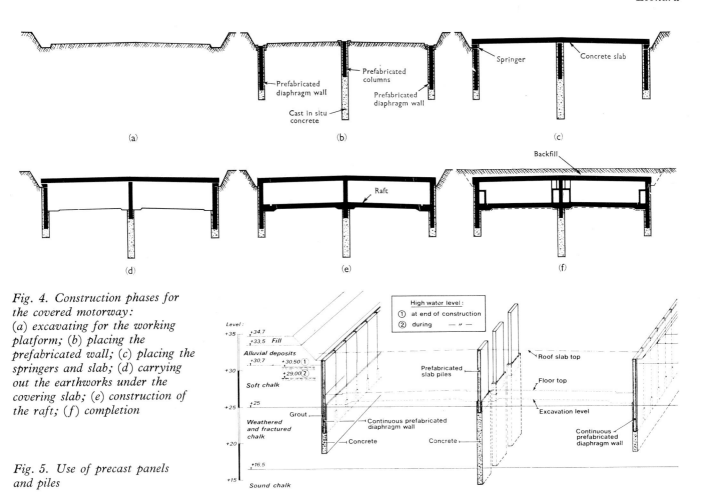

Fig. 4. Construction phases for the covered motorway: (a) excavating for the working platform; (b) placing the prefabricated wall; (c) placing the springers and slab; (d) carrying out the earthworks under the covering slab; (e) construction of the raft; (f) completion

Fig. 5. Use of precast panels and piles

The walls are made up of a continuous line of precast concrete panels averaging 12 m × 2·40 m × 0·40 m.

9. The design vertical loading on the outside walls varied from 390 kN/m to 780 kN/m. These loads are low compared with those of the centre wall. This would have permitted the founding of the precast elements on the cement-bentonite grout used during the excavation of the trenches but, in order to provide a homogeneous foundation, cast in situ concrete was again employed. The average total depth of the diaphragm wall from the top of the panel to the bottom of the concrete foundation is 15 m (Fig. 5).

CONSTRUCTION METHODS

10. At the extreme west of the site were 34 pouring platforms to cast the slabs. Production varied from 10 to 15 slabs per day, with the first lifting at least 48 hours after pouring. The total volume of concrete used for casting the slabs was 8500 m³. The steelwork was assembled before being placed on the casting platforms.

11. Along the centreline of each wall a walled guide trench was prepared, upon which the position and level of each slab was marked. The guide walls served as guides for the kelly grab and formed sound bases for suspension of the panels (see Fig. 7). The trenches were cut with a kelly rig equipped with a grab of width 0·60 m.

Fig. 6. Core through a panel, cast in situ concrete and chalk

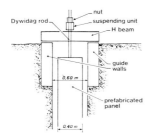

Fig. 7. Method of suspending the panels

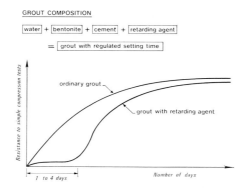

Fig. 8. Grout characteristics

Fig. 9. View of the site showing excavating and placing of the panels

The excavation was carried out under a special grout which had an imposed setting time. Grout characteristics in the short term were as follows.

(*a*) The grout had to remain sufficiently liquid (<50 s Marsh) during excavation and placing of the panel slab. The grout maintained the open trench stable until the concrete was poured and the panel put in place.

(*b*) The resistance of the grout was required to increase in order to allow the trench to be continued without disturbing the panels in place, but at the same time the grout could not be too hard as this would have reduced the production of the kelly rig.

In the long term, the resistance was required to increase progressively in order to transmit the lateral ground loads to the slab, the hardened grout reconstituting the ground between the slabs and the undisturbed ground. Fig. 8 shows the characteristics of the grout, which was composed of water, bentonite, cement and retarding agents.

12. Each panel was coated with an anti-sticking fluid to facilitate removal of the grout when the soil inside the tunnels was excavated. Where necessary, the joint of each panel already in place at the end of a trench was scraped with a special tool to ensure that the abutting joints were free from sediment.

13. Once the excavation was complete, concrete was poured in the trench with a tremie tube to 1 m above the future base of the precast panels. Each panel was lowered into the trench and hung in position as shown in Fig. 7. The positioning of the panels was a delicate operation: each panel weighed 28 t. The displaced grout was pumped back to the grout station where it was regenerated.

Table 2. Plant used on site

6	kelly rigs
4	tyred concreting cranes
5	track cranes for lifting and placing the panels
6	lorries for removal of excavated material
4	earth movers
1	articulated lorry for transport of panels
3	250 kVA generators
3	grout stations
5	km of pipeline

14. The plant used on site is listed in Table 2. Fig. 9 shows the plant and the method of placing the panels. The total number of staff on site was 87 for single shift work.

15. A particular feature of the plant was an automatic grout preparation station. The grout was very strictly controlled with changes in composition following ground conditions. Automatic batching and mixing ensured a high standard of consistency and economy on the grouts.

16. Temporary anti-noise screens were placed around the site to limit the disturbance to the neighbourhood.

FINAL RESULT

17. The edges of each slab were so formed that a joint had the section shown in Fig. 10. Once the roof and floor were constructed, the wider part of the joint was opened to a depth of 2 cm. (The remaining grout was quite sufficient to maintain the joint watertight during this phase.) In order to perfect the permanent water-tightness between panels, a special resin was placed in the joint which was then filled with mortar.

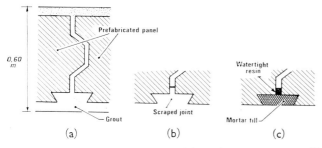

Fig. 10. Joint between panels: (a) before treatment; (b) cleaned; (c) treated

Fig. 11. View of the covered motorway after excavation

18. A very high standard was obtained in the finish of the job (Fig. 11). The method allowed openings in the panels for galleries, pipes, etc., and precise placing of starter bars in the steelwork of the panels before the concrete was poured.

19. The tolerances imposed were that vertical and horizontal deviations should be less than 1%, and no abutting panels should be offset more than 2 cm one from the other in the vertical plane. The wide opening of the joint on the open side of the panel enabled any joint between badly offset panels to be improved to avoid excessive discontinuities.

COMPARISON BETWEEN PRECAST AND CAST IN SITU WALLS

20. The clear difference between precast and cast in situ walls is the precision that can be obtained in the final result. A precast wall can be placed with a precision far greater than that generally accepted for a cast in situ wall.

21. Secondly, for precast walls the conditions of pouring and reinforcement are known and controlled to a far greater extent than with in situ walls. One can use higher stresses in the concrete which allows the walls to be up to 30% thinner than cast in situ walls. The gain in space, which may seem small at first, enabled 800 000 F* to be saved in the buying of land for the A13 site.

22. Starter bars can be incorporated without difficulty and a much higher degree of precision in the final positioning is possible. There are none of the frequent bumps encountered with traditional walls, and there is no cutting

---

*£70 000 approx.

off of excess poor concrete from the top. There is no need to place a facing in front of the wall to improve its aspect, and there are none of the surprises that occasionally occur when a cast in situ wall is uncovered.

23. The excavation is carried out under a grout and not a bentonite slurry for the following reason. If the excavation is carried out under bentonite, which is later replaced by a grout, the percentage of sand fines in the bentonite has to be kept to a very low figure to avoid any mixing with the grout at the moment of its displacement by the grout. When a continuous open trench can be used, the use of two fluids (bentonite then grout) cannot be accepted as the two fluids will mix and the final grout quality will be poor.

24. It is important to distinguish between the degree of watertightness obtained by the two types of wall. In the short term the grout in the joints is enough to ensure that the excavations are carried out in dry conditions. In the long term the grout may shrink with drying out and small movements of the panels may occur. With time damp patches occur unless the joints are treated. A precast wall is far easier to treat as the form of the joint can be designed to accept various widths of watertight resins, and the joint is easier to clean. The concrete of a precast panel can be less permeable than in a cast in situ wall, as vibrated concrete can be used with a lower water/cement ratio than cast in situ concrete. Therefore for permanent watertightness a precast wall gives the best results.

ECONOMIC ASPECTS

25. At first glance one would think that a precast wall is more expensive than a cast in situ wall. By considering only the cost of materials (cost of one square metre of concrete) this is true, but when the subsequent time-consuming jobs of preparing starter bars, removing excess concrete, treating leaking joints, etc., are taken into account, the final cost shows that a precast wall is more economical.

26. For example, a precast concrete wall is 20% more expensive than a cast in situ wall if the wall is to provide only temporary watertightness and the internal aspect is unimportant. As soon as the wall has to be watertight and well-finished the precast wall is generally cheaper for a better end result. The general trend is towards the precast wall for economic reasons.

CONCLUSION

27. The large-scale use of precast walls gives a finished result superior to a cast in situ wall with respect to quality, time and unit cost. The method has been accepted by the Administration and they are most satisfied with the result.

ACKNOWLEDGEMENTS

28. The Paper was prepared with the help and experience of Messrs Pertusier and Namy of Soletanche Entreprise. The client was La Direction Départementale de l'Equipement des Hauts-de-Seine, and the contract was carried out as a joint venture with SA Entreprise L. Ballot and Capag–Cetra responsible for civil engineering and earth moving, and Soletanche Entreprise responsible for diaphragm walls.

Temporary retaining wall constructed by Berlinoise system at Centre Beaubourg, Paris

B. O. CORBETT, BSc, FICE, MIStructE, *Geotechnics Division, Ove Arup & Partners*

M. A. STROUD, MA, PhD, *Geotechnics Division, Ove Arup & Partners*

The basement at Centre Beaubourg is approximately 154 m × 122 m in plan by 16·50 m deep, extending locally to a depth of 20·50 m. The Paper discusses the design, construction and performance of the temporary retaining wall at the eastern side of the excavation along the Rue du Renard. Beneath this street, metro line 11 runs parallel to the excavation and above the base of it. The wall is constructed by the Berlinoise system, consisting of king piles and anchors. It is shown that the method was economical and quick to construct. It is demonstrated that the technique is ideally suited to conditions in Paris but that it could have severe limitations elsewhere.

INTRODUCTION

The French Government held an International Competition during 1970 for architectural and engineering proposals for the creation of the Centre Culturel Beaubourg, to be situated in the heart of Paris close to Les Halles. The result of this competition led to the construction of the scheme put forward by the architects Piano and Rogers, assisted by Ove Arup & Partners. Centre Beaubourg includes under one roof an extensive information centre with a large reference library, a cinema and a museum of contemporary art. Much emphasis has been placed on flexibility and the provision of large areas of uninterrupted floor space. The centre is novel both in its function and in its architecture.

2. The main building covers a plan area of 60 m × 167 m and is situated over the eastern half of an area of basement and car parks, which extend over an area approximately 122 m × 154 m (see Figs 1 and 2). The basement is generally 16·50 m below street level (+ 36·00 m NGF*), but deepens to 20·50 m over the area of the cuvelage (a sub-basement for mechanical services). The structure above ground level forming the main exhibition centre is being built in steel and the loads are supported by twin columns 6 m apart grouped at 12·8 m centres along the longer sides of the building. The loads on these columns, the inner one of which carries a compressive force of 50 000 kN and the outer one a tensile force of 10 000 kN, are transferred to reinforced concrete shear walls at + 27·00 m NGF. These shear walls in turn are supported by barrettes, the tops of which are at a level of + 21·00 m NGF. The outer edge of the eastern barrettes is some 5 m from the perimeter wall around the basement.

* NGF Niveau Général de France.

3. The architectural requirements for the permanent retaining wall called for watertightness and a humidity barrier.

4. In this Paper, the design and construction of the temporary retaining wall, bounded by Rue du Renard, is discussed (see Figs 1 and 3). This project was complicated by the presence of metro line 11 running under Rue du Renard behind the retaining wall (see Figs 2 and 4).

5. The Conditions of Contract for the work were based on an Appel d'Offres.[1] Construction followed the practice laid down by DTU no. 12.[2]

GROUND CONDITIONS

6. The geological sequence below the site consists of alluvial deposits of the River Seine overlying soils and rocks of Eocene age. The sequence of the Tertiary rocks at Centre Beaubourg can be compared with the general sequence in the Belgian, London and Hampshire basins. The most recent of the Eocene deposits at Beaubourg are the Marnes et Caillasses, corresponding in age to the Bracklesham Beds. The general profile is shown in Table 1.

7. A site investigation into the properties of these different soils was carried out.[3,4] The parameters used in the design calculations are given in Table 1.

8. Over the whole area, there were one or two levels of basement beneath the old buildings. Many of these had been collapsed during demolition and backfilled with masonry rubble. Underneath the streets, the filling around the cellars consisted of a mixture of plaster, masonry and sand. The base level of the fill found in the boreholes varied between + 27·90 and + 32·00 m NGF.

9. It was very difficult to differentiate the alluvions modernes from the alluvions anciennes, since they both occur as dense, gravelly sands. Both materials are called alluvions anciennes throughout this Paper. N values from the results of the standard penetration tests were always greater than 40 and generally greater than 100. The underside of the alluvium was generally at about + 24·00 m NGF.

10. The thickness of the Marnes et Caillasses varies between 16·8 m in the north-east of the site and 5·3 m at the south-west. About 5 m above the top of the Calcaire Grossier there is a thin fossiliferous bed and this marker band effectively divides the Marnes et Caillasses into two types of material. The upper beds consist of a very stiff to hard chalk marl, which has been very disturbed by weathering and leaching of gypsum. Localized

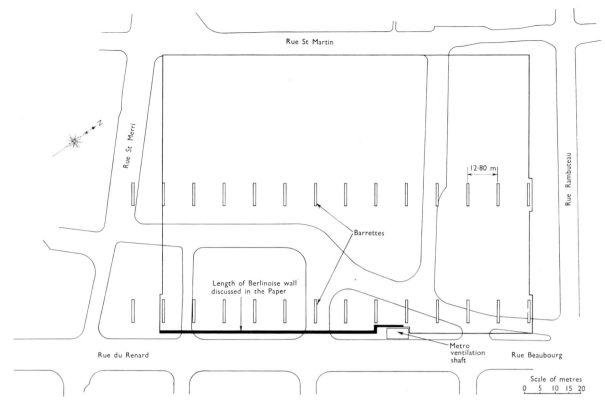

Fig. 1. Plan of site

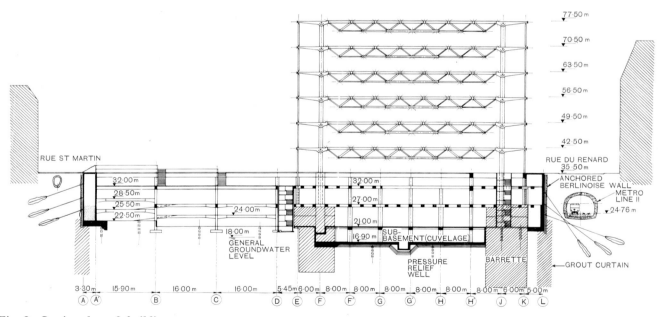

Fig. 2. Section through building

bands of strong, redeposited limestone occur frequently and these are almost horizontal and up to about 0·4 m in thickness. In this zone, solution cavities occasionally occur filled with Seine alluvium or the Sables de Beauchamp, a grey-green clayey sand found above the Marnes et Caillasses on other parts of the site.

11. The carbonate content of the Marnes et Caillasses is almost 100% and chemical analysis indicates that the carbonate is approximately two-thirds dolomite and one-third calcite. The water content in situ is in general about 33%, but can vary from 25 to 47%. The plastic limit is about 35% and the liquid limit averages 63% (varying between 35 and 88%).

12. The lower part of the Marnes et Caillasses is stratified and consists of alternating beds of calcareous mudstone and highly fractured, open jointed limestone.

Fig. 3. General view of wall

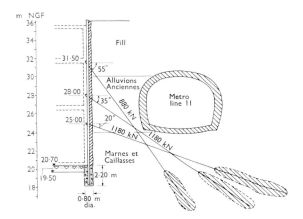

Fig. 4. Cross-section through wall

Table 1. Soil profile and properties

Type of soil or rock	Thickness, m	Properties in terms of effective stress	
		Cohesion c', kPa	Angle of internal friction ϕ', degrees
Remblais (fill, old foundations)	6	0	35
Alluvions modernes et anciennes (alluvium)	6	0	35
Marnes et Caillasses (marls and stony deposits)	5–17	50	28
Calcaire Grossier (massive limestone)	15	—	—

The strength of the limestone increases with depth and about 2 m above the top of the Calcaire Grossier there is a bed of strong, relatively massive, fossiliferous limestone (La Rochette).

13. The stratum immediately beneath the Marnes et Caillasses is the Calcaire Grossier, a moderately strong, fossiliferous limestone extensively quarried in other parts of Paris for building stone. Although this rock and the deeper, uncemented deposits have a bearing on the design of the basement and the behaviour of the structure, they are not relevant to the design and construction of the temporary retaining wall.

14. Over the whole site, the groundwater level in the Marnes et Caillasses and the Calcaire Grossier is at about +18·00 m NGF, i.e. 18 m below ground level. This level is lower than that in the River Seine about 1 km to the south, which in normal conditions is at +26·00 m NGF. The groundwater level is held down by pumping in the northern parts of Paris. Under conditions of flood, water can back up the sewers and the extreme design condition is for a groundwater level at +32·00 m NGF.

THE NEED FOR THE RUE DU RENARD RETAINING WALL

15. It had at first been thought that the nature of the soil and rock would permit a large part of the foundation works to be constructed in an open excavation. However, it was necessary to abandon this idea for several reasons.

16. Firstly, the usable space at the bottom of the excavation would have been very much reduced, particularly restricting the lower basement (or cuvelage) along the Rue du Renard.

17. Secondly, it was essential that the installation and positioning of the barrettes were carried out with great accuracy and for that reason it was necessary that they be built from the bottom of the excavation. Space had thus to be provided close to the perimeter of the basement for the cranes, grabs and ancillary equipment used for constructing the barrettes.

18. Thirdly, the form of the superstructure was critical to the success of the project and thus had an impact on all aspects of the work that preceded the erection of the steelwork. The lowest level, for instance, from which the steelwork could be stored and then erected was at +27·00 m NGF. One of the principal considerations in the construction of the substructure was thus the necessity of forming as extensive a working platform as possible at this level.

19. It was therefore necessary to adopt as the solution a vertical temporary retaining wall extending over the whole depth of the excavation without any interior support.

Along the Rue du Renard the temporary wall had to be particularly secure, since it was necessary to support not only the main thoroughfare running parallel to the wall but also the twin track tunnel of metro line 11 (see Fig. 4). There are buildings along the east side of the road that would have suffered from excessive movement.

TEMPORARY RETAINING WALLS

20. Several methods of construction of the temporary retaining walls were studied. Four of these received detailed consideration: puits blindé, diaphragm wall, la Parisienne and la Berlinoise.

Berlinoise method

21. In the Berlinoise method, steel joists are inserted vertically into borings (which are either empty or have been drilled under bentonite mud) along the perimeter of the future basement. The toes of these king piles are concreted up to basement excavation level; in other words, they are supported on short concrete piles.

22. As excavation proceeds, horizontal planks are inserted between the king piles and supported on their flanges. The planking can be simply of timber, precast concrete or vertical slabs cast in situ. It is essential in using this method of support that the soil can stand temporarily over a height of about 1 m to be able to place the planking. After excavation to an intermediate level in successive stages, the stability of the wall is ensured by means of ground anchors. Each horizontal bed of anchors is installed and then stressed progressively; these operations take about one week for each level of anchors. The excavation proceeds in stages as soon as successive beds of anchors have been installed.

23. In most countries where the Berlinoise method has been employed, heavy walings are used to spread the loads between the king piles. However, at Centre Beaubourg, as is usual in Paris, the king piles were anchored individually without any lateral continuity. The temporary wall provides a back shutter when placing the concrete for the permanent wall.

Other methods rejected

24. Puits blindé has been the traditional method in France of building basement retaining walls on urban sites. It was the only method used until the development of the diaphragm wall about 15 years ago. The method consists of digging a series of shafts by hand around the perimeter of the excavation at regular intervals apart. Sections of wall are built inside the shafts over the full height. The soil between the short lengths of wall is then excavated in a second series of pits and a continuous wall completed. Traditionally this type of wall was supported by struts from within the excavation, but in recent years these have been replaced by ground anchors.

25. The main inconvenience of this method is that a large labour force is necessary and construction is slow. This was not acceptable on this project.

26. The development of the diaphragm wall process has been adequately described elsewhere.[5] This method of construction was rejected because the depth of the guide walls through the old basements would have meant extensive preliminary excavations within a strutted trench.

27. The Parisienne method has many similarities to the Berlinoise; the principal difference is that the king piles are of precast prestressed concrete. The whole depth of the boring is backfilled with a bentonite–cement slurry.

28. As with the Berlinoise, panels are constructed between the king piles. These are of in situ concrete and continuity is provided between the piles as horizontal starter bars are fixed to the reinforcement in the piles and are bent out into the infilling panels prior to placing the concrete. The construction of ground anchors to support the wall as excavation proceeds is similar to the Berlinoise method.

29. After a preliminary design had been prepared for a Parisienne wall, it was clear that the main problem was that the weight of the 20 m long sections would be about 180 kN. This would have made the placing of them difficult when compared with the weight of the Berlinoise king piles at about 40 kN. Ancillary problems would have been the space required to cast the sections, to move them about and stock them on the site. In the end, the limited space available swayed the decision towards the Berlinoise wall at the expense of the Parisienne.

PERMANENT RETAINING WALL

30. The method adopted for construction of the permanent retaining walls consisted of building the walls in situ. The wall is cast against the face of the Berlinoise; the anchors are destressed as the wall reaches them and the remainder of the temporary wall buried.

31. A bituminous membrane is provided between the permanent and temporary construction to provide a vapour barrier.

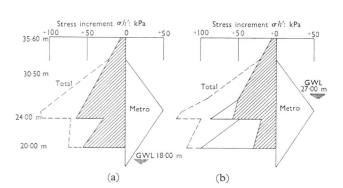 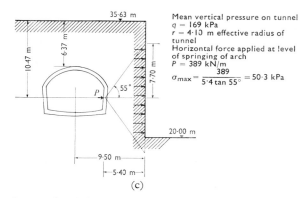

Fig. 5. Earth pressure diagrams: (a) temporary condition (groundwater level 18·00 m NGF), Berlinoise wall; (b) permanent state (groundwater level 27·00 m NGF), permanent retaining wall; (c) section

Fig. 6 (above left). Backfilling of old cellars

Fig. 7 (left). Drilling for king piles

Fig. 8 (above right). Drilling for ground anchors

Fig. 9 (right). Procedure in under-cutting toe of wall

32. The structure of the wall is of vertical columns and horizontal ribs, with infilling concrete panels between. In places, panels are omitted to make space for ducts and other services.

33. A grout curtain cut-off was installed below the retaining wall and pressure relief drainage wells constructed within the area of the basement to relieve the uplift pressure on the base slab.

DESIGN AND CONSTRUCTION

34. The earth pressure diagrams used for the design of the wall are shown in Fig. 5. These were based on classical theory.[6] It was assumed that the wall moves sufficiently to mobilize the total active pressure and 50% of the total passive pressure, and that there is no friction on the wall, for in the temporary state the wall tends to move downwards relative to the soil under the vertical component of the anchor force, and in the permanent state there is a bituminous layer between the reinforced concrete wall and the sheeting of the Berlinoise wall.

35. The arch of the metro tunnel was assumed to be rigid and the thrust at the springing was calculated to be 389 kN/m. The distribution of pressure on the wall taken for design purposes is shown in Fig. 5.

36. The king piles were at 2·5 m centres. Three levels of multi-stand prestressed anchors were installed; these were inclined so as to avoid the metro tunnel The design forces and inclinations of the anchors are shown in Table 2. The vertical component on each king pile was 1800 kN.

37. The contract for the excavation and construction of the temporary wall was let to a consortium comprising H. Coutant (Ivry), Etudes et Travaux de Fondations (ETF) and Intrafor-Cofor. Coutant were responsible for the excavation and for constructing the infilling panels between the profiles. ETF carried out the calculations and installed the king piles. Intrafor-Cofor constructed the ground anchors. The contract was let on the basis of an Appel d'Offres.[1]

38. With street level at +36·00 m NGF, the top level

Table 2. Design forces and inclinations of anchors

Level	NGF, m	Force, kN	Inclination to horizontal, degrees
Top	31·50	880	55
2nd	28·00	1180	35
3rd	25·00	1180	20

of anchors was installed at +31·50, the second level at +28·00 and the third level at +25·00 (Fig. 4).

39. By 9 August, 1972, bulk excavation had proceeded from the original ground level of +36·00 m NGF down to +26·00, except for a berm that was left around the edges of the excavation to provide a working platform for the plant installing the king piles. Along the Rue du Renard at this time the old cellars were full of debris from the demolition of the buildings above them. Starting from the southern end, the debris was being cleared out and the cellars were backfilled with sand and gravel (Fig. 6).

40. The auger arrived on site about 30 August and the first series of king piles was being installed by the metro ventilation shaft during the first week. By 15 September work had proceeded some half way along the wall on the installation of the king piles (Fig. 7).

41. Between 15 September and 12 October, excavation started in the area of the metro ventilation shaft at the northern end of the first phase and was taken down to a depth of 2–3 m (+34·00 to +33·00 m NGF). At the same time, the top level anchors were drilled and stressed and the excavation was deepened a further 2 m below the anchors at the southern end of the wall and approximately 1 m at the northern end. Towards the end of this period of three weeks, drilling started on the second level anchors, working from north to south. By 12 October, all the top level anchors had been stressed and the excavation taken down to +28·50 m NGF at the northern end.

42. During the next period, up to 2 November, the second level anchors were installed and the majority of these stressed, again working from north to south. All second level anchors were stressed by 2 November.

43. Shortly after this, drilling for the third level anchors proceeded as before (Fig. 8) and the anchors at the northern end by the metro ventilation shaft were stressed.

44. By early February 1973, the excavation was at its full depth at +19·50 m NGF. By the end of this month, excavation for the cuvelage away from the wall had started. By mid-March, the whole cuvelage excavation was open down to a level of +15·50. At a late stage in the design, it had proved necessary to enlarge the cuvelage and extend it eastwards up to the Berlinoise wall. This involved undercutting the toe of the wall and the work for this was proceeding in cofferdam by this time (Fig. 9). The whole of this undercutting was completed by May, and during June the bottom slab was placed and the lower part of the wall reinforcement put in position.

45. By November 1973, the permanent retaining wall had been brought up to the level of the second level anchors along most of the length south of the metro ventilation shaft. The second phase of the excavation in the northern part of the site had been completed down to formation level by this time.

MONITORING

46. The horizontal movements of the steel profiles were measured at the top of the wall. These are shown in Fig. 10. The movements ranged from −6 to +20 mm with an average of +5·8 mm. (Minus is away from the excavation, plus towards the excavation.) Using Henkel's method,[7] the ratio of modulus of elasticity E to coefficient

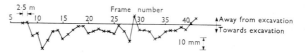

Fig. 10. Horizontal movement at top of king piles between 26.10.72 and 27.3.73

of earth pressure at rest K_{ot} is calculated to be $E/K_{ot} = 26·8 \times 10^4$ kPa. (This compares with the deformation of London Clay at a number of sites in London where on average this ratio is $E/K_{ot} = 3·2 \times 10^4$ kPa.)

47. Each ground anchor was proof tested on initial stressing. Some were fitted with stress measuring devices. The changes were recorded initially at one month intervals and the changes of force in the anchors with time were insignificant.

48. At a distance of some 18 m back from the wall on the eastern side of the Rue du Renard, the heave measured ranged from 0 to 4·0 mm, generally in the range 3–4 mm. This movement is attributed to the relief of stress following excavation and, to some extent, by claquage following the grouting for the impermeable cut-off.

PERFORMANCE

49. The method adopted for the construction of the temporary retaining wall was fast and economical.

50. The Berlinoise wall has proved successful at Centre Beaubourg for the following reasons.

(*a*) The soil and rocks are sufficiently strong to permit the wall to be undercut in order to construct the next infill panel.

(*b*) The vertical loads on the king piles can be safely carried at the base of the excavation with little risk of bearing failure or excessive settlement.

(*c*) The deformation characteristics of the soils and rocks are sufficiently high to keep movements in the zone behind the wall within acceptable limits. This is very important in a case such as this where a metro tunnel runs behind the wall.

(*d*) The anchors provide the majority of the passive restraint, so that little passive pressure is mobilized at the toe of the wall and consequently the inwards movement of the toe is small.

(*e*) The groundwater level is generally below the base of the excavation. The Berlinoise is a type of retaining wall difficult to make waterproof; had the groundwater table been high, seepage through the wall would have made construction and waterproofing of the permanent retaining wall extremely difficult. In the final stages of the excavation, where it was necessary to excavate the cuvelage down to a level of +15·50 m NGF, some 2·5 m below the existing phreatic surface, groundwater lowering was satisfactorily achieved by a system of filter wells.

51. The forces in the inclined ground anchors impose high vertical loads on the king piles. This could be catastrophic where the ultimate bearing capacity of the ground in which the toes of the king piles were founded was approached or exceeded. In this situation, the wall would tend to move downwards relative to the soil, mobilizing wall friction. The lateral earth pressure on

the upper part of the wall in this case would be much greater than that calculated by Rankine theory and the the anchors might well become overstressed.

52. The system could nevertheless be designed to accommodate a certain amount of wall friction, provided the anchor forces are appropriately increased and careful consideration is given to the frictional bond between king piles and infill panels.

53. In the situation at Centre Beaubourg where the cuvelage was locally extended up to the retaining wall (see Fig. 9), the base support for a number of king piles was substantially reduced, although the passive restraint at the toe was maintained by strutting. However, it was estimated that a large proportion of the remaining vertical forces would be transferred from the unsupported king piles to adjacent ones by the infill concrete panels acting as shear walls. To ensure stability, it was specified that the additional deepening be carried out by opening up the excavation adjacent to the wall in short lengths at a time. The operation was performed with complete success.

54. While the indications are that the Berlinoise wall is stiff in shear and may possibly accommodate the removal of base support from the occasional king pile, it is less capable of withstanding the local removal of lateral support. It is not difficult to envisage that the failure of a small number of anchors in one section, for example, could cause a cascade failure and rapid collapse of the whole support system. Where there is any doubt about the performance of the anchors, waling beams would seem to be essential.

CONCLUSIONS

55. The Berlinoise method of construction has provided a quick, cheap and successful solution for the temporary retaining walls at Centre Beaubourg.

56. The method permitted the site within the retaining walls to be kept clear of obstructions so that other operations could proceed unhindered.

57. Movements at the top of the wall due to installation have been shown to be small. Horizontal displacement ranged from -6 to $+20$ mm (minus away from the excavation, plus towards the excavation), with an average of about $+6$ mm.

58. The method is particularly suited to sites where the water table is no more than a few metres above the bottom of excavation level.

ACKNOWLEDGEMENTS

59. The client is l'Etablissement Public du Centre Beaubourg and the management contractor l'Entreprise Générale GTM Bâtiment et Travaux Publics. The Authors would like to thank Ove Arup & Partners, for permission to publish this Paper, and especially their colleagues John Morrison, Rob Peirce, Peter Rice and Rens Stigter.

REFERENCES
1. BARCLAY M. *et al.* Working in France. *Struct. Engr*, 1974 **52**, Jan., 3.
2. CENTRE SCIENTIFIQUE ET TECHNIQUE DU BÂTIMENT. *DTU No. 12: Travaux de terrassement pour le bâtiment.* Centre Scientifique et Technique du Bâtiment, Paris, 1964.
3. OVE ARUP & PARTNERS. *Centre Beaubourg, avant projet detaillé: rapport su la géotechnique.* 1972. Unpublished.
4. SIMECSOL. *Etablissement public pour la réalisation du Centre Beaubourg: etude des sols.* 1972. Unpublished.
5. FENOUX Y. La réalisation des fouilles en site urbain. *Travaux*, 1971, Aug.–Sept.
6. COULOMB C. A. Essai sur une application des règles des maximis et minimis a quelques problèmes de statique relatifs à l'architecture. *Mem. Acad. R. Sci. Paris*, 1776, **3**, 38.
7. OVE ARUP & PARTNERS. *Report on the geotechnical aspects of the investigations for the Barbican Arts Centre.* 1970. Unpublished.

Discussion on Papers 10–13

Reported by F. A. SHARMAN

The four Papers selected for this session might have provoked some controversy among the Authors and their disciples, since each extolled the merits of a different system. Rather than querying or attacking rival solutions, however, contributors were mainly concerned to elaborate their own cases, either with a respectful belief that the others must equally be right for their own different circumstances, or with an insufficiency of time to make a judgement in any situation but their own. Despite this lack of debate, or of expressions of scepticism, there was lively interest in the processes described, and the contributions to the discussion covered an even wider range of topics than the Papers themselves.

2. It was Mr Fuchsberger's description of his practical experience over many years of diaphragm wall construction (Paper 10) which drew most comment. He had, for example, mentioned the tendency to overwidth in trench digging in gravel zones. Professor Veder remarked that this meant that the bentonite cake was not able to keep the width the same as that of the working tool: the excess might be as much as 50%. In some cases this overbreak did not occur, and he had identified the difference in one such comparison as due to the greater roughness of the individual stones in the gravel in the case of the more stable material. The relative appearance of the rough and smooth stones is illustrated in Figs 1 and 2, and

Fig. 1. Stone (limestone) with a very smooth surface

Fig. 3. Thin cut (30 μm) of stone of Fig. 1: enlargement ×35

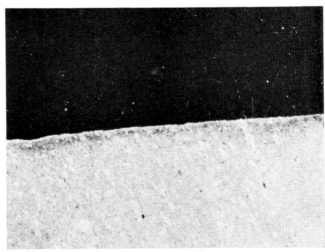

Fig. 2. Stone (limestone) with a normal rough surface

Fig. 4. Thin cut (30 μm) of stone of Fig. 2: enlargement ×35

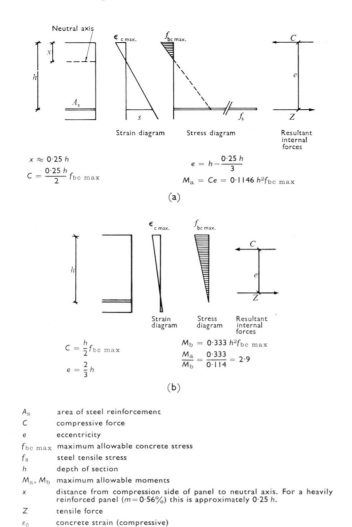

$x \approx 0.25 \, h$

$C = \dfrac{0.25 \, h}{2} f_{bc \, max}$

$e = h - \dfrac{0.25 \, h}{3}$

$M_a = Ce = 0.1146 \, h^2 f_{bc \, max}$

(a)

$C = \dfrac{h}{2} f_{bc \, max}$

$e = \dfrac{2}{3} h$

$M_b = 0.333 \, h^2 f_{bc \, max}$

$\dfrac{M_a}{M_b} = \dfrac{0.333}{0.114} = 2.9$

(b)

A_s	area of steel reinforcement
C	compressive force
e	eccentricity
$f_{bc \, max}$	maximum allowable concrete stress
f_s	steel tensile stress
h	depth of section
M_a, M_b	maximum allowable moments
x	distance from compression side of panel to neutral axis. For a heavily reinforced panel ($m = 0.56\%$) this is approximately $0.25 \, h$.
Z	tensile force
ε_c	concrete strain (compressive)
ε_s	steel strain

Fig. 5. (a) Conventionally reinforced panel; (b) prestressed panel

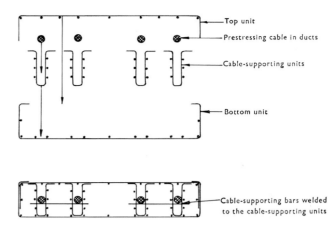

Fig. 6. Construction of reinforcement cage

Fig. 7. Construction of reinforcement cage

the surfaces as shown up by thin (30 μm) sections at ×35 magnification are seen in Figs 3 and 4. Mr Fuchsberger agreed that this interesting revelation would explain the phenomenon: stones of the size illustrated could act like roller bearings in some circumstances.

3. Another contributor referred to the possibility of reinforcement cages rising during the placing of tremie concrete. In one case a 5 t cage in a 19 m deep wall had risen 250 mm in one movement. The speaker had subsequently always secured the cages to struts wedged across

Table 1

Conventionally reinforced wall allowing bending moment 80 mt		Prestressed wall allowing bending moment 160 mt	
Structural reinforcement	66 kg/m²	Weight of prestressing cable	21 kg/m²
Construction reinforcement	24 kg/m²	Construction reinforcement	24 kg/m²
Total	90 kg/m²	Total	45 kg/m²

the guide trench. Mr Fuchsberger agreed that the problem of cage movement was frequently encountered, and was best solved in the way described.

4. The same speaker suggested that the lean concrete backfill which had been mentioned as suitable for sealing off leaks due to old pipes or for making good trench collapses should have a cement content of 50 kg/m³, with a slump of 150–225 mm: it would set in less than 24 h and produce a diggable material strong enough to stand without bentonite support for a height of 2–3 m.

5. The effect of moving groundwater was also discussed, and cases were described in which a narrowing of the permeable gap during wall construction had increased the velocity or pressure gradient to such an extent that trenches had collapsed.

6. Mr Gysi of Lugano gave some interesting facts and opinions about the use of prestressed concrete walls constructed in slurry-filled trenches. Some 30 000 m² of such walls had been successfully completed in Switzerland and Italy, and a further 15 000 m² were currently under construction. Table 1 shows his comparison between

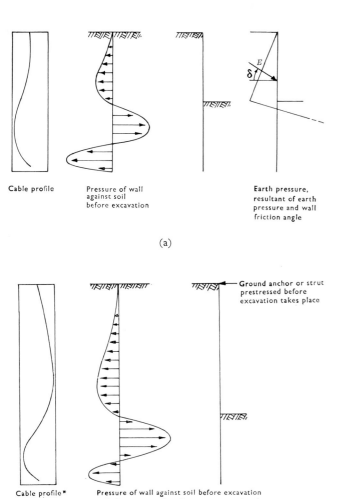

Cable profile   Pressure of wall against soil before excavation   Earth pressure, resultant of earth pressure and wall friction angle

(a)

Cable profile*   Pressure of wall against soil before excavation and after prestressing of strut or ground anchor   Ground anchor or strut prestressed before excavation takes place

* Cable profile to a distorted scale

(b)

Fig. 8. (a) Cantilever wall; (b) horizontally supported wall

a normally reinforced and a prestressed wall, assuming a thickness of 0·8 m.

7. He claimed that since the prestressed wall would nowhere be in tension, it would only be vulnerable to cracking and consequential water seepage at panel joints, which would be sealed in the usual way. To substantiate the stress calculation on which the above steel content comparison was based, Mr Gysi showed the diagrams of Fig. 5 and the reinforcement cage arrangement, Figs 6 and 7. He remarked that the prestressing eccentricity causes bending in the wall, but that this gives a favourable interaction with the soil, as illustrated in Fig. 8.

8. The meeting was hardly able to scrutinize these claims, but speakers did point out that the whole stress picture was critically dependent on the exact positioning within the slab of the prestressing cable, and the relationship between the theoretical and the actual concrete in cross section.

9. Another contributor said that literature and recent experience on the subject of diaphragm walling had revealed a number of trench collapses in saturated silts,

in circumstances in which conventional analysis would have indicated an expectation of stability. One explanation was the incidence of vibration from plant or the disturbing action of the grab. These influences would cause the silt to behave like a viscous liquid. Silt could of course be dewatered, and this remedy was likely to be more effective than the raising of the working platform level in order to increase the hydrostatic pressure in the trench.

10. Another possible explanation for the collapses was that the filter cake might form rather slowly on the walls of the trench in silt. In silts of permeability 10^{-5}– 10^{-7} cm/s and with a differential head of about 1·25 m between bentonite and groundwater levels, the loss of fluid to the ground would be very slow and correspondingly the formation of the filter cake would be slow. On an overall basis, the seepage stresses should be sufficient to stabilize the excavation, but it was possible that the material immediately adjacent to the walls of the trench might not be sufficiently stabilized before formation of the cake, and small sections of the silt might slump off, leading to a progressive collapse. The remedy was to increase the rate of filter cake formation by either raising the bentonite level or lowering the groundwater level, or by some combination of these acts.

11. If collapses or serious cavitations in silt were to be avoided, this contributor suggested, the differential head between the bentonite and the groundwater ought to be substantially greater than that recommended by the Federation of Piling Specialists in their *Specification for cast in place concrete diaphragm walling*. A figure of 3–5 m might be suitable, but this was not to be taken as a dogmatic statement. A very detailed site investigation was suggested when diaphragm walling was to be installed in silt.

12. An Israeli contributor widened the discussion to include a third member of the diaphragm wall family, which trio he considered to consist of the in situ diaphragm, the prefabricated diaphragm wall and the load bearing element. The first two members were, he felt, under reasonable technical control, the first because of its exposure after excavation, and the second because of its factory quality. Attention should be directed to the load bearing element, which required reliable control during construction.

13. In some important projects in Israel, he said, foundations in water-bearing sand had been inspected by a gamma ray test method. In this, a gamma ray source with detector is lowered down a 2 in. dia. steel pipe forming part of the reinforcement. Two to four elements can be inspected by this nuclear method for the cost of one core drilling, it was claimed.

14. Mr Colas des Francs emphasized a point implicit in his Paper: prefabricated wall panels could be given so good a finish before installation that much money could be saved in facing work when the wall was exposed. In this way precast walls could be justified financially as well as by the other advantages.

15. Mr Leonard enlarged on the reasons for adopting precast walling for the work described in Paper 12. He claimed that in addition to ensuring cleaner and quieter work near the hospital, the method had proved more economical than an in situ wall could have been. This allegation was not disputed at the meeting.

16. Mr Corbett spoke of an interesting discrepancy

between the rules normally applied to permissable loads on ground anchors and some actual experiences with tests to failure.

CONTRIBUTORS

Mr D. David, Consulting Engineer, Israel.

Mr H. J. Gysi, A. Linder, Zurich.
Dr S. A. Jefferis, King's College, University of London.
Mr N. A. Sadleir, Twickenham College of Technology.
Mr F. A. Sharman, Sir William Halcrow and Partners.
Professor C. Veder, Technische Hochschule, Graz, Austria.

Measured performance of a rigid concrete wall at the World Trade Center

S. K. SAXENA. Phd, MCASE, *Senior Engineer, Dames and Moore (formerly Soils Engineer, Port Authority of New York and New Jersey)*

This Paper documents the measured behaviour of the few instrumented panels of the World Trade Center's perimeter wall constructed by the slurry trench method. The bending moment and consequently the pressures along the rigid, tie-back concrete wall are computed. It is hoped that the information will provide guidelines to designers for walls constructed by the slurry trench method. The field data consists of continuous records of the loads and the slope of the wall throughout its height during the construction period and for a considerable time afterwards. The load history is reviewed along with the progress of construction, and the instrumentation is briefly described. Profiles of curvature (at various stages of construction), axial loads, bending moments and pressures are presented. The estimated moments are compared with the moments computed on the basis of dynamometer readings (these dynamometers were welded with the reinforcements in pairs on either side). The tie loads are studied with time and explanations for the reduction in tie loads are presented.

INTRODUCTION

One of the most important applications of the construction of a rigid concrete wall by the slurry trench method has been at the Port Authority of New York and New Jersey's World Trade Center in New York. The construction has been described by Kapp.[1] The World Trade Center is located in New York City's downtown area occupying eleven acres. A concrete wall was constructed in 1968 and 1969 to allow the excavation and construction of six basement floors about 70 ft below the ground elevation. The concrete wall is generally 3 ft thick and was supported by tie-back rock anchors during the construction stage. Final support is provided by the floor system and the ties are distressed. Ten panels were instrumented with electric resistivity load cells, slope inclinometers and electric strain meters attached to the reinforcement bars of the concrete walls. In addition piezometers and well points for observation of water pressures on the wall were

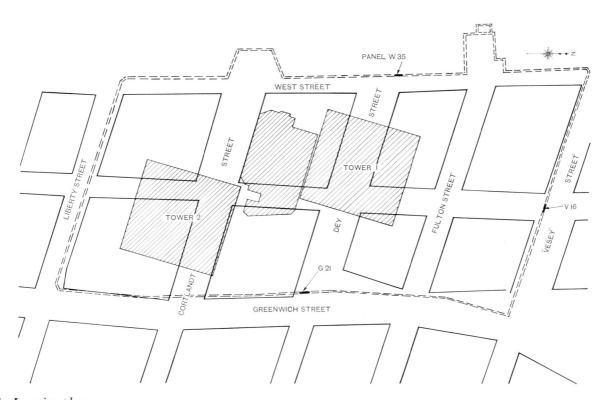

Fig. 1. Location plan

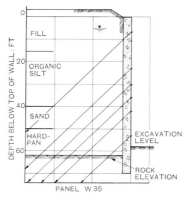
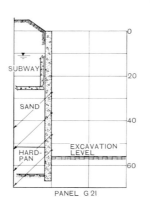
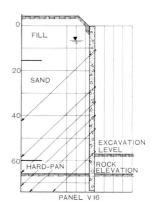

Fig. 2. Panel sections

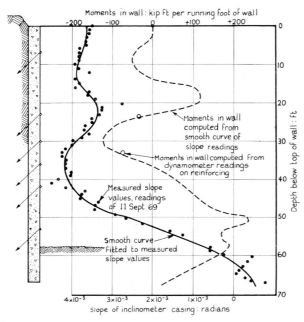

Fig. 3. Panel W35: typical slope curve and corresponding bending moment curve

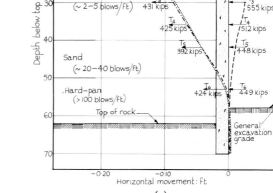

(a)

Fig. 4 (above right and below). Panel W35: (a) horizontal movements; (b) bending moments; (c) horizontal pressures

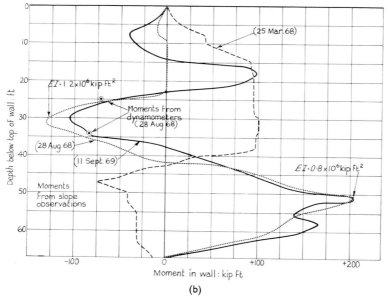

(b)

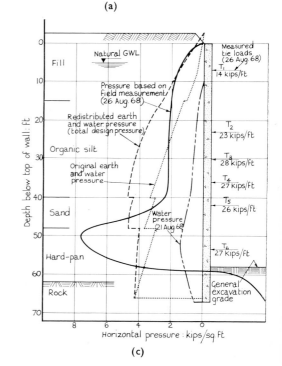

(c)

also installed. This Paper documents the measured behaviour of a few instrumented panels.

WALL PANELS

2. The wall panels were 65–70 ft high with a nominal thickness of 3 ft. The rigidity of the wall (EI) in the uncracked stage was 1.5×10^6 kip ft². Each panel was supported by two rows of four to six prestressed tie-backs carried through the soil overburden and penetrating 30–35 ft into bedrock. The ties were inclined at an angle of 45° and were prestressed to design loads of as much as 600 kips each. Of the ten instrumented wall panels, only four are provided with load cells on tie-backs. Three instrumented panels are selected for presentation in this Paper—W35, G21, and V16. Figure 1 shows the location of these panels and Fig. 2 shows the panel sections with positions of ties and different soil types.

3. As far as field measurements are concerned, the essential data were the readings of the slopes. A long pendulum slope indicator was used to obtain slopes at every foot along the depths of the wall. The slope indicator casing was placed within a steel casing, which was installed at the time the concrete wall was cast, and the annular space between the two casings was filled with cement grout. Extensive care was exercised to ensure the accuracy of readings, by checking the algebric sum of 180° readings which should be constant within limits, and by holding the pendulum reading device under water for about 15 min to ensure constant temperature.

4. The inclinometer readings provided a direct measure of the slope of the wall at any elevation. Differences in slope between two elevations (usually every foot or half foot) furnished an estimate of curvature between the elevations. Bending moments were computed from the curvatures. A smooth curve for the estimated profile of slopes was hand drawn through the observed points. Gould *et al.*[2] observed that nine out of ten readings fell within a band of $\pm 2 \times 10^4$ radians on either side of a curve for panel W35. A typical slope curve and the corresponding bending moment curve for the same panel are shown in Fig. 3.

5. Dynamometers (strain meters), mounted on opposite sides of the reinforcing cage, provided an independent check at sporadic points.

Panel W35

6. Panel W35 had six ties evenly distributed over the height of the wall, with the upper tie about 7 ft below the top of the wall. The ties were installed at 100% of their design loads, except for T_1, which was installed at 90% of its design loading. The observations show that the wall moved continuously into the soil as excavation proceeded and lower ties were installed. The maximum deflexion was about 0.2 ft and this was observed about a year and a half after the start of excavation (see Fig. 4(a)). After lock-off, the tie loads decreased continuously as indicated in Fig. 4(a). Bending moments and horizontal pressures acting on the wall have been developed and are shown in Figs 4(b) and 4(c) for panel W35. The horizontal pressures presented differ from those presented by Gould.[3] The pressure diagram is for the completion of excavation stage. The earth pressures used in design are

also shown in the same figure. The Appendix describes the procedure of obtaining the bending moments from slope readings.

7. Figure 5(a) shows the decrease in load of various ties of panel W35 with time. A look at Fig. 5(b) would indicate that loss in the tie-back can be entirely explained by the elastic movement of the wall into the soil with the accompanying reduction in tie-back elongation. It appears that creep in the grout rock contributed almost nothing to tie load decrease especially because the bond stresses had been nearly constant for a year after excavation.

Panel G21

8. The performance of panel G21 has been just the opposite of panel W35 (see Figs 6(a) and 6(b)). Because of the location of a subway structure adjoining this panel, the first tie was installed at a lower elevation than most panels (25 ft below the top of the wall). Furthermore, to avoid overstressing the wall of the subway structure, the tie was locked off at 40% of its design load. A temporary brace held the wall until the first tie was in place. The remaining three ties were installed at 100% of their design load. Figure 6(a) shows that the wall moved continuously towards the excavation during construction. The maximum deflexions were slightly more than 0.2 ft.

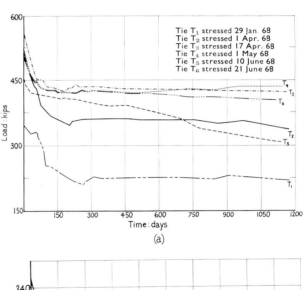

(a)

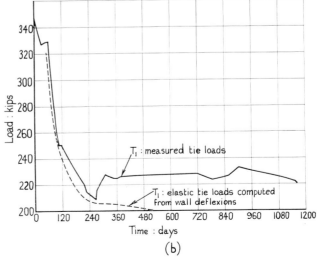

(b)

Fig. 5. Panel W35: tie-back loads versus time

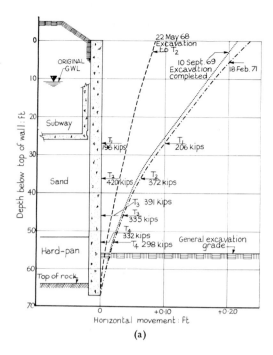

(a)

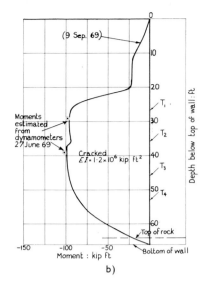

b)

Fig. 6 (above and below left). Panel G21: (a) horizontal movements; (b) bending moments; (c) horizontal pressures

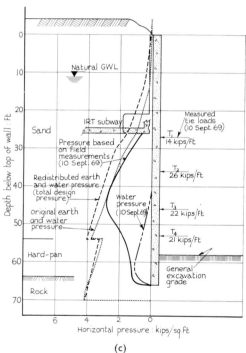

(c)

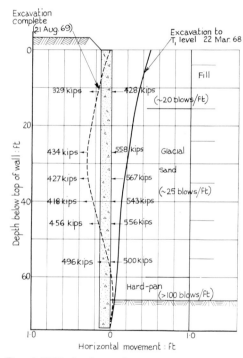

Fig. 7. Panel V16: horizontal movements

Bending moments and pressure diagrams for the completion of excavation stage are shown in Figs 6(b) and 6(c). The slight increase of load at T_1 can be fully explained by elastic elongation. However, the remaining three ties experienced a loss of load, even though the movements of the wall exhibit elastic elongation. Either slippage between tie and grout or creep (or a combination of both) therefore is indicated at anchorage.

Panel V16

9. This panel, like panel W35, had six ties. The wall leaned towards excavation initially as the excavation to the elevation of first tie was achieved. Then as the other ties

were installed it pulled away from excavation. At completion of excavation, the wall had its tip almost at its original position before excavation proceeded (Fig. 7). The deformation behaviour of this panel is therefore somewhere in between panels W35 and G21. Figure 8 shows some interesting features. Figure 8(a) is a slope diagram of the panel at the end of excavation stage. Very near the elevation of tie T_6 there is an abrupt change of slope indicating a plastic hinge. At the time these readings were taken, the wall was being cut for the floor beams. The bending moment diagram reflects the existence of a large moment at that point. Figure 8(b) also shows the design bending moments and the bending moments obtained by numerical analysis of a beam on

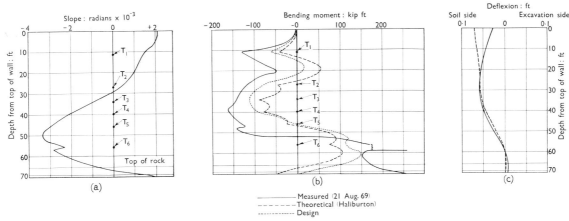

Fig. 8. Panel V16 at end of excavation: (a) slope; (b) bending moments; (c) deflexions

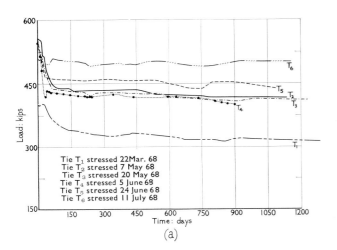

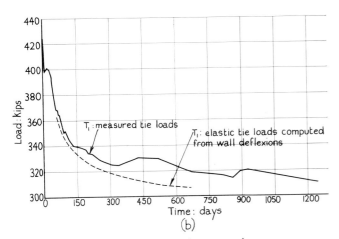

Fig. 9. Panel V16: tie-back loads versus time

non-linear foundation proposed by Haliburton.[4] Figure 8(c) shows the observed deflexion together with the values obtained from the numerical analysis method. The agreement between the deflexions measured from the slope indicator and those computed by the numerical method is good except at the top.

10. Figure 9(a) shows the measured loads in the ties

of panel V16 with time and Fig. 9(b) confirms the reduction in load due to reduction in elastic elongation.

DISCUSSION

11. Field data presented here shows that movement of a cast in place concrete wall due to bending of the wall can be limited to small values. The most influential factors affecting movement were the position and magnitude of lateral support, and the depth of embedment of the panel into rock.

12. In all the cases presented, the wall panels acted as semi-rigid members rotating around a point near the base of the wall. The location of the point of rotation was very much dictated by the depth of embedment into rock, while the direction of rotation (either into the excavation or into the soil) was governed by the positioning of the ties and the magnitude of the tie loads.

13. The Appendix shows that all computations of bending moments are based on the empirical relationship proposed by Branson.[5] Branson proposed his equations for beams under pure flexure. Since there is no evidence in the literature that the equations can be extended to the case where in the beam is prestressed, the effect of prestress has not been presented. The effect of prestress would be to increase the cracking moment if the principle of superposition of stress at the outermost tension fibre were applied. Panel W35 was studied with the effect of prestress and the results[6] indicate that the bending moments at peak points can be reduced by as much as 30%.

14. Apart from the above, there are many uncertainties in the analysis of a structure like this and most of these are introduced as a result of the slurry trench method of construction. The variations in wall thickness, the locations of reinforcements and the conditions of restraint assumed at the rock embedment level make the analysis tedious.

15. Finally, the use of plane strain analysis for a thick plate structure may have its shortcomings, though it is not clear that thick plate theory could provide an exact analysis due to lack of precise knowledge of the end boundary conditions.

16. Nonetheless, the simplified method used within the context of beam theory and non-linear soil spring does provide a reasonable insight of the problem.

ACKNOWLEDGEMENTS

17. The Author wishes to acknowledge the guidance and support of his colleague Mr D. L. York of the Port Authority of New York and New Jersey in this work.

APPENDIX

18. The inclinometer readings provide the slope along any elevation of the wall. If the slope be denoted as S and distance from top of wall be termed as y, the moment M can be computed using the well known equation

$$M = E_c I_{ef} \frac{dS}{dy} \quad \ldots \ldots \quad (1)$$

where E_c = modulus of elasticity of concrete
$\quad I_{eff}$ = effective moment of inertia of the wall section.

19. The effective moment of inertia will vary with the degree of cracking at the section. As long as there are no cracks

$$I_{eff} = I_{gs} \quad \text{for } M < M_{cr}$$

where I_{gs} = gross moment of inertia of the wall section.
The cracking moment M_{cr} is the moment which corresponds to a maximum tensile stress equal to the modulus of rupture of concrete.

$$M_{cr} = f_t I_{gs}/y \quad \ldots \ldots \quad (2)$$

where f_t = tensile stress in concrete.
For $M < M_{cr}$ an empirical formula recommended by Branson[5] is used:

$$I_{eff} = \bar{A}^4 I_{gs} + (1 - \bar{A}^4) I_{cr} \quad \ldots \ldots \quad (3)$$

where $\bar{A} = M_{cr}/M$
$\quad I_{cr}$ = moment of inertia of cracked section.

A table can be made for different values of $\bar{A}$ relating I_{eff} with $\bar{A}$.

20. Equation (1) can be rewritten as follows:

$$S' = M/(E_c I_{eff}) \quad \ldots \ldots \quad (4)$$

where S' = change in slope per unit length.
When $M = M_{cr}$, $I_{eff} = I_{gs}$,

$$S'_{cr} = M_{cr}/(E_c I_{gs})$$

Hence

$$\frac{S}{S'_{cr}} = \frac{M}{M_{cr}} \cdot \frac{I_{gs}}{I_{eff}} = B/\bar{A} \quad \ldots \ldots \quad (5)$$

where $B = I_{gs}/I_{eff}$.

Using values of M_{cr} from equation (2) and S'_{cr} from equation (5), the M–S' relationship can be obtained, giving the value of M for any value of S'. The values of S' are derived from the inclinometer readings as differences between values of slope at points one foot apart.

21. The above analysis is valid for $S' > S'_{cr}$.

22. For $S' < S'_{cr}$:

$$M = E_c I_{eff} S'$$

Slope angle from dynamometer data

23. The change in the slope per unit length is

$$S' = \frac{f_i - f_o}{D E_s}$$

where f_i = net stress in inside steel
$\quad f_o$ = net stress in outside steel
$\quad D$ = distance between outside and inside steel
$\quad E_s$ = modulus of elasticity of steel.

24. Knowing the slope angle from the above equation, the moment is computed exactly as described for the inclinometer readings.

REFERENCES

1. KAPP M. S. Slurry-trench construction for basement wall of World Trade Center. *Civ. Engng*, 1969, Apr., 36–39.
2. GOULD J. P. and DUNNICLIFF D. J. Accuracy of field deformation measurements. *Proc. 4th Pan American Conf. Soil Mech. and Foundation Engng, San Juan, Puerto Rico*, 1971, American Society of Civil Engineers, **I**, 313–366.
3. GOULD J. P. Lateral pressures on rigid permanent structures. *ASCE Speciality Conference on Lateral Stresses in the Ground and Design of Earth Retaining Structures*, American Society of Civil Engineers, 1970, 219–269.
4. HALIBURTON T. A. *Soil structure interaction: numerical analysis of beams and beam-columns.* School of Engineering, Oklahoma, 1971, Tech. Pub. No. 14.
5. BRANSON D. E. Deflections of reinforced concrete flexural members. *J. Am. Concr. Inst.* 1966, June, 637–674.
6. KERR W. Private communication, 1974.

A case history study of multi-tied diaphragm walls

G. S. LITTLEJOHN, BSc, PhD, MICE, FGS, *Lecturer in Geotechnics, Department of Engineering, University of Aberdeen*

I. M. MACFARLANE, BSc, FICE, FGS, *Director, Trocoll Cementation Engineering Ltd*

The Paper is concerned with full-scale performance of multi-tied diaphragm walls, designed according to an empirical method of analysis first introduced in 1971. At the present time this design method appears unique in that it takes account of the continuous wall construction and excavation stages and the procedure is amenable to varying soil strata. Problems encountered and lessons learned are detailed, and where possible records of wall movement are analysed in relation to design assumptions. The Paper indicates that wall deflexions and bending moments, which occur as excavation proceeds, follow a pattern similar to those predicted by the new method. In addition the results suggest a triangular pressure distribution for a semi-rigid diaphragm wall, as distinct from the trapezoidal distribution commonly assumed. In all the case studies presented, no significant vertical or horizontal wall movements have been monitored during the construction period. The Authors submit that the empirical design method results in economical structures exhibiting satisfactory performance in the field.

INTRODUCTION

An empirical repetitive single-tied wall design method has recently been introduced[1] for the analysis of multi-tied diaphragm walls. This method may be used in the design office and the theoretical background, together with results from small-scale tests, are discussed in detail by James and Jack in Paper 6 of this Conference.

2. At the present time the method appears unique in that it takes account of the continuous wall construction, excavation and anchoring stages, and the procedure is applicable to varying soil strata.

3. Certain basic assumptions are made in the new method, e.g. that the soil pressure distribution is of triangular form and the wall yields progressively as excavation proceeds. In addition, a point of contraflexure in the wall is assumed to occur at a point where the factor of safety is unity against overturning. The purpose of the full-scale monitoring studies described is to check the validity of the basic assumptions with particular regard to bending moment profiles, and where possible to monitor overall movements in order to confirm satisfactory performance of this anchored diaphragm wall system in the field.

4. Case histories have been chosen which are representative with respect to depth of excavation, number of anchor levels and variation in ground conditions. Prob-lems encountered and lessons learned are detailed, and where wall deflexions have been monitored these are analysed in relation to the design assumptions. The diaphragm wall at Victoria Street, London, which is described by Hodgson in Paper 7 of this Conference, represents an additional case study for this design method.

CASE STUDY NO. I: GUILDHALL PRECINCTS REDEVELOPMENT, LONDON

5. For the Guildhall precincts redevelopment, the main contractor was Trollope and Colls Ltd and specialist contractors were Cementation Piling and Foundations Ltd and Cementation Ground Engineering Ltd.

6. A plan of the site is shown in Fig. 1. The ground conditions comprised 6–8.8 m of gravel overlying London Clay. The effective excavation depth was 10.4 m and the soil was retained by a diaphragm wall 0.51 m thick and anchored at two levels into the gravels (see Figs 2 and 3).

7. To facilitate the study of full-scale results and comparison with design assumptions it was considered that information was required on contact pressures, wall displacements and anchor loads at various stages during the excavation.

8. The installation of in situ strain gauges and earth pressure cells in diaphragm walls formed under bentonite is expensive and difficult, and since contact pressures and wall displacements are interrelated it was decided that the most convenient and robust approach would be to record the displacement profile of the wall at each construction stage using a sensitive inclinometer. Designed to operate inside an aluminium extruded duct measuring 44.5 mm square internally, the inclinometer for this investigation had a gauge length of 152 mm, sensitivity of 10″ and a

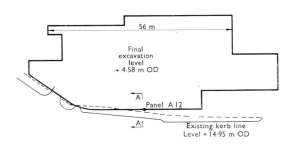

Fig. 1. Site plan—Guildhall

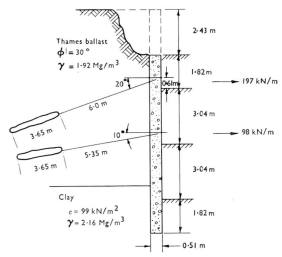

Fig. 2. Section AA through panel A12—Guildhall

Fig. 3. General view of anchored diaphragm wall—Guildhall

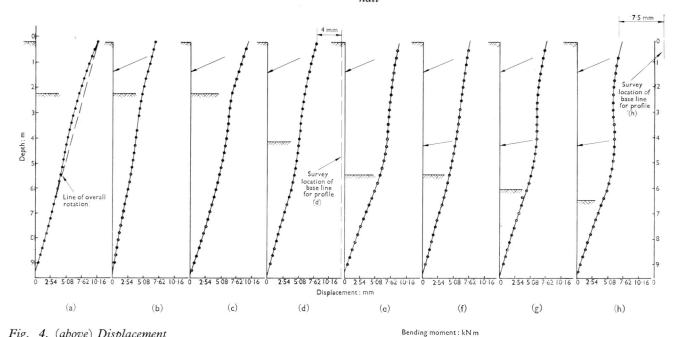

Fig. 4. (above) Displacement profiles of panel A12 at various construction stages

Fig. 5((a)–(h)). Bending moment profiles of panel A12 at various construction stages (moments based on 0·30 m strip)

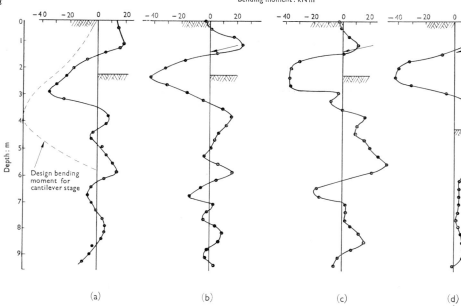

range of $\pm 3°$. Recording errors would tend to cancel over the 9·7 m length of the duct, but for the unlikely extreme case of summation of errors a total error of 1 mm was estimated.

9. To monitor anchor prestress, load cells consisting of steel annuli of 76 mm internal diameter, 9 mm wall thickness and 76 mm length were constructed, and located between the load bearing plate and the anchor stressing head. To measure axial strain four electrical resistance gauges (two axial and two circumferential) were fixed onto the outer surface, diametrically opposite each other to eliminate bending effects.

10. With regard to overall movements of the wall a permanent base line, measured accurately by invar tape, was established at a remote distance from the excavation and was transferred by triangulation to a temporary base line on site. A theodolite reading to 1″ was employed throughout. By further triangulation wall stations were located. A geodetic level was used to measure any wall settlements. Three general surveys were carried out during this investigation thus enabling the overall horizontal and vertical movements at two stages to be computed.

11. Panel A 12 (Fig. 1) was chosen to be monitored to ensure minimal effects from corners, and displacement profiles were measured along the neutral axis of the wall at points between two vertical rows of anchors. All displacement profiles (see Fig. 4) were plotted relative to the toe of the wall, no account having been taken of the overall displacement of the wall, except where this movement was measured during a general survey (see profiles (d) and (h) of Fig. 4). It should be emphasized that full explanations for the sequence of profiles obtained cannot be given authoritatively without accurate information on the absolute location of each profile, but valuable data on bending moments can be obtained by graphical differentiation of the displacement profile gradients (see Fig. 5).

Discussion of results

12. In general the displacement profiles given in Fig. 4 display a logical progression and give rise to confidence in the inclinometer data. Profile (a) shows an overall rotation towards the excavation together with a superimposed cantilever action above the excavated depth. The maximum differential displacement (10 mm) between the crest and toe of the wall occurred during this initial cantilever stage of construction, giving an overall rotation of about 3·5 minutes of arc i.e. a slope of 1/970. The corresponding moment diagram curve ((a) of Fig. 5) appears to have been unduly affected by the stiffening effect at crest level due to the guide wall, and the peak of the curve occurs at the same elevation as the base of the guide wall.

13. The effect of stressing the top anchors is shown in profile (b) of Fig. 4, where the wall has been drawn back towards its original profile and the cantilever action has been reduced. The difference between profiles (b) and (c) corresponds to a time lapse of four days and indicates the magnitude of the movement set up within the gravels to mobilize the necessary restraint. During this period there has been an overall movement towards the excavation and a bulging deflexion below excavation level. This is further accentuated in profile (d).

14. Bending moment profiles (e)–(h) of Fig. 5 follow a pattern similar to that predicted by the design method. In particular (g) and (h) appear to confirm conclusively the bending moment profile that is assumed.

15. It was considered mathematically unsatisfactory to further differentiate the bending moment curve twice in order to obtain the resultant pressure distribution, but use of a sixth order polynomial resulted in a distribution which was approximately triangular. Together with the good agreement between design and measured bending moment profiles at the later construction stages, this reinforces the assumption of triangular pressure as opposed to the trapezoidal distribution normally assumed.

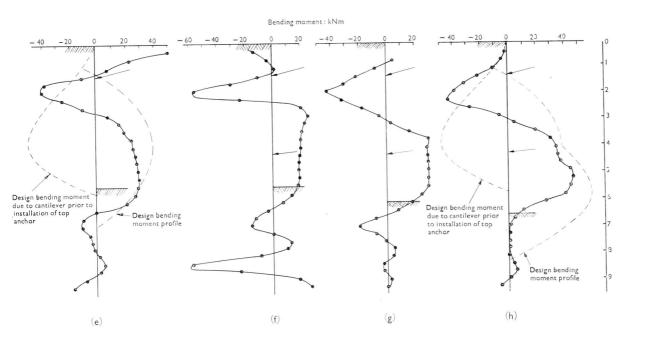

Bending moment: kNm

(e)   (f)   (g)   (h)

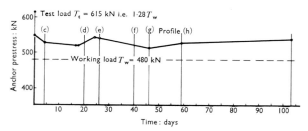

Fig. 6. Long term behaviour of anchor A12/T2

16. From the general survey the overall displacements monitored (profiles (d) and (h) of Fig. 4) show a displacement into the retained soil mass and some conflict appears to exist between these wall movements and the inclinometer profiles. Although the results were checked the apparent error was probably due to movement of the base line joining the fixed stations. In future the use of deep boreholes as fixed datum points is recommended to facilitate the monitoring of overall wall displacements at each construction stage.[2]

17. Vertical settlements measured showed no appreciable changes throughout the project, changes being scattered and of the order of 0·25 mm.

18. The upper anchors on panel A12 were monitored over a period of 103 days, and Fig. 6 shows that the prestress force remained remarkably constant throughout. These results suggest that there were no significant relative movements between the fixed anchor zone and the wall.

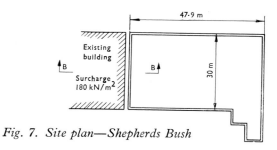

Fig. 7. Site plan—Shepherds Bush

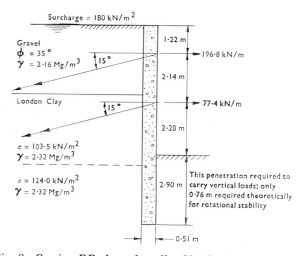

Fig. 8. Section BB through wall—Shepherds Bush

CASE STUDY NO. 2: NEW TELEPHONE EXCHANGE, SHEPHERDS BUSH, LONDON

19. For the new telephone exchange at Shepherds Bush, London, the main contractor was F. G. Minter and specialist contractors were Cementation Piling and Foundations Ltd and Cementation Ground Engineering Ltd.

20. Site conditions and soil parameters are shown in Figs 7 and 8. The site plan indicates a heavy surcharge from the existing multi-storey telephone exchange immediately behind the wall, which necessitated the adoption of panel lengths ranging from 4·5 m down to 2·2 m at this site.

21. For this diaphragm wall the total excavated depth was 10 m and generally the upper anchors were taken into the gravel and the lower ones into the clay. In spite of the close proximity of the existing exchange (see Fig. 9) no movement or distress was observed within this building during the construction period. However, the external re-entrant corner exhibited small lateral displacements (10–20 mm) into the excavation, and this led to groundwater seepage at local panel joints. The joints were subsequently sealed by cement grout but it is clear that the stability of external re-entrant corners is more critical than a normal straight section of diaphragm walling and special measures pertaining to wall design and the location and testing of anchors are recommended as follows.

(*a*) Reinforcement should be continuous in the diaphragm wall around the corner of re-entrant panels.

(*b*) A temporary capping beam should be cast if practicable around the top of the re-entrant corner and extended for two panels on either side, and should remain in position until the permanent floors or supports are installed.

(*c*) The grouted or fixed anchor zone of one row of cables must be outside the zone of influence of the active wedge of the wall parallel to this row of cables.

Fig. 9. Close-up view of anchored diaphragm wall —Shepherds Bush

(*d*) When an individual 24 hour check is carried out on an anchor supporting a re-entrant corner, there is a possibility of interaction with adjacent anchors still to be stressed. In this situation all anchors concerned (usually within three panels on either side of the corner) should be tested during the same period, preferably on one day, even if the 24 hour check has already been carried out on some of the anchors in question.

(*e*) Until more information is available on full-scale behaviour of external re-entrant corners, all anchors within three panels on either side should remain re-stressable until the permanent floors or supports are installed.

CASE STUDY NO. 3: CPF BUILDING, SINGAPORE

22. The CPF building was the first diaphragm walling contract to be executed in Singapore, and the case history is included in this Paper simply to illustrate a successful application overseas in difficult ground conditions. The anchorage contractor was Cementation Ground Engineering Ltd.

23. The site plan and panel section for this excavation are shown in Figs 10 and 11. Typical of the waterfront area of Singapore, ground conditions were highly variable (loose sands, soft marine clays, soft to stiff clay, sandstone and shales) with many old river inlet channels.

24. In such a situation it is rare for any anchored diaphragm wall enquiry to be accompanied by a comprehen-

sive site investigation report. For example it is unlikely that the boreholes will be adequate in number, depth or disposition over the site area, and rarely are samples available for inspection. The designer is therefore faced with making engineering judgements on the stratification

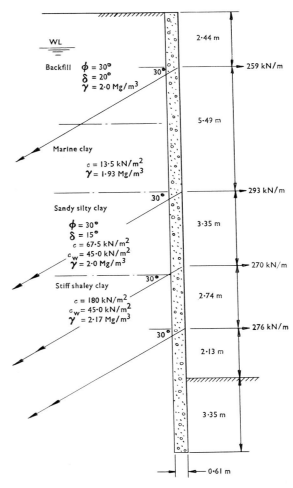

Fig. 11. *Section CC through wall—Singapore*

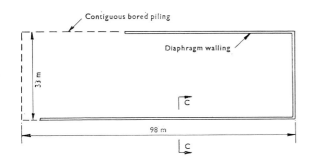

Fig. 10. *Site plan—Singapore*

Fig. 12. *General view of anchored diaphragm wall—Singapore*

characteristics and strength parameters of the ground mass not just where results are available from boreholes, but also between boreholes spaced a hundred or more metres apart. The economics of the anchored wall construction and its field problems are intimately and inextricably tied to these fundamental judgements, which must be made before any design calculations can proceed.

25. For this type of project a construction investigation is necessary, in which continued observations are used to check information previously obtained, so that as the picture of the ground conditions becomes better defined anchor positions and forces may be varied if required. For this purpose it is necessary for a rapid design method to be available, and it is noteworthy that a computer program has been produced which enables design amendments to be made quickly to take account of significant differences in strata and soil conditions as these become exposed on site. This facility proved invaluable at Singapore. However, bearing in mind that the program is now readily available to UK engineers for local and overseas contracts, care must be taken to ensure that site investigations yield appropriate or relevant data of adequate quality before the computer program is invoked to execute the calculations.

26. With regard to wall construction, the panel thickness was 0·61 m and up to four rows of anchors were installed as excavation progressed downwards to a maximum depth of 16·8 m. The presence of soft clays and a high water table led to a reduction in length of wall panels from 4·5 to 3·5 m in certain areas, these panels in general being excavated down into stiff or very stiff clay and sometimes founded on the hard shale. Where the hard shale was above final formation level the wall had to be underpinned in short sections and in these areas the contact zone between the underside of the wall and the shale was found to be quite clean, the bentonite having been swept clear by the tremied concrete.

27. Throughout the construction period no distress was observed on adjacent roads and footpaths although these were heavily laden with contractor's plant, supply lorries and the dense Singapore traffic (see Fig. 12).

CASE STUDY NO. 4: KEYBRIDGE HOUSE, VAUXHALL, LONDON

28. Figure 13 shows a plan of a large double basement excavation for Keybridge House, a new telecommunications centre situated in a built-up area, most of the surrounding buildings being old with shallow foundations. Of particular concern was the close proximity of the Waterloo line, supported on viaduct arches, and St Anne's Church, where the diaphragm wall passed adjacent to the church foundations along two sides of the old building.

29. The main contractor for this project was Taylor Woodrow Ltd and specialist contractors were Cementation Piling and Foundations Ltd and Cementation Ground Engineering Ltd.

30. A diaphragm wall, 0·61 m thick and anchored at three levels, was adopted to retain 14·5 m of soil as shown in Fig. 14. Wall panel 9 was chosen to be monitored because it was fairly remote from corners and on a straight section of wall. It became evident during the excavation that the top of the clay dipped sharply in the region of panel 9, possibly indicating an old river channel. The

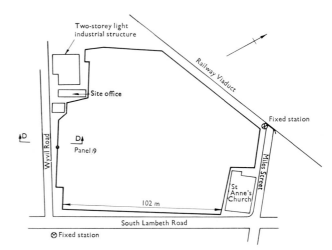

Fig. 13. Site plan—Vauxhall

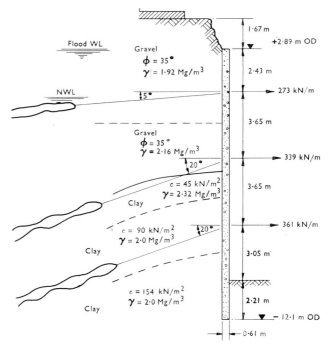

Fig. 14. Section DD through panel 9—Vauxhall

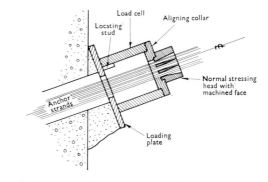

Fig. 15. Location of anchor load cell

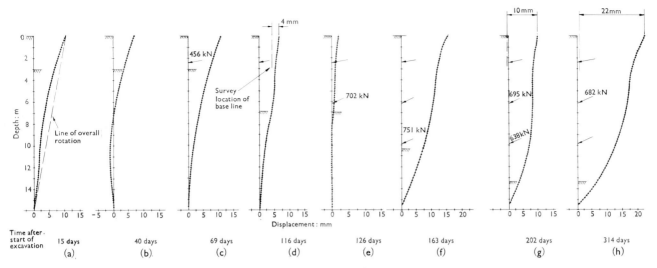

Fig. 16. Displacement profiles of panel 9 at various construction stages

presence of such channels, which appear to prevail in abundance in this district of London, necessitated considerable lengthening of certain cables in order to reach suitable strata for the grouted fixed anchor zone.

31. At panel 9, wall deformations were measured midway between two vertical rows of anchors. The method was similar to that described for Guildhall, except that an improved inclinometer[3] was employed (range ±5°, gauge length 200 mm and sensitivity 15″) and the base line distances from fixed stations to one wall station (see Fig. 13) were measured using a Tellurometer MA100 or Geodimeter in conjunction with a Kern DKM3 optional theodolite. These general surveys were supplemented by measurements of the vertical position of each station using a Kern GK23 level and staff, and closing errors indicated an accuracy of about 0·3 mm.

32. Anchor loads were measured using improved load cells fitted with eight electrical resistance strain gauges (four axial and four circumferential) to reduce effects of load eccentricity, and improved alignment of the loading system was also obtained by the use of a machined collar between the stressing head and load cell (see Fig. 15).

Discussion of results

33. Figure 16 illustrates the wall displacement profiles relative to the toe of the wall, the overall displacements being included only at the time of a general survey.

34. Profile (a) of Fig. 16 shows that an overall rotation of the wall occurred at an excavation depth of 3·05 m, similar to that of panel A12 at Guildhall. Slight bending is indicated between the excavation depth and 9·0 m, i.e. ±3 m about the upper level of the London Clay.

35. It is clear from both panels monitored over the cantilever stage (Figs 4 and 16, profile (a)) that the displacement and overall rotation of the wall represent a large proportion (about 50%) of the corresponding movements at full excavation, thus illustrating the need for early support if wall movements are to be kept to a minimum.

36. Movement of diaphragm walls and associated basement heave are closely related to the method of

support and manner of excavation. With anchored diaphragm walls the position of the top row of cables is governed by a balance between increase of the initial cantilever moments in the wall and limitation of the resultant inward movement towards the excavation. In the terraced areas of London where gravels overlie London Clay, the first row of anchors is usually located 3–4 m below ground level and the resulting inward deflexion of the wall is reduced to some extent when subsequent anchors are installed and tensioned. However, if movement of the wall must be kept to an absolute minimum, the first row of anchors must be located as close to ground level as possible; the limiting depth is usually 1·5 m, as with shallower anchorages there is a risk of local ground failure occurring behind the wall on tensioning the cables, with associated wall damage.

37. Profile (b) of Fig. 16 was monitored when all anchors had been installed and stressed to 450 kN for one week, except for one anchor immediately adjoining the inclinometer duct. Nevertheless this wall panel has been drawn back with apparent toe rotation, although the bending in the region below 12 m indicates resistance by the stiffer clay at toe level to the upper displacement of the wall by the anchors.

38. After a further 29 days when all upper level anchors were stressed to 450 kN, the wall deflexion (profile (c)) had reverted to the shape of profile (a). No major change in prestress load was monitored.

39. Following excavation to 6·8 m it was observed in profile (d) that below a depth of 6·5 m the deflexions were identical to those of profile (c), and an inward toe displacement towards the excavation of at least 2·25 mm occurred due to consolidation of the clay on the cut side. With the development of beam action below anchor level, the cantilever action within the upper 6 m almost disappeared. The degree of bending at this stage was very low, the maximum deflexion being only 0·6 mm with respect to the line of overall rotation.

40. With two levels stressed a further wall displacement towards the excavation occurred (profile (e)) and the deflexions within the upper 6 m were identical to those of profile (d), leading again to low bending moments.

41. Following excavation to 10·4 m and the stressing of all anchors the differential displacement between the upper anchor levels increased from 0·77 mm to 2·04 mm. Thus transition from profile (e) to profile (f) involved a further overall rotation of about 5 minutes of arc.

42. At the final excavation stage further rotation was indicated (profile (g)) and also toe displacement (about 1·5 mm) into the excavation. It would appear from this profile that further excavation caused outward bulging at and below lower anchor level and rotation about the toe and some point between the two upper anchors. This latter effect led to a reduction of differential displacement between crest and toe.

43. At this time the general survey indicated overall displacements of 10 mm and 0·5 mm into the excavation for the crest and toe respectively.

44. Vertical settlement readings indicated that the crest of the panel had moved down 12·2 mm, probably due mainly to the total vertical load component of the three levels of inclined anchors, equivalent to 432 kN/m².

45. After the final design stage was reached a delay of three months occurred in the construction programme, and profile (h) of Fig. 16 shows that differential displacement between crest and toe doubled, although the central anchor load exhibited only a slight loss of prestress. This indicates possible consolidation of the highly stressed soil surrounding the fixed anchor, or more likely that overall movement of the retained soil mass containing the anchors occurred. The total displacement of the crest was estimated to be 22·0 mm.

46. For the final two profiles the calculated bending moment curves are given together with the design values for the corresponding stage of excavation (see Fig. 17). Design values for the initial cantilever condition are also shown, where a modulus of 24 500 MN/m² ($3·5 \times 10^6$ lb/in²) and the full width of the wall (0·61 m) was used in calculating the second moment of area.

47. It can be observed that the magnitudes of the bending moment maxima are in good agreement with the design values although some variations in moment distribution occur. The magnitudes of peak bending moments measured are less than the design values by about 23%, although the magnitude of the design cantilever moment is similar to that created by the stressing of the upper row of anchors. However, it should be noted that the measured moments relate to the normal groundwater level whereas the design curves have been established on the basis of flood level (see Fig. 17).

48. Bending moment discontinuities at anchors occurred at slightly lower levels than the design elevations, the divergence being up to one metre. This anomaly is thought to be a function of anchor inclination and overdig before anchor installation. The moment curve above the upper anchor level is similar to one period of a sinusoidal waveform. The probable explanation for this phenomenon, as in the case at Guildhall, is the influence of the guide wall at the rear of the crest which had a depth of 1·3 m.

49. Comparison of the deformation profiles recorded up to final excavation stage indicates that bending moments were at maximum values at this stage.

50. With regard to overall behaviour the profiles indicate that more efficient anchoring was obtained with the gravel anchors and the panel exhibited rotation about the upper anchor regions. For the two anchors successfully monitored on the panel, a drop in load occurred over a period of six months. These reductions were small, being 20 kN (2·8%) for level 2 and 100 kN (12·7%) for level 3.

51. In general, crest displacements did not exceed 10 mm during the construction stage, and even at a later date the displacement was limited to 21·5 mm which corresponds to a ratio of displacement to excavated height of 1/620, and hence no noticeable settlements are likely. There is a dearth of information on acceptable movements associated with anchored walls but Jennings[4] has produced some interesting observations (see Table 1) for some large

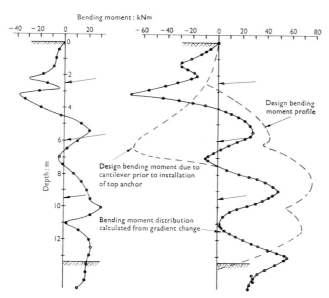

Fig. 17. Bending moment profiles of panel 9 at final excavation stage (moments based on 0·30 m strip)

Fig. 18. General view of anchored diaphragm wall—Vauxhall

Table 1

Predominant soil supported	Depth of exca-vation H, m	Crest dis-placement Δ, mm	$\dfrac{\Delta}{H}$	Remarks
Firm fissured clay	14·7	76	1:194	Damage to ser-vices in the street and buildings across the street
Firm fissured clay	14·7	38	1:388	Acceptable movement
Firm fissured clay	22·9	38	1:600	Acceptable movement
Very stiff fissured clay	14·7	19	1:773	Acceptable movement
Soft jointed rock	18·3	25	1:730	Acceptable movement

deep cuts in South Africa. From the results it would appear that crest displacements should be limited to 40 mm otherwise severe damage to adjacent services or structures may result. At Vauxhall, although the movements were acceptable, it must be emphasized that the total crest displacement more than doubled in the four month period immediately following the final excavation stage, and clearly the consolidation of the clay on the excavation side was aided by the much reduced drainage path length.

CONCLUSIONS

52. Full-scale monitoring studies indicate that wall deflexions and bending moments, which occur as excavation proceeds, follow a similar pattern to those predicted by the empirical design method. In addition the results suggest a triangular pressure distribution which is contrary to the trapezoidal distribution assumed in the design of deep strutted excavations.

53. In view of the logical progression displayed by the displacement profiles formed using inclinometer data, it is recommended that monitoring studies of the type described should continue but investigations should also be carried out on a variety of similar construction techniques, e.g. on strutted diaphragm walls and multi-tied contiguous bored piled walls.

54. Finally, with regard to all the case studies presented no significant vertical or horizontal movements of walls have been monitored during the construction period. The Authors therefore submit that the empirical design method results in economical structures exhibiting satisfactory performance in the field, always provided that those design procedures are supported by an adequate site investigation programme.

REFERENCES

1. LITTLEJOHN G. S. *et al.* Anchored diaphragm walls in sand. *Ground Engng*, 1971, **4**, Sept., 14–17, Nov., 18–21; 1972, **5**, Jan., 12–17.
2. BURLAND J. B. and MOORE J. F. A. The measurement of ground displacement around deep excavations. *Field Instrumentation in Geotechnical Engineering*. Butterworths, London, 1974, 70–84.
3. PHILLIPS S. H. E. and JAMES E. L. An inclinometer for measuring the deformation of buried structures with reference to multi-tied diaphragm walls. *Field Instrumentation in Geotechnical Engineering*. Butterworths, London, 1974, 359–369.
4. JENNINGS J. E. Discussion on Deep excavations and tunnelling in soft ground. *Proc. 7th Int. Conf. Soil Mech., Mexico*, **3**, 1969, 331–335.

High anchored wall in Genoa

Prof. Ing. G. BARLA, *Politecnico, Torino, Italy*

Dott. Ing. C. MASCARDI, *Studio Geotecnico Italiano, Milano, Italy*

The construction and behaviour of an anchored wall in Genoa is described. The main features of the work are maximum excavated height 34 m; total length 147 m; minimum distance from existing buildings 3 m; total number of anchorages 658. The crest of the wall was kept constantly under observation, measurements being taken of horizontal and vertical displacements during construction. When the excavation reached approximately mid-depth, a finite element analysis of its behaviour was performed. This confirmed assumptions made in a design revision concerning soil parameters and natural stresses in the soil mass. The displacements measured during the progress and completion of work fairly matched those predicted by use of the finite element method.

INTRODUCTION

The design of a new tall building in Genoa required a vertical cut 34 m deep in the side of a hill (Fig. 1), at a minimum distance of only 3 m from existing old houses.

These buildings have shallow foundations and uncertain statical conditions as they were built long ago, partially rebuilt and modified during the centuries and finally damaged during the second world war; it was not possible to demolish them as they represent one of the last examples of typical architecture of the ancient town (see Figs 2 and 3).

2. Subsoil conditions were explored by eight boreholes prior to wall design, and subsequently by eleven supplementary boreholes, drilled during wall construction in order to get a more precise and detailed knowledge of soil nature and parameters. The final result of soil investigations is shown in Fig. 4.

3. Choice of the retaining structure finally adopted was based on the need for varying as little as possible the existing stresses in the soil, in order to minimize related strains in the existing buildings around the excavation; and the need for flexibility of wall design, in order to allow significant variations in the retaining structure, as

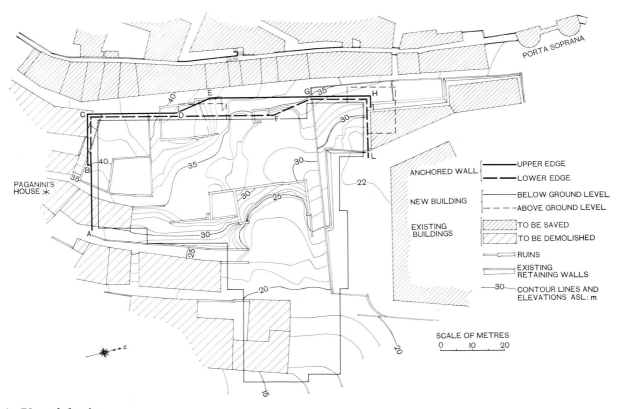

Fig. 1. Plan of the site

Fig. 2. Southern corner of the wall, with excavation in phase 5 at elevation 12·00 m, and the southern wing of the new building under construction

additional information about the final state of equilibrium of the soil was to be provided by repeated measurements during construction. When the excavation reached elevation 21 m, the behaviour of the old houses around the pit did in fact suggest a revision of the initial design.

MAIN FEATURES OF THE WORK

4. The retaining structure comprises the following main parts, arranged as shown in Fig. 5:

358 vertical bored piles, spaced 0·60–0·80 m, reinforced with steel H beams, for a total length of 5848 m;

658 grouted anchorages, Tirsol IRP type, inclined 20° to the horizontal, having nominal service loads of 569 and 853 kN, for a total length of 16 457 m;

14 rows of horizontal steel beams, linking the anchor heads with the vertical piles;

a reinforced concrete wall, 0·50 m thick, spanning from one horizontal beam to the next, able to carry 50% of the horizontal pressure on the wall, thus accounting for a reduction in pile efficiency due to corrosion.

5. The vertical piles were bored from ground level and from an intermediate elevation, with the purpose of limiting deviations at lower depths and consequent difficulties for the correct placement of horizontal beams: the wall has an upper edge and a lower one, where the second series of vertical piles was bored.

6. The excavation proceeded step by step as anchors and horizontal beams were installed: from ground level down to elevation 21 m the whole area was sliced horizontally; from elevation 21 m to the base the wall was divided into smaller zones and each was excavated by slices corresponding to the horizontal steel beams (Fig. 4).

DESIGN ASSUMPTIONS AND SITE MEASUREMENTS

7. The initial design was based on the assumption of a triangular distribution of horizontal pressure with depth, referring to an ideal active state of stress, affected by a factor of safety. The design was checked according to the limit equilibrium method of Ranke and Ostermayer.[1]

8. From the start of excavation, vertical and horizontal displacements of the crest of the wall were measured to an accuracy of 1 mm.

9. When the excavation reached elevation 21 m some fissures appeared in the ancient buildings along the longer side of the excavation. A check of the tension in the cables of the anchorages, installed later, showed increases of about 10%. Work was stopped and the supplementary boreholes were drilled.

10. These boreholes showed the presence of a heavily overconsolidated clay which was not found in the previous investigation. The measurement of overconsolidation pressure by CHG consolidation tests, and the use of the diagram of K_0 versus OCR given by Brooker and Ireland,[2] allowed an evaluation of the original horizontal pressure in the overconsolidated clay. (Deviation from K_0 conditions due to the original slope was neglected in evaluating the initial stresses in the soil.)

11. As work proceeded below elevation 21 m, the displacements of the wall were measured also at a number of vertical sections by means of pendulums, installed with the hinge at the top and the weight at various elevations. The tension in 12 anchorages was measured by means of disc-type dynamometers permanently installed: they showed a slight decrease in the load (4–5%) due to creep.

Fig. 3. Northern corner of the wall photographed at the same time as Fig. 2: in the background are the towers of the Porta Soprana

FINITE ELEMENT ANALYSIS

12. A finite element study of the anchored wall was carried out by Ordisor, Paris, using their GFPA program. This allowed simulation of the excavation process by slices and accounted for the placing of the anchorages in the soil. Each anchorage was represented in the model by a bar element. The soil–anchor interaction was accounted for by distribution of the anchor load between points situated at the two extremes, avoiding anomalous stress concentrations in the soil.

13. The numerical results obtained for the section at G (Fig. 5) are of particular interest because the available deformability and strength parameters of the soil refer to

Table 1. Soil parameters used in finite element analysis and safety factor evaluation

Soil	Deformation modulus E, MN/m^2	Poisson's ratio ν	Cohesion c, MN/m^2	Angle of friction ϕ, degrees
Silty clay	19·6	0·40	0·0196	27
Overconsolidated clay	58·8	0·40	0·0588	31·5
Limestone boulders in silty clay matrix	78·5	0·30	0·098	46
Fractured limestone or sandstone	1177	0·15	0·098	46

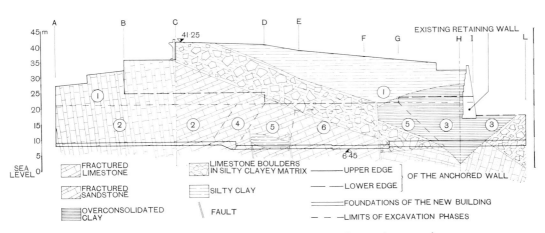

Fig. 4. Vertical section along the wall: numerals indicate the order of main phases of excavation

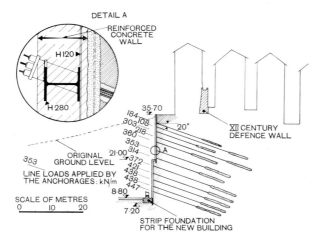

Paper 16

Fig. 5. *Vertical cross-section on vertex G. (Bottom elevation was changed from 9·40 to 8·80 m during construction.)*

samples taken from boreholes located in the near vicinity of this section. The finite element model used to represent the section is shown in Fig. 6. It consisted of 201 nodal points and 174 elements. Plane strain conditions were assumed. The soil was taken as linearly elastic and isotropic, with material parameters as shown in Table 1. The nodal points along the vertical outer boundaries were set as free to move in the vertical direction. The nodal points along the horizontal boundary were taken as fixed.

14. The initial state of stress in the soil was assumed to be given by a vertical stress due simply to the gravity load and a horizontal stress distribution as shown in Fig. 7. The horizontal stress in the overconsolidated clay layer was evaluated on the basis of a K_0 value assigned according to the K_0 versus OCR relationship.[2]

15. Eight different excavation steps were considered in the finite element analysis. The actual construction sequence for placing the anchorages in the soil was

reproduced as closely as possible. The loads applied to the lines of anchors were set equal to those shown in Fig. 5. No account was taken of the influence of the vertical piles and horizontal steel beams.

STRESS DISTRIBUTION IN THE SOIL

16. Figure 7 shows the distribution of stresses in the soil adjacent to the anchored wall for excavation level at elevation 9·4 m. Also shown are the elements subjected to tensile stress. Tensile stresses develop in the silty clay as the excavation proceeds (Fig. 8), first arising close to the wall and propagating towards the right vertical boundary of the model.

17. Table 1 shows the shear strength parameters c and ϕ of the soil, derived from laboratory tests performed on samples taken from boreholes adjacent to the section under consideration. Using these parameters a stability analysis was made of the anchored wall on the basis of stress level only.

18. A safety factor F_{SL}, defined as the ratio of the deviator stress at failure to the mobilized deviator stress,[3] was evaluated for each element of the structure in its final configuration. Figure 8 shows that F_{SL} is less than 1·5 only in the silty clay, and generally reaches extremely high values in the remaining areas, indicating effective anchorages and appropriate design.

19. The use of the linearly elastic analysis of the structure is justified on the same basis: it can therefore be anticipated that the application of a nonlinearly elastic scheme would not mean significant changes in the displacement values discussed below.

COMPUTED AND MEASURED DISPLACEMENTS

20. Figure 9 shows a comparison between measured and computed values for the vertical and horizontal displacements at the crest of the wall, at each stage of excavation.

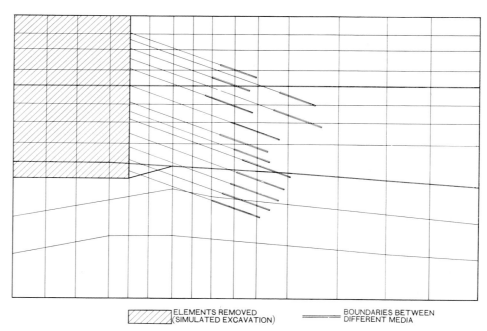

Fig. 6. *Finite element model of section on vertex G*

126

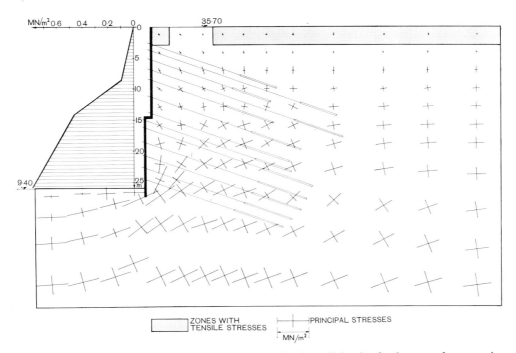

Fig. 7. Initial horizontal stress distribution, and principal stresses in the soil in the final stage of excavation

21. For the horizontal displacements of point A in the soil, the values computed by means of the finite element analysis match field observations made during wall construction and at work completion. The horizontal displacement measured near to the crest of section H (only the value at work completion is shown in Fig. 9), where similar subsoil conditions occur (Fig. 5), confirms the good agreement between computed and measured values.

22. For the vertical displacements, a distinction is made between displacements for point A, pertaining to

the soil, and B, considered as jointed to the vertical piles. The displacement of B is evaluated under the assumption that this point follows the behaviour of a point located on the same vertical at the base of the wall. Values predicted in this way for point B agree remarkably well with the measured vertical displacements at the crest of the wall.

23. The agreement between field observations and numerical predictions for displacement values confirms that an appropriate choice was made for the material parameters and natural stress distribution in the soil. The finite element analysis was carried out when the

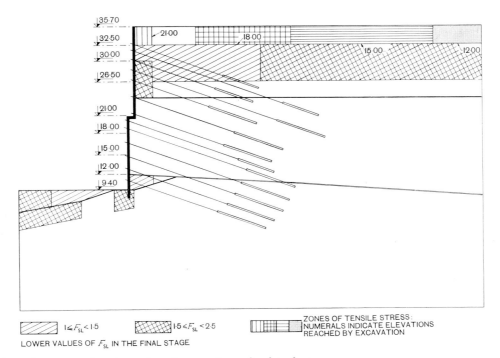

Fig. 8. Zones of tensile stress and regions where lower values of safety factor occur

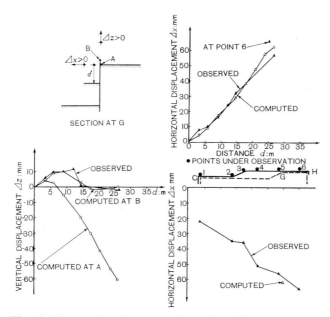

Fig. 9. Comparison between observed and computed displacements

excavation reached elevation 21 m, allowing the values of the final displacements in the structure to be inferred correctly in advance, as was subsequently checked by means of measurements.

ACKNOWLEDGEMENTS

24. For their permission to publish the data reported in the present Paper, the Authors are grateful to Immobiliare San Gallo, Genova, proprietors; Ing. Giovanni Rodio, SpA, Milano, contractors for special works; and Ordisor, SA, Paris.

REFERENCES

1. RANKE A. and OSTERMAYER H. Beitrag zur Stabilitätsuntersuchung mehrfach verankerter Baugrubenumschließungen. *Bautechnik*, 1968, **45**, Oct., No. 10, 341–50.
2. BROOKER E. W. and IRELAND H. O. Earth pressure at rest related to stress history. *Can. Geotechnical J.*, 1965, **2**, Feb., No. 1, 1–15.
3. KULHAWY F. H. *et al. Finite element analyses of stresses and movements in embankments during construction.* University of California, Berkeley, 1969, Report TE 69–4, Off. of Res. Serv.

Some experiences with ground anchors in London

J. M. MITCHELL, MIStructE, *Geotechnics Division, Ove Arup & Partners*

Experience with ground anchors on five sites in London was not satisfactory. No details of the sites are given, but aspects of developing a suitable method to assess the performance of anchors are discussed, together with acceptability criteria. Comment is made on contractual procedures and responsibility for design is discussed.

INTRODUCTION

Some form of support system is normally required during the construction of deep basements. Since props or berms may hinder the works, the use of ground anchors to allow unobstructed excavation has obvious merit.

2. The Author's field experience with ground anchors, employed to retain the sides of basement excavations, began in 1968 and now includes five sites in the London area. Anchors were installed in the terrace gravels and the underlying London Clay to support either steel sheet piles or concrete diaphragm walls.

3. The experiences on these sites were far from satisfactory. Details of these sites are not given, but this Paper attempts to describe some of the problems that were encountered, and the testing procedure that was evolved to assess the performance of ground anchors. Criteria for the acceptance of ground anchors were developed and typical test results are given to illustrate the approach.

4. Drawing from this experience some suggestions are made which it is felt might provide a more satisfactory contract procedure when ground anchors are used. Such procedure must involve responsibility for design and this is also briefly discussed.

FIELD EXPERIENCE

5. The difficulties that occurred indicated that a realistic testing procedure to assess the performance of ground anchors was the principal requirement for site quality control. Some of the more important points which emerged as necessary for an acceptable testing procedure were that sufficient strand should be provided in the tendon so that any test load does not exceed 70% of the ultimate tensile strength of the strand, that the development of the 'free' anchorage length should be ensured, and that acceptability criteria should be established. In addition, an appropriate sequence should be employed to load the tendon, if it is formed with several strands.

Loading sequence

6. Ideally the jack used to apply the load to the tendon should pull all the strands simultaneously. This was not possible on four of the five sites; on the fifth site single Macalloy bars were used for the anchor tendon.

7. An example shows the difficulties that can occur. On one of the sites, each of the two strands comprising the anchor tendons was loaded separately to the test load. No direct readings of load were made but strain gauge readings indicated that the load on the first strand relaxed during the loading of the second strand. This behaviour is shown in Fig. 1 and clearly indicates that the loading procedure was unsatisfactory. Because the stressing systems employed for the succeeding sites again only allowed the tensioning of one strand at a time, the following procedure was developed to avoid a repetition of the behaviour illustrated in Fig. 1.

8. A small initial load was applied to bed each strand of the tendon. The remaining load was then applied in four or five equal increments with each strand being loaded in turn in a specific sequence, such as is shown in Fig. 2, adopted in an attempt to equalize the load applied to the 'fixed' anchorage length during stressing. After every increment, the actual load on each strand was checked using the jack in conjunction with a loading bridge. The collet and wedges securing the strand were lifted clear of the thrust plate, and the separation was measured by the use of a feeler gauge. A typical arrangement for this procedure is also shown in Fig. 2.

9. In a tendon consisting of five or six strands, the greatest reduction of load always occurred in the first strand loaded, and the least reduction of load occurred in the last strand. With the use of the procedure described above, the difference in load between the first and the last strand after a complete loading cycle was generally found to be small. To illustrate the method, Table 1 gives details of the record of a typical test, including the proposed load on each strand and the results of load checks. In Fig. 3 the results recorded in Table 1 are plotted as a load–extension diagram. Average values for both load and extension have been taken to obtain the points plotted on the figure.

Providing sufficient strand

10. Since the consequence of the failure of ground anchors can be severe, it is prudent to provide the facility to test any contract anchor to 150% working load.[1] This provision also enables random testing of contract anchors, which is a vital part of quality control. Also, should an anchor fail the test, similar tests on further anchors can be carried out. It is suggested that prestressed concrete practice should be followed, and the stress at 150% working load in the anchor tendon should be limited to 70% of its ultimate tensile strength. The application of this limit should prevent any non-recoverable or 'plastic' extension of the strand or bar.

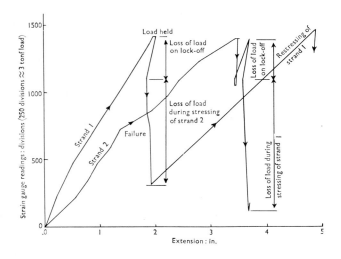

Fig. 1. Tendon loading sequence

11. The provision of sufficient strand was not adopted as a general rule on four of the five sites, and several alternative testing methods were used. For example, on the site relating to Fig. 1 the anchors were tested to 133% working load with the strand stressed to 75% of its ultimate tensile strength. On another site, for anchors installed in terrace gravels, the procedure employed was to stress the tendon to 80% of its ultimate tensile strength, and hold the load for five minutes, to give a test load of 138% working load. This was then reduced to 70% of the ultimate tensile strength of the strand, to give a test load of 120% working load. Loss of load was then monitored for 48 h. These tests are also discussed in §§ 12–14.

Development of the 'free' anchorage length

12. In the UK it is common practice to completely fill the anchor hole with cement grout. Anchor tendons with a short required life are usually bound with grease-impregnated tape to allow movement of the tendon within

Table 1. *Anchor test record*

Increment no.	Pump gauge value	Strand no.	Initial load, lb/in.²	Load check,* lb/in.²	Jack ram movement before loading, in.	Jack ram movement before lock-off, in.	Length of strand after lock-off, in.	Second load, lb/in.²	Second load check, lb/in.²	Final strand extension, in.
1	1100 lb/in.² (20 tons)	1	1200	—	$18\frac{11}{16}$	$18\frac{7}{8}$	$30\frac{13}{16}$	—	—	—
		2	1200	—	$18\frac{14}{16}$	$18\frac{15}{16}$	$26\frac{15}{16}$	—	—	—
		3	1200	—	$18\frac{5}{8}$	$18\frac{3}{4}$	$26\frac{13}{16}$	—	—	—
		4	1200	—	$18\frac{11}{16}$	$18\frac{13}{16}$	30	—	—	—
		5	1200	—	$18\frac{11}{16}$	$18\frac{13}{16}$	$29\frac{11}{16}$	—	—	—
		6	1200	—	$18\frac{3}{4}$	$18\frac{7}{8}$	$28\frac{1}{8}$	—	—	—
2	2500 lb/in.² (40 tons)	1	2700	1100	$18\frac{11}{16}$ ($18\frac{9}{16}$†)	$19\frac{1}{8}$ ($18\frac{7}{8}$†)	31	—	—	—
		2	2600	1300	$18\frac{5}{8}$ ($18\frac{9}{16}$†)	$18\frac{15}{16}$ ($18\frac{13}{16}$†)	$27\frac{1}{8}$	—	—	—
		3	2600	2000	$18\frac{9}{16}$	$18\frac{13}{16}$	$26\frac{15}{16}$	—	—	—
		4	2600	—	$18\frac{5}{8}$	$18\frac{7}{8}$	$30\frac{3}{16}$	—	—	—
		5	2600	—	$18\frac{5}{8}$	$18\frac{15}{16}$	$29\frac{15}{16}$	—	—	—
		6	2600	—	$18\frac{9}{16}$	$18\frac{7}{8}$	$28\frac{3}{16}$	—	—	—
3	3600 lb/in.² (60 tons)	1	3800	2700	$18\frac{9}{16}$	19	$31\frac{5}{16}$	—	—	—
		2	3800	2700	$18\frac{7}{8}$	$19\frac{1}{4}$	$27\frac{3}{8}$	—	—	—
		3	3800	3000	$18\frac{5}{8}$	$19\frac{1}{16}$	$27\frac{7}{8}$	—	—	—
		4	3800	2600	$18\frac{3}{4}$	$19\frac{11}{16}$	$30\frac{9}{16}$	—	—	—
		5	4000	3000	$18\frac{5}{8}$	$19\frac{1}{16}$	$30\frac{1}{4}$	—	—	—
		6	4000	3000	$18\frac{5}{8}$	$19\frac{1}{16}$	$28\frac{5}{8}$	—	—	—
4	4700 lb/in.² (80 tons)	1	4900	3300	$18\frac{11}{16}$	$19\frac{3}{16}$	$31\frac{5}{8}$	5200	3600	$31\frac{15}{16}$
		2	4800	3400	$18\frac{15}{16}$	$19\frac{3}{8}$	$27\frac{7}{8}$	5000	4200	$27\frac{7}{8}$
		3	4900	3300	$18\frac{5}{8}$	$19\frac{1}{8}$	$27\frac{1}{2}$	5000	4000	$27\frac{13}{16}$
		4	4800	4000	$18\frac{5}{8}$	$19\frac{1}{8}$	$30\frac{7}{8}$	5000	4100	$31\frac{1}{16}$
		5	4900	4000	$18\frac{5}{8}$	$19\frac{3}{16}$	$30\frac{11}{16}$	4900	4800	$30\frac{15}{16}$
		6	4800	3800	$18\frac{9}{16}$	$19\frac{1}{8}$	29	4800	4700	$29\frac{3}{16}$
5	5800 lb/in.² (94 tons)	1	5900	4400	$18\frac{15}{16}$	$19\frac{3}{8}$	$32\frac{3}{8}$	5800	4800	$32\frac{3}{8}$
		2	5900	4500	$18\frac{7}{8}$	$19\frac{5}{16}$	$28\frac{1}{16}$	6000	4300	$28\frac{3}{16}$
		3	5800	4500	$18\frac{7}{8}$	$19\frac{3}{8}$	$28\frac{1}{16}$	6000	4700	$28\frac{1}{4}$
		4	5900	5000	$19\frac{3}{16}$	$19\frac{11}{16}$	$31\frac{3}{8}$	6000	4700	$31\frac{1}{2}$
		5	5800	4600	$18\frac{13}{16}$	$19\frac{5}{16}$	$31\frac{1}{8}$	6000	5200	$31\frac{7}{16}$
		6	5900	4500	$18\frac{15}{16}$	$19\frac{7}{16}$	$29\frac{1}{2}$	6000	6200	$29\frac{3}{4}$
6	6300 lb/in.² (110 tons)	1	6300	5000	$19\frac{1}{4}$	$19\frac{9}{16}$	$32\frac{9}{16}$	6400	4900	$32\frac{3}{4}$
		2	6300	4600	$19\frac{1}{4}$	$19\frac{5}{8}$	$28\frac{7}{16}$	6300	4800	$28\frac{9}{16}$
		3	6300	4600	$19\frac{1}{16}$	$19\frac{1}{2}$	$28\frac{3}{8}$	6300	5000	$28\frac{1}{2}$
		4	6200	5000	$19\frac{11}{16}$	$19\frac{7}{16}$	$31\frac{11}{16}$	6300	5200	$31\frac{3}{16}$
		5	6200	5000	$19\frac{1}{16}$	$19\frac{7}{16}$	$31\frac{1}{2}$	6200	5500	$31\frac{5}{8}$
		6	5200	4700	19	$19\frac{5}{16}$	$29\frac{13}{16}$	6300	5800	$29\frac{15}{16}$

★ Load checked after development of cracks behind wall and sudden loss of load.
† Check value.

the free anchorage length. If this movement is not allowed to take place, the use of a load–extension plot to assess the performance of an anchor becomes more difficult.

13. For example, Fig. 4 gives details of the load–extension curves of four anchors installed in terrace gravels, and loaded using the procedure described in § 11. The theoretical elastic extension of the free anchorage length of 26 ft has also been plotted, and an examination of the extensions indicates that the effective length of the strand was shorter than the free length. Grout may have escaped into the loose fill behind the concrete diaphragm wall during the final stages of grouting, and it is possible that the shaft of grout was in fact bearing against the wall, and acted as a strut during the test.

14. To ensure complete freedom of movement of the anchor tendon, it is recommended that the practice should be adopted as in the USA[2] of backfilling the free anchorage length with sand, sand and gravel, weak grout, or stone chippings.

Anchor test method

15. The test method chosen should allow the selection of realistic acceptability criteria.

16. The creep behaviour of the anchor should be investigated. The test load of 150% working load is therefore applied, and the load remeasured after 24 h.

17. To expedite the testing only one cycle of loading is used. The loading sequence is described in § 8. Measurements are made of the extension of the strand during tensioning, using a steel tape with the thrust plate as a datum; the load on each strand, by means of the calibrated jack pump gauge; and the movement of the walings and steel sheet piles during loading.

18. Table 1 gives details of the record of a typical test. The jack ram movement was measured before and after loading to give the full extension of each strand. This movement should be compared with the change in length of strand which has passed through the jack (see Fig. 2). Any difference in the values is due to the slight relaxation of load that occurs when the wedges are hammered home to lock off the strand. An examination of the extensions recorded in Table 1 indicates that these losses decreased to negligible amounts as the test proceeded.

19. Anchor failure is generally denoted during the loading sequence by excessive extensions and loss of load during tensioning, as illustrated in Fig. 3.

20. The test load on anchors that appear to be satisfactory during the loading stages is measured again after 24 h, using the loading bridge in conjunction with the jack pump gauge.

21. Insignificant movements of the walings and steel sheet piles were recorded on the three sites where measurements were made.

ACCEPTABILITY CRITERIA

22. The criteria developed are
(*a*) the load–extension plot should compare reasonably with the theoretical elastic extension of the free anchorage length;
(*b*) there should be no more than 5% loss of load after 24 h, after initially loading to 150% working load.

23. In order to assess whether an anchor is likely to give a factor of safety of at least 2·0 against failure, the load–extension curve above the value of the test load is

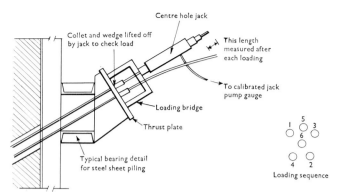

Fig. 2. *Load measuring arrangement and strand loading sequence*

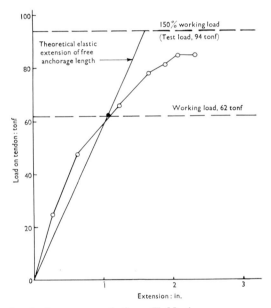

Fig. 3. *Anchor test result from Table 1*

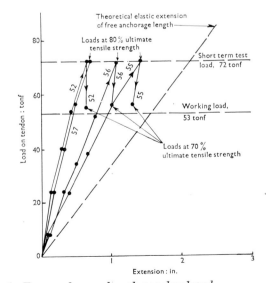

Fig. 4. *Free anchorage length not developed*

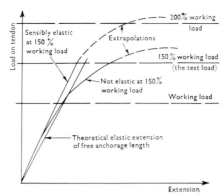

Fig. 5. Concept of extrapolating test results

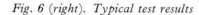

Fig. 6 (right). Typical test results

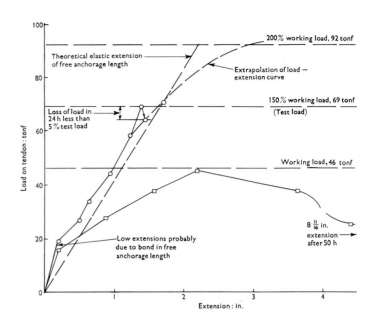

sketched in as illustrated in Fig. 5. A similar approach has been put forward,[3] but is based on a range of extensions measured for a large number of anchors tested to failure. It is believed, however, that the criterion regarding loss of load in 24 h is a better indication of the likely long term behaviour of the anchor.

24. Figure 6 illustrates the use of the two criteria for tests on two typical anchors. One test result is obviously a failure, while the second result satisfies both criteria. The extrapolation of the load–extension curve also indicates that the second anchor is adequate. When the results of a test satisfy criterion (*a*) but there is more than the 5% permissible loss of load in 24 h, the test load is again applied, and the load measured after a further 24 h. This is repeated until the second criterion is satisfied, or a reduced working load is estimated and agreed.

SOME PROPOSALS ON THE BASIS OF FIELD EXPERIENCE

25. The unsatisfactory nature of the field experiences included two sites where anchors had to be replaced by strutting. On the other sites the anchor loads were downgraded to 50%, following a series of poor test results. No

Table 2. Anchor record

Anchor reference number
Diameter of borehole
Direction and dip of borehole
Date of boring
Date of grouting
Date of stressing
Length and diameter of temporary lining used during boring
Consistency, colour, structure and type of various soil strata penetrated
Details of obstructions encountered
Details of tendon installed
Length of free anchorage
Length of fixed anchorage
Size and spacing of any enlargements in fixed anchorage length
Grout quantity injected and pressures employed
Grout test results
Tendon extension during loading (including any test results)
Maximum and minimum daily site temperature on each day that grouting loading and monitoring is carried out
Measured loads

trial anchor tests were carried out on any of the five sites, and it is suggested that this is one of the reasons for the difficulties that occurred.

26. It is fair to say that the understanding of the behaviour of ground anchors is at present incomplete. Therefore only anchor subcontractors who have extensive technical support and the ability to deal with the unexpected should be employed. The provision of sufficient strand to permit testing of all anchors, the proper creation of the free anchorage length, and a preliminary programme of trial anchors all require added expenditure. Such costs are small when set against the cost of failure and the replacement of anchors by struts. Certain anchor subcontractors instrument their anchors to monitor load, and this practice should be encouraged.

27. It is not suggested that the expenditure of more money will necessarily overcome all problems with ground anchors, but if it is spent in the manner outlined above it should go a long way towards minimizing problems.

28. A recommended contract procedure incorporating the above suggestions might be as follows.

(*a*) Install and complete testing of trial anchors. These tests serve three main purposes: to confirm the suitability of the proposed method of installation of the contract anchors; to confirm the design of the fixed anchorage; and to develop performance criteria to control the quality of the contract anchors. Trial anchor tests should be carried out on anchors installed in each stratum sustaining the anchor load, and at the same angle of inclination as that proposed for the contract anchors. Consideration should also be given to the excavation of some trial anchors as a fundamental test.[4]

(*b*) Interpret the test results and agree modifications if necessary.

(*c*) Commence construction of contract anchors, all with the facility to be tested to 150% working load. Each anchor should have a complete record of load–extension as given in Table 1.

(*d*) Carry out random tests on the contract anchors to 150% working load with the test load being remeasured after 24 h.

(*e*) Monitor the performance of the ground support system during the progress of the works: include checking the loads on selected anchors, with load cells if possible; line and level surveys of the tops of the anchored walls; visual, and if possible line and level, surveys of adjacent buildings and roadways.

29. Good records are essential. The information required by the Author's firm is listed in Table 2. The requirements are quoted in full to illustrate that the anchor subcontractor should provide a full-time supervisor for the contract. The designer (see §§ 30–33) should also have a representative present to ensure that the contract procedure given in (*a*)–(*e*) of § 28 is being correctly implemented. With such personnel present any variations in either the gound conditions or the construction of individual anchors[5] can be quickly assessed and appropriate action taken.

RESPONSIBILITY FOR DESIGN

30. Anchors often form part of temporary works and the assignment of responsibility for these works is not always clear, particularly when the contract is under the Royal Institute of British Architects (RIBA) Standard Form of Contract. Two situations have to be considered: firstly, when the Engineer proposes that ground anchors should be used as part of the ground support system; secondly, when the Contractor makes the proposal. In an attempt to clarify the position regarding responsibility for design of temporary works the following approach is suggested.

31. When the Engineer proposes the use of ground anchors:

(*a*) the design of the ground support system using anchors shall be the responsibility of the Engineer;

(*b*) the implementation of the design shall be the responsibility of the Contractor;

(*c*) individual ground anchors shall be installed and tested in accordance with the Engineer's specification.

32. When the Contractor proposes the use of ground anchors:

(*a*) the design of the ground support system shall be the responsibility of the Contractor and shall be to the approval of the Engineer;

(*b*) the implementation of the design shall be the responsibility of the Contractor;

(*c*) the individual anchors shall be installed and tested in accordance with a specification to the approval of the Engineer.

33. In the first of the cases described above, the Engineer's specification would be embodied in the contract documents and be mandatory on contractors tendering for the contract.

CONCLUDING REMARKS

34. Field experience of Ove Arup & Partners with ground anchors in London has been far from satisfactory. However, much has been learned and this has led to the development of a simple testing procedure for assessing anchor performance.

35. Recommendations are given on the contract procedure which should be adopted, together with some thoughts regarding the assignment of the responsibility for the design of ground anchors when they form part of the temporary works. It is hoped that these suggestions will assist in creating an improved technical and contractual basis for the use of ground anchors.

ACKNOWLEDGEMENTS

36. The Author wishes to thank the resident engineers, contractors and subcontractors at the sites concerned for their co-operation and assistance with the development of a testing procedure. The Author would like to thank his colleagues K. W. Cole for his guidance during the works described and F. G. Coffin, Dr J. A. Lord, and D. F. T. Nash for their helpful comments during the drafting of the Paper.

REFERENCES

1. LITTLEJOHN G. S. Ground anchors today—a foreword. *Ground Engng*, 1973, **6**, Nov., 20–22.
2. WHITE R. E. Anchorage practice in the United States. *Consult. Engr*, 1970, May, Suppl. *Ground Anchors*, 32–37.
3. BASSETT R. H. Discussion on Soil anchors. *Ground Engineering*. Institution of Civil Engineers, London, 1970, 89–94.
4. OSTERMAYER H. Discussion. *Proc. 5th European Conf. Soil Mech. Foundn Engng*. Sociedad Espanola de Mec. del Suelo y Cimentaciones, Madrid, 1972, **2**, 334–336.
5. BASSETT R. H. Discussion. *Proc. 5th European Conf. Soil Mech. Foundn Engng*. Sociedad Espanola de Mec. del Suelo y Cimentaciones, Madrid, 1972, **2**, 330–334.

Discussion on Papers 14–17

Reported by J. C. McKENZIE

The Chairman, Dr T. Whitaker, opened the session by pointing out that, in the section of geotechnics covered by the Conference, all those involved, dealing as they were with a very complex material, had learned to use caution when applying the recognized mathematical formulae. These had generally been designed to obtain solutions by over-simplifying the conditions. Advances from this state must be by painstaking measurement of what happens in practice. By such work, the over-simplification could be corrected next time. The Papers of this session were directed towards this end.

2. The monitoring involved not only helped future projects but also was of unquestioned value as a check on the design and safety of any project.

3. The Authors generally introduced their Papers by highlighting the main points. In the case of Paper 17, it was stressed that the design of anchors is empirical and that the Paper was based on results achieved on five sites in one area, indicating the need for more precise rules for procedure and checking of the installation as it proceeds.

ANALYSIS OF WALL BEHAVIOUR

4. Dr P. L. Bransby spoke about the relationship between the settlements behind a diaphragm wall and the lateral displacements of the wall. The topic is touched on in Paper 15 by Littlejohn and MacFarlane, who remark that there is little published data available.

5. He presented a simple analysis giving some insight into the behaviour of the soil behind the wall. The analysis relates the immediate (i.e. not consolidation) settlements of the ground to the lateral displacements of the wall.[1] Figure 1 shows both the initial (AB) and the deflected (CB) shape of a cantilever wall. The slip lines[1] which control the deformations in the soil behind the wall are inclined at $\pi/4 - \nu/2$ to the vertical, where ν is the angle of dilation and gives a measure of the amount by which the soil expands during shear. The displacement vectors for all points along each slip line such as GH are all of constant magnitude, fixed by the displacement of the wall at G, and are inclined at ν to GH. Thus, the deflexion of the wall fixes the magnitude of the displacement vector for each slip line, and the displacement, and hence settlement, of the surface of the soil behind the wall can easily be determined.

6. The predictions given by the analysis for the case of a model flexible cantilever wall in dense sand are compared[2] with the observed settlements in Fig. 2. The agreement between prediction and observation is excellent.

7. The analysis is even simpler when the soil deforms at constant volume (i.e. $\nu = 0$), as might be the case for undrained clay or loose sand. The slip lines are inclined at $\pi/4$ to the vertical (Fig. 3) and in this case the displacement vectors are directed along the slip lines, though again the magnitude of the displacement along any slip line is fixed by the movement of the wall at the point

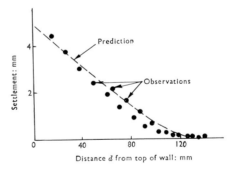

Fig. 2. Predicted and observed settlements[2] behind a model sheet pile wall in dense sand. (The distance d is measured from the point A shown in Fig. 1.)

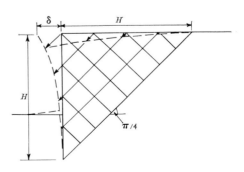

Fig. 1. Slip line field behind a cantilever wall when $\nu \neq 0$

Fig. 3. Slip line field behind a cantilever wall for a material which deforms at constant volume

Diaphragm walls and anchorages. Institution of Civil Engineers, London, 1975, 135–140.

135

where the wall intersects the slip line. The horizontal and vertical components of any displacement vector are now equal, and so the vertical settlement of a point at the surface of the soil is the same as the lateral movement of the point on the wall on the same slip line. The curve of settlement of the surface against horizontal distance from the top of the wall therefore has exactly the same shape and magnitude as the curve of the lateral displacement of the wall against vertical distance from the top of the wall.

8. Figure 4, obtained from Peck,[3] shows surface settlements near a braced sheet pile wall and lateral deflexions of the wall, plotted to the same scale for different stages of excavation. The most direct comparison of settlement and lateral displacement is given by the data for days 44 and 45, when the similarity between the shapes of the settlement and lateral displacement is striking, the maximum movement being about 10 cm in each case.

9. In summary, Dr Bransby said that the maximum settlement is equal to the maximum movement of the wall; the ratio of deflexion of the top of the wall to the height of the wall gives a rough measure of the angular distortion of the ground surface; and the deflected shape of the wall controls the ground settlements.

10. He emphasized that the analysis applied for immediate settlements only, and that consolidation settlements (or swelling), including those due to groundwater changes, must be added.

11. Dr Saxena pointed out in his reply that the point of rotation is often unknown and so the drawing of slip lines is very difficult. This is particularly true after the ties have been placed. He also thought that the lateral displacement of the wall determined the displacement behind the wall. An investigation was required in which load cells were used in the vicinity of the wall. In most

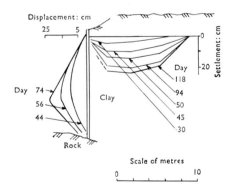

Fig. 4. Observed settlements behind and lateral displacements of a braced sheet pile wall in soft clay, at various stages of excavation of the cut[3]

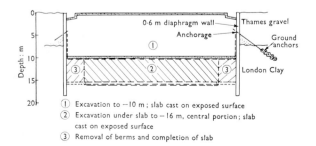

Fig. 5. Central London YMCA: excavation sequence

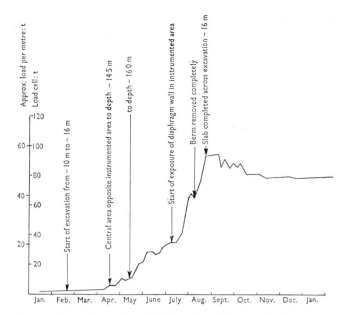

Fig. 6. Central London YMCA: load changes during excavation (load cell −10·2)

cases this had not been done; a knowledge of the actual pressures involved would be useful.

CASE HISTORIES OF WALL BEHAVIOUR
Central London YMCA

12. Mr H. D. St John presented some results from an investigation carried out by the Building Research Station of the behaviour of the ground around the foundations for the new Central London YMCA in Tottenham Court Road. The basement is 16 m deep and is bounded by 0·6 m diaphragm walls of depth 18–19 m over a plan area of approximately 40 m × 70 m. The surrounding ground consists of 7 m of gravel and fill overlying a 17 m bed of London Clay.

13. Figure 5 shows a cross section of the basement. The diaphragm wall was constructed during demolition of the existing building, and excavation to −10 m was carried out with support provided by a slab at ground floor level, and a single row of ground anchors at −4 m. A 300 mm slab was cast on blinding on the exposed clay surface at −10 m. Excavation of the 6 m below this slab was the most critical stage, largely because of the very limited toe to the wall (2−3 m). When the basement is excavated to depth adjacent to the wall, the immediate effect is to induce high horizontal stresses in the soil in front of the wall, and to reduce vertical stresses in this region to virtually zero. Subsequently, the effect of drainage and time is to reduce the resistance provided by the soil by a considerable amount. This means that, adjacent to the wall, it is important to ensure that low vertical stresses exist for as short a time as possible before adequate support can be provided for the wall.

14. The method of excavation adopted was, therefore, to remove the central portion, leaving only a surcharge in the form of a high berm 4−5 m wide around the edge; to cast a heavy slab in this area; then to remove the berm in sections, propping the wall temporarily from the slab before completing the slab to the wall.

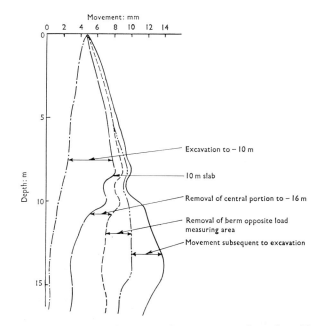

Fig. 7. Central London YMCA: movement of south wall

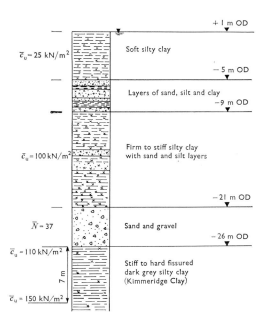

Fig. 8. Humber Bridge: typical borehole log

15. A number of instruments including inclinometers were installed, numerous levelling points were fixed around the structure, and a network of survey stations was used for measuring movement of the ground outside the site, and as a control to internal measurements. Over a section of the wall perimeter all horizontal loads transmitted from the wall to the structure were measured by load cells, inserted before the props were loaded.

16. During the initial stages of excavation small load changes only were measured at the ground floor level and on the ground anchors. Figure 6 shows the build-up of load on a typical load cell at the main propping slab (−10 m). Only a small portion of the total load was developed during excavation over the central area from −10 m to −16 m. As the operation of removing the berm approached the monitored area, the load started to rise more rapidly, and a brief lull occurred while this area was passed before a rapid and dramatic increase which was curbed immediately by the pouring of the slab to the wall. Since that time, the load had fluctuated but shown no marked trends.

17. Figure 7 shows the measurements of wall movement at the load measuring section, taken using an inclinometer. During the final stage the toe has kicked in, and the curvature of the wall has increased in the vicinity of the propping slab at −10 m.

18. The observations illustrate the effectiveness of leaving a fairly limited berm in this situation, and emphasize the care required when excavating close to the toe of a wall at this sort of depth, where drainage paths are short, and drainage could considerably reduce the available toe resistance.

Humber Bridge

19. Mr J. R. Busbridge's contribution concerned the use of diaphragm walls in the construction of the south anchorage of the Humber Bridge.

20. Figure 8 shows a typical bore log for the ground

at the anchorage site. The Kimmeridge Clay is a stiff to hard overconsolidated and severely fissured clay, extremely susceptible to softening. If immersed in water, it tends to completely disintegrate within a few hours. This property greatly influenced the design of the anchorage foundation. Plate tests carried out in the Kimmeridge Clay gave the relationship shown in the figure for undrained shear strength and depth.

21. The south anchorage was designed to resist a mainly horizontal pull of 38 000 t. After consideration of various alternatives it was decided that the best solution was to construct a diaphragm wall gravity anchorage at a founding level of −29 m OD, i.e. 3 m into the Kimmeridge Clay.

22. The anchorage is shown in Fig. 9. The overall design concept was to prevent softening of the clay due to water entering fissures which would open on the removal of overburden. The diaphragm walls assisted in various ways: the excavation could be carried out in the dry; five discrete trenches could be excavated individually, so localizing stress relief in the clay; during their construction, high lateral at rest pressures were relieved in the clay; the water present in the sand and gravel was cut off; and heave was reduced by the penetration of a rigid diaphragm below excavation level.

23. An initial excavation was carried out to −4·5 m OD. From this level, 800 mm thick diaphragm walls were constructed to a depth of 28·5 m. At the same time the phreatic surface for the sand and gravel was reduced to about −15 m OD.

24. In view of various uncertainties in the design, particularly as to the rate of softening of fissured clay, it was decided to monitor the excavation.

25. The first trenches to be excavated were the central trench and the west outer trench. Various instruments were installed, including inclinometers, piezometers and borehole extensometers for measuring heave. Two vertical sections of struts were instrumented for load.

26. A typical profile given by one of the inclinometers

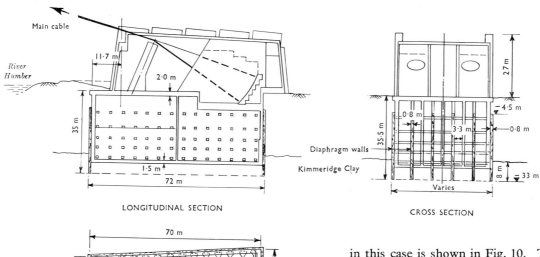

LONGITUDINAL SECTION

CROSS SECTION

PLAN OF CELLULAR BLOCK

Fig. 9. Humber Bridge: Barton anchorage

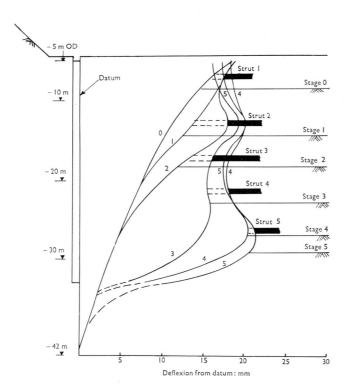

Fig. 10. Humber Bridge: deflexion profiles as given by averaged inclinometer readings

in this case is shown in Fig. 10. The wall is strutted and no preload was applied.

27. The lines shown in Fig. 10 are smooth lines drawn through average deflexions from the datum for each excavation stage. The datum is taken from only one reading but other lines represent the average of at least six readings.

28. The inclinometer was positioned about 1 m from the back of the outer diaphragm wall down to a level of −42 m OD. The instrument used was the standard Soil Instruments model with readings taken at 1 m intervals. Certain observations can be made as to general deflexions of the wall, with reference to the bottom of the inclinometer.

(a) Initial rotation corresponds to 1·6 minutes for a 4 m cantilever condition (stage 0).
(b) When the excavation reached −22·5 m (stage 3) with three props in position, a small movement was detected at the bottom of the wall, with rotation about some point near the top.
(c) Further excavation caused progressive deflexion of the wall towards the excavation until at the final level (stage 5) the wall had moved in by 21 mm at the excavation level and 10 mm at the bottom of the wall. At this stage slight movement was measured in the soil beneath the wall.

GROUND ANCHOR TESTS

Wandsworth Bridge southern approach

29. Mr J. Bundred spoke on some GLC experience with trial anchors in London Clay.

30. An 82 m length of diaphragm retaining wall, maximum height 8·8 m and anchored at high level, was constructed in 1968 for the GLC Wandsworth Bridge southern approach highway underpass scheme (Fig. 11).

31. A single, longitudinal row of multi-strand ground anchors, 18 m long, inclined at 30° below the horizontal, and acting 2 m below the wall's crest at 1·15 m centres, was formed in fissured London Clay. These anchors have a resin-protected 12 m free length and a 6 m far end grouted by the system of tube à manchette (depending on a regrouting technique to achieve load capacity).

32. Before commencing anchor installation, tests to destruction were carried out on two 12 wire trial anchors, the 12 wires having a minimum nominal capacity of 1000 kN. Later, these two anchors were exposed and examined.

Fig. 11. Wandsworth Bridge southern approach: anchored diaphragm wall (right); view looking north-east at anchoring stage

33. The trial anchors failed to reach their anticipated ultimate load of approximately 800 kN because of wire bond failure; study of their fixed ends showed that there were clay inclusions trapped in the wires and that part of the wires had apparently coatings of grout and clay slurry (Fig. 12). Clearly this could have affected the bond between the wires and the grout; the tested anchors probably failed through inadequate bond between steel and grout rather than via block movement of the fixed anchor relative to the surrounding clay, but the time factor from grouting to testing was also involved.

34. The test anchors had been wash bored and then washed 'clean' with water and air. The lack of bond between wires and grout revealed a difficulty of cleaning out the bores properly. This was overcome by washing out with a weak grout solution. This problem would not have been present to the same degree with an augered hole cleaned out by air pressure only.

35. The presence of clay inclusions raised the question of permanent protection against corrosion of the 6 m fixed anchor length. The use of a weak grout to wash out the bores adequately reduced this risk. Alternatively, the risk could be reduced by using auger methods. Another suggestion was to use a polypropylene stocking over that 6 m far end length of the wires as protection when the cable was homed, but which would not prevent the grout from bonding with the surrounding clay

36. An intriguing feature was the presence of grout layers adjacent to trial anchor 1. These varied in thickness up to 38 mm. All the layers were approximately horizontal and provided proof of overstressing of the London Clay; it would appear that there were fissures expanded by high secondary grout pressures. The fact that no vertical fissures were grouted is probably because the clay is overconsolidated with a retained horizontal stress. Elsewhere, overstressing of the clay resulted in heave at the surface which was first noticed after regrouting to produce the 800 kN anchor blocks.

37. Six measured anchor loads had remained remarkably constant since early 1970, four months after underpass road slab completion, giving a mean anchor load of 136 kN with a variation of $+1\frac{1}{2}\%$ throughout the subse-

quent $4\frac{1}{2}$ years. These wall ground anchors appeared to be entirely satisfactory six years after installation, with a working factor of safety of 5 on anchor block capacity.

38. Although the Soletanche anchorages at Wandsworth had proved to be satisfactory overall, the wire bond failure induced in the trial anchors showed that these anchors should have been left for a longer time before load testing.

39. The grouting technique was very versatile with the tube à manchette system; the capacity of any anchor could be readily varied and the testing of its load capacity expedited.

40. There was an upper pressure and quantity limit to the regrouting technique as used for 800 kN anchors at the Wandsworth site and grout pressures in excess of 1650 kN/m² could overstress the ground immediately surrounding a 6 m fixed anchor length. Mr Bundred suggested that, to provide an anchorage of higher capacity than 800 kN in similar fissured London Clay, use of a

Fig. 12. Grout body showing two distinct layers formed on successive days; a grout layer running into a clay fissure is apparent (bottom right)

fixed anchor length greater than 6 m was preferable to increasing either the grout quantity or the secondary grout pressure above 1650 kN/m². An alternative would be to provide the equivalent by using a greater number of lower capacity anchors.

Thames tidal defences

41. Mr Brulois contributed some points on ground anchor tests in connexion with Paper 17.

42. He agreed that all strands of a tendon must be pulled out simultaneously, which is perfectly possible with several prestressing systems using, for example, a hollow centre jack.

43. The tendon displacement must be carefully and accurately measured and it must be an absolute measurement not taking into account the settlement of the reaction block especially when the top soil is poor and very compressive.

44. For a complete anchor test, a group of at least two (preferably three) similar anchors must be tested. By a direct and quick pulling out the first one gives a rough idea of the anchorage capacity, enabling convenient and adequate pulling sequences to be planned for the other anchors of the same group. These sequences must provide for measuring the rate of creeping under load, and in soft soils a long duration creep test is advisable.

45. On experimental work carried out for the Thames tidal defences, anchor tests were performed using a group of three. The ground consisted of Thames sand and gravels with overlying silty sands, which are a very poor material in a loose state with some standard penetration test values under 10.

46. For these anchors the Bachy TMD system was used, which separates the anchorage operation into two phases. First, a special tube à manchette is anchored into the ground by stage pressure, grouting being performed

as in alluvial deposits with the tube-à-manchette technique; both grouting and pressure can be easily controlled and regrouting can take place as many times as necessary. Later, the tendon is sealed into the tube à manchette.

47. The anchors used the VSL Losinger prestressing system. Tendons were 12 half-inch strands giving a total yield strength of 1800 kN. Anchorage lengths were 6 m in Thames sand and gravels, and 6 m and 9 m in silty sands. Free lengths were 28 m in Thames sand and gravels, and 17 m and 24 m in silty sands.

48. The test procedure consisted in successive cycles of loading and unloading by stages, and also long duration creep measurements.

49. Figure 13 shows the method and equipment used for accurately plotting the stress–strain diagram. The system is direct reading by displacement transducers.

50. In sand and gravels, a load of about 90% of the tendon yield strength (1570 kN) was reached without anchorage failure. The critical stability load was found to be around 1400 kN. Under a 1000 kN load, creep for 12 days was 3·9 mm and was stabilizing.

51. In silty sands, with a 6 m anchorage, failure occurred under 800 kN, giving a critical stability load in the vicinity of 500 kN. But under a 300 kN load, creep for 12 days was 2·9 mm without any evidence of stabilization.

52. With a 9 m anchorage length, results were generally improved by more than 60% and, moreover, under a 550 kN load, creep for 12 days was 2·5 mm with evident stabilization.

Test requirements and problems

53. Mr M. J. Vanner drew attention to the increasing emphasis, as shown in Papers 17 and 18, on the necessity for proof testing of every anchor. Because of the general variability of soil, this was the ideal aim. If anchors were formed either below the water table or in an area where the ground conditions were not likely to vary greatly over the years, then this procedure would give safe results.

54. A problem arose if anchors, constructed and tested in 'dry' conditions, were later subjected to conditions when the ground became saturated. There was a possibility that apparently safe anchors could fail. This problem required further investigation and consideration.

CONTRIBUTORS

Dr P. L. Bransby, University of Cambridge.
Mr J. Brulois, SIF Entreprise Bachy.
Mr J. Bundred, Greater London Council.
Mr J. R. Busbridge, Freeman Fox and Partners.
Mr J. C. McKenzie, Edmund Nuttall Ltd.
Mr H. D. St John, Building Research Station.
Mr M. J. Vanner, Balfour Beatty and Co. Ltd.
Dr T. Whitaker, Consultant.

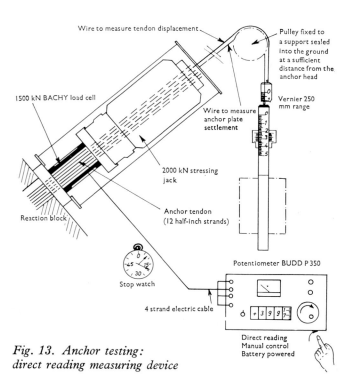

Fig. 13. Anchor testing: direct reading measuring device

Wire to measure tendon displacement

Pulley fixed to a support sealed into the ground at a sufficient distance from the anchor head

1500 kN BACHY load cell

Vernier 250 mm range

Wire to measure anchor plate settlement

2000 kN stressing jack

Anchor tendon (12 half-inch strands)

Reaction block

Stop watch

4 strand electric cable

Potentiometer BUDD P 350

Direct reading
Manual control
Battery powered

REFERENCES

1. BRANSBY, P. L. Contribution to discussion on General theories of earth pressure. *Proc. 5th European Conf. on Soil Mechanics, Madrid*, 1972, **2**, 75–78.
2. MILLIGAN, G. W. E. *The behaviour of rigid and flexible retaining walls in sand.* PhD Thesis, Cambridge University, 1974.
3. PECK, R. B. Deep excavation and tunnelling in soft ground. *Proc. 7th Int. Conf. Soil Mech., Mexico*, 1969, State of the Art Volume, 225–290.

Construction, carrying behaviour and creep characteristics of ground anchors

H. OSTERMAYER, Dipl Ing, *Technical University of Munich, Germany*

In Germany, ground anchors have been used for the last 15 years for temporary purposes and now are being installed in greater numbers as permanent anchors. For instance, in the construction of the Olympic tent-type roof a large number of permanent anchors were used. The Paper describes the experience gained from fundamental tests in which about 300 anchors were dug out and precisely inspected in order to check the suitability of the anchor systems. In addition, different types of anchors and corrosion protection systems are discussed. It is important that suitability tests should be carried out at each construction site in order to check the load carrying behaviour of the anchor in special soil conditions. On the basis of such tests at different sites, coupled with a research programme, the influence of diameter and bond length on skin friction is given. Diagrams are given to help estimate the load carrying capacity of anchors for different soil conditions. For permanent anchors the two important considerations are long term stability and creep characteristics. A test is necessary in which the tensile load is held constant over a certain period so as to permit determination of the allowable working load and estimation of future creep displacements and load reduction. In conclusion, measurements of anchor forces and wall displacements for some anchor problems are given.

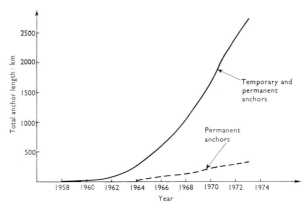

Fig. 1. *Total length of grouted anchors installed by Bauer since 1958*

INTRODUCTION

It was in Germany in 1958 that Bauer[1,2] for the first time demonstrated that steel rod can be directly anchored in a borehole of 6 cm dia. when cement suspension is injected into the hole. Since then the development of such grouted anchors has steadily gained momentum, as can be seen from Fig. 1. Permanent anchors as well as temporary anchors show this trend.

2. The economy of grouted anchors is due to the use of borehole diameters of only 8–14 cm to carry into the soil relatively large working loads, 30–50 tons in cohesive soils and 40 to over 100 tons in non-cohesive soils. In Germany, therefore, bracings are seldom used for enclosure walls in built-up areas of a city. For instance, in subway construction in Munich, for excavation sites over 12 m wide anchored walls are regarded as more economical than braced walls.

3. Figure 2 shows construction of an anchored wall involving relatively deep excavation for a subway station in Hannover. The H pile walls and bored pile walls were supported by five rows of anchors with a maximum length of 27 m. During and after construction no damage occurred in the adjoining property.

4. In addition to the extensive use of temporary anchors for enclosure walls, permanent anchors can be used wherever tensile loads are to be permanently carried into the soil, for example to ensure the safety of embankment and retaining walls (30 m high bored pile wall in Stuttgart[3]), to take uplift pressure (Tivoli, Munich, Fig. 3(a)), to anchor tension ropes of suspension bridges or tent-type roofs (Olympic tent-type roof, Munich,[4] Fig. 3(b)), and to ensure safety against overturning for tall structures (ski-jumping structure, Oberstdorf,[5] Fig. 3(c)).

5. In all these cases, expensive dead-weight foundations have been replaced by ground anchors where dead weight of soil instead of concrete is utilized. For the Olympic tent-type roof at Munich the cost for the anchored construction was about DM 150 per ton of tensile load compared to DM 180 for dead-weight construction.

6. The trend is to use tendons of still higher tensile steel so that for the same size of borehole greater loads can be carried by the anchor. When such a steel is used, the steel/grout interface and the soil at the grout/soil interface are highly stressed. It is therefore essential that the anchor system, as well as each anchor used, be examined carefully in order to make sure that the load is safely carried over a long period of time.

7. The Paper deals with the construction and testing of anchors. Important data collected from experience and research are presented in the hope that those concerned with anchor problems find this useful from the practical point of view. Actual measurements are reported in order to demonstrate possibilities and limitations of using grouted anchors.

CONSTRUCTION OF ANCHORS

8. In Germany about 300 anchors were dug out after the fundamental test and scrutinized carefully.[6,7] These

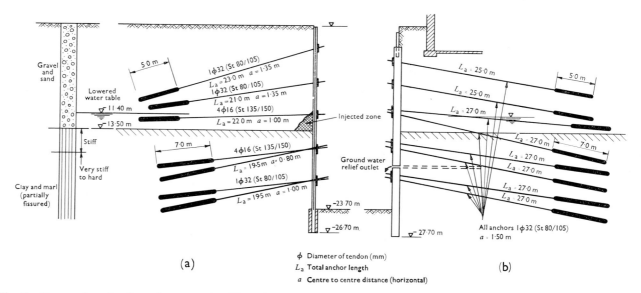

Fig. 2. *Excavation site for subway station Kröpcke, Hannover:* (a) *H piles and lagging construction;* (b) *bored pile wall*

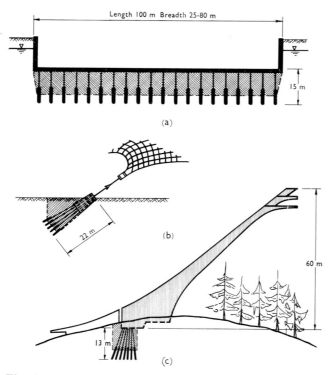

Fig. 3. *Examples of use of permanent anchors:* (a) *safety against uplift, Tivoli Munich;* (b) *anchorage of Olympic tent-type roof, Munich;* (c) *ski-jumping structure in Oberstdorf*

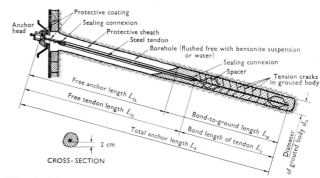

Fig. 4. *Temporary anchor construction of type A*

an outer sheath sealed tightly to the steel tendon and the bond length of the tendon is protected by a cement-grout covering of at least 2 cm. The requirement for a covering of 2 cm is based on the assumption that radially only hair cracks develop due to elongation of the tendon and that there are no longitudinal cracks. The corrosion protection at the head is also of great importance, because of the possibility of hostile liquid agents seeping into the anchor head. It is therefore essential that at least a coat of paint is applied to the anchor head and the tendon portion above the protective sheath.

Permanent anchors

11. As permanent anchors (Fig. 5) both type A (grouted body under tension) and type B (grouted body under pressure) are used. In contrast to temporary anchors the permanent anchors must have a double system of corrosion protection, protection against mechanical damage being provided in addition to uninterrupted corrosion protection.

12. In the case of type B anchors the corrosion protection can be applied and tested under factory-controlled conditions without any difficulty. In this type the force is passed on to the grouted body from the rear end via the pressure pipe (Fig. 5(c)). The corrosion protection covering over the whole length of the tendon is examined elec-

tests later formed the basis for standards which temporary and permanent anchors must meet.[8-10]

Temporary anchors

9. As temporary anchors, type A (Fig. 4) are generally employed. The main feature is that the grouted body is subjected to tension due to elongation of the steel tendon.

10. Only one corrosion protection is deemed necessary, this being that the free tendon length is protected by

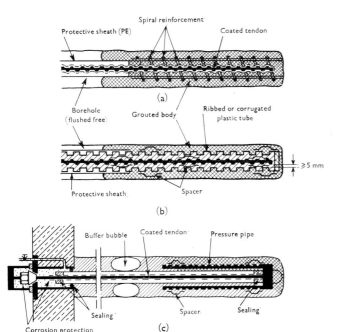

Fig. 5. Permanent anchor constructions: (a) type A with coated tendon; (b) type A with tendon in a ribbed or corrugated plastic tube; (c) type B with pressure pipe (Bauer/Stump)

trically and is then covered with a sheath which acts as a mechanical protection. As the corrosion protection does not carry any bond stresses a relatively elastic material can be used. Protection pastes or grease pressed into the annular space between the sheath and the tendon will also serve the purpose.

13. In the case of type A anchors the application of corrosion protection which remains undamaged during construction and stressing of anchors is indeed a difficult task. For the anchor shown in Fig. 5(a) the difficulty lies in finding a synthetic coating which not only should have an excellent bond with steel but, in addition, must also be flexible and strong enough to carry high bond stresses over a long period of time. When the coating is thick the danger exists that the grouted body will be subjected to high bursting stresses. A spiral reinforcement is therefore provided to take up these stresses.

14. In the type A anchor of Fig. 5(b) the tendon is inside a ribbed plastic sheath. The annular space is filled with cement. Although this cement is cracked, as in all type A anchors, the criterion of corrosion protection is fulfilled when the cement in the annular space is at least 5 mm thick and the PVC or PE tube is at least 1 mm thick and remains tightly sealed for a period of more than 50 years. When a ribbed plastic sheath is used the danger of material fatigue is less than in the case where protective coating has been directly applied to the tendon, as in Fig. 5(a).

15. The requirements of double corrosion protection must also be met at the anchor head, as shown in Fig. 5(c).

TESTING

Fundamental test of anchor systems

16. An anchor system is allowed to be installed only when the system is subjected to a fundamental test in which three anchors are stressed to at least 1·5 times the allowable working load and then dug out and carefully examined. The purpose of such a test is to make certain that the anchor system performs as planned and to detect any shortcomings that may affect the load carrying behaviour of the anchor over a longer period of time.

17. Points particularly noted[11] are the form and quality of the grouted body, its strength and possible defects; any discrepancies between the planned and actual bond-to-ground lengths; the position of the tendon (whether or not central) to check if cement covering of the prescribed thickness has been provided; the formation and distance between cracks for calculating the width of cracks under stressed conditions; and the quality of the corrosion protection coating especially in the case of permanent anchors.

Suitability test at site

18. Before the construction work is started, three or four anchors ought to be subjected to a suitability test. In this test the anchor is subjected to loading and unloading cycles so that elastic and plastic deformations can be ascertained for loads up to 1·5 times the working load. The elastic deformations allow calculation (and hence check) of the free length of tendon; the plastic deformations correspond roughly to the displacements of the grouted body. These deformations therefore depict the carrying capacity of an anchor.

19. In addition, the time–deformation relationships under constant load ought to be observed for various loading steps. The creep, so measured, forms an essential test criterion for permanent anchors (§§ 35–45). (For cohesive soils the minimum observation time is 24 h for 1·5 times the working load.)

Acceptance test at site

20. Variations in soil conditions and installation processes can cause the carrying capacities of anchors to differ widely, and it is essential that each anchor at site be subjected to test. To obtain the allowable working load of the tendon a safety factor of 1·75 with respect to the yield stress of steel has been fixed. The test load T_t for temporary anchors is 1·2 times the working load T_w, with 5% of the anchors being tested for 1·5 times the working load. In case of permanent anchors, each anchor is subjected to 1·5 times the working load.

Experience gained from tests

21. The experience gained from such tests (primarily from fundamental tests) for non-cohesive and cohesive soils is outlined in §§ 22–25.

22. *Non-cohesive soils.* For non-cohesive soils the boring technique (ramming or boring with or without flush water) has no decisive influence on the quality and carrying capacity of anchors. It is, however, essential that the cement suspension be injected with a pressure of at least 10 atm so that any soil zones that may have become loose during boring are recompressed. Under this pressure, water is filtered out of the cement suspension to achieve a better quality of grouted body and, in addition, the value of skin friction at the grout/soil interface is increased. In order that the bond-to-ground length does not exceed the planned length, use of water (or better still

bentonite suspension) to flush out remaining cement from the borehole is found to be very effective.

23. *Cohesive soils.* For cohesive soils the boring technique has a decisive influence on the quality and carrying capacity of the anchors. Boring without casing or with casing combined with the use of flushing water generally results in lower values of ultimate failure load. From the working and economical point of view flush water is used in cohesive soils. For such cases it is important that immediately after boring the hole is cleaned and, starting from the rear end, the hole is grouted with a cement suspension of water/cement ratio about 0·4. The carrying capacity is greatly enhanced with post-grouting (§§ 32, 33). The process consists of providing one or several post-grouting tubes in the grouted body. These tubes are fitted with sleeve valves thereby allowing one or more post-groutings as required. During the process of post-grouting the grouted body bursts open, the cracks get

Fig. 6. Cross-section of grouted bodies showing 7 steel tendons (dia. 16 mm) along with centrally placed post-grouting tube: the white lines are the cracks filled with post-grout material

simultaneously filled with the post-grout and this enlarged monolithic mass abuts tightly against the surrounding soil. Care must be taken when the anchors lie close to ground surface as the post-grouting may cause upheaving. Figure 6 shows multi-tendon anchors after post-grouting.

24. Multi-underreamed anchors[12,13] were also tried in Germany but produced no positive results. In spite of the fact that the enlarged cavities or bells of the under-reams were thoroughly cleaned, relatively high plastic deformations and creep displacements took place. The method of post-grouting was found to be cheaper and more effective than underreaming for increasing the carrying capacity of anchors.

25. Although for anchors in non-cohesive soils spacers may not be necessary, it is essential that spacers be provided for exact centering of tendons in cohesive soils (Figs 4 and 5). In non-cohesive soils the bond-to-ground length can be kept within its planned limit by the simple process of flushing out any remaining cement from the borehole (Figs 5(a) and 5(b)). Another method that can be employed in the case of cohesive soils is by pumping up a buffer rubber bubble (Fig. 5(c)) or by installing an elastic packer.

CARRYING CAPACITY

Experiments to determine the load carrying behaviour

26. Loads up to 80 tons in cohesive soils and over 150 tons in non-cohesive soils, for anchors of 10–15 cm dia. and 4–8 m length, cannot be accounted for by means of the classic laws of soil mechanics. It is therefore necessary to collect all the empirical data on soil conditions, anchor dimensions and installation methods. In a research programme the results of all fundamental tests were evaluated, and additional tests were performed in which anchor diameter and length were the variables. The anchors were laid open after the tests for careful examination.

27. Model tests carried out by Werner[14] in sand show

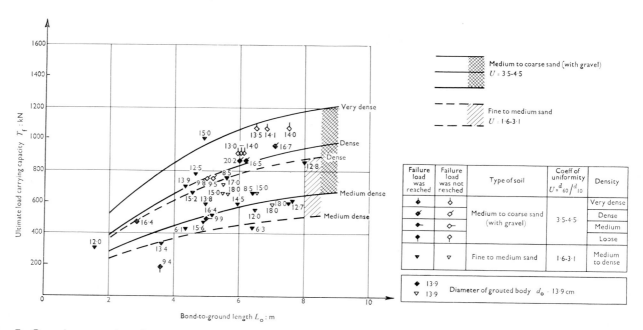

Fig. 7. Carrying capacity of anchors in sand and gravelly sand

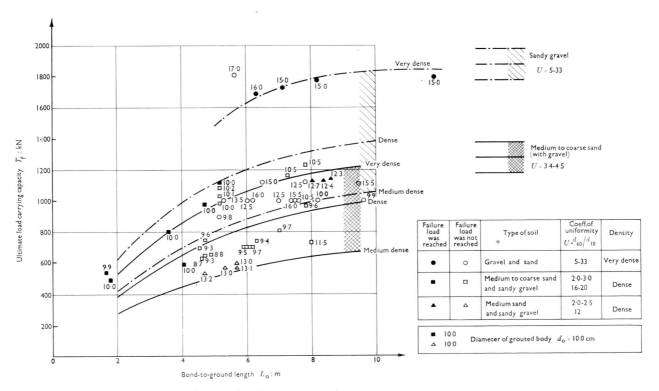

Fig. 8. Carrying capacity of anchors in gravel and gravelly sand

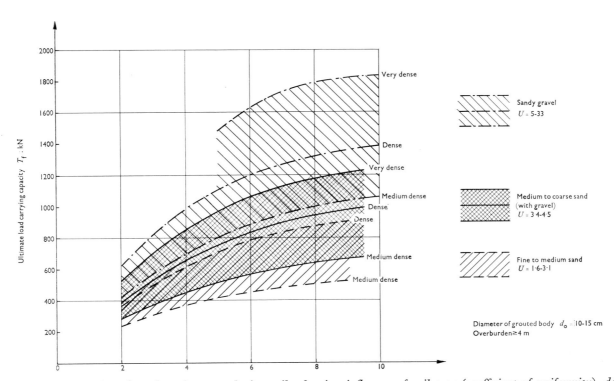

Fig. 9. Carrying capacity of anchors in non-cohesive soils showing influence of soil type (coefficient of uniformity), density and bond-to-ground length

close agreement with the in situ tests. In these model tests the influence of relative density and the size of grouted bodies were examined. Wernick[15] also carried out model tests in sand in which radial stresses in the soil were measured. These measurements confirm the hypothesis that high carrying capacity of an anchor is in part due to the locking-in effect of the soil.

Carrying capacity in non-cohesive soils

28. *In situ tests.* From Figs 7–9 the following results are of importance.

(*a*) For a given soil the carrying capacity increases rapidly with increase in soil density. For example, in the case of sandy gravel, when the soil is very dense the values are 80% higher than when it is medium dense.

(*b*) For the same value of relative density the carrying capacity increases with an increase in coefficient of uniformity. For example, in a dense sandy gravel soil with $U = 5–33$ the carrying capacity is 50% higher than in a dense sand with $U = 1·6–3·1$.

(*c*) The increase in carrying capacity of an anchor tapers off steadily with the length, so that in general a length of 6–7 m is optimal from the economical point of view.

(*d*) Up to 10 cm dia. a slight increase in carrying capacity is obtained with an increase in diameter of the grouted shaft. In the range 10–15 cm dia. the points are scattered and no definite conclusion can be drawn.

(*e*) The boring technique (ramming with non-recoverable bit or boring) has no influence on the carrying capacity.

29. For the example in Fig. 10 the calculations of skin friction are based on the assumption that at failure the shear stresses are uniformly distributed over the surface of the grouted shaft. In spite of the scatter it can be stated that the skin friction decreases as the shaft diameter increases. When the diameter is increased beyond 10 cm no substantial increase in the carrying capacity is obtained. In addition one sees that the average skin friction decreases as the anchor shaft length (bond-to-ground length) increases. The reason is that larger elastic deformations of the anchor result in a progressive failure at the grout/soil interface. In order to determine the exact distribution of skin friction, tests are being made with strain gauges fixed along the bond length of the tendon.

30. The calculated value of skin friction is as high as 500 kN/m² for sands and more than 1000 kN/m² for sandy gravel. This can only be explained as due to the locking-in effect of the soil when the anchor is given a pull. The affect of this phenomenon is to increase the normal stress at the grout/soil interface. This comes out to be 2–10 times the effective overburden pressure. The locking-in effect, which increases as the relative density and uniformity coefficient increase, can in part be traced back to dilatation of dense soils when sheared, the net result being an increase in radial stress.

31. *Model tests.* The radial stresses mentioned above were measured in the soil by Wernick[15] in model tests. As expected the radial stress and therefore the skin friction decreased as the shaft diameter was increased. The influence of anchor shaft dimensions and relative density of soil were investigated by Werner[14] also in model tests. The results of 150 tests showed that the measured skin friction (as a result of the locking-in effect) was greater than values calculated on the basis of overburden pressure.

In Fig. 11 the ratio of these two values is defined as the lock-in factor. These tests also confirmed the tendency shown by in situ tests that the lock-in factor increases with increasing relative density and with decreasing anchor diameter (Fig. 11(a)). In the model tests the effect of bond-to-ground length on the average skin friction was not shown (Fig. 11(b)), as the process of progressive failure as a result of elastic deformation of steel tendons could not be duplicated in the model tests.

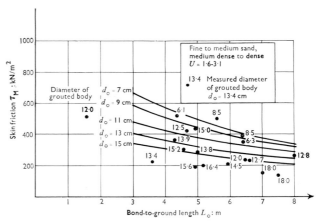

Fig. 10. Influence on skin friction of diameter and length of grouted body

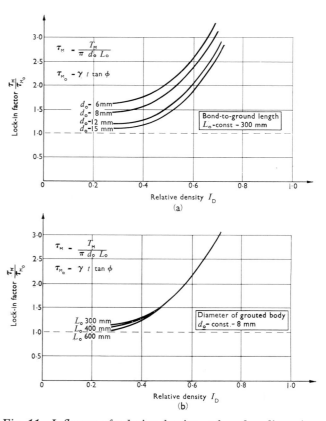

Fig. 11. Influence of relative density and anchor dimensions on lock-in factor (after Werner): (a) variable diameter of grouted body; (b) variable length of grouted body

Carrying capacity in cohesive soils

32. In cohesive soils the influence of the length and diameter of the grouted body on the carrying capacity of anchors has not yet been investigated systematically. Figure 12, however, shows the results of a large number of fundamental tests, from which an idea can be formed about the carrying capacity of anchors in a certain soil. The following conclusions can be drawn from Fig. 12.

(a) The average skin friction per unit area of anchor shaft is independent of measured diameter in the range 9–16 cm, i.e. the carrying capacity increases proportionally with diameter.

(b) The skin friction is independent of bond-to-ground length up to a value of 100 kN/m². For values higher than this a slight decrease is to be expected with increase in bond-to-ground length, but for rough calculations the value can be taken as constant.

(c) The skin friction increases with increasing consistency and with decreasing plasticity. In stiff clays (I_c 0·8–

1·0) with medium to high plasticity the value of skin friction of 30–80 kN/m² is the lowest. Sandy silts of medium plasticity and very stiff to hard consistency (I_c 1·25) give the highest value of more than 400 kN/m².

(d) The skin friction can be significantly increased with the help of post-grouting. Using this method the value increased from about 120 kN/m² to about 300 kN/m² in the case of a stiff clay of medium to high plasticity.

33. From Fig. 13 it can be seen that the carrying capacity of anchors in cohesive soils can be substantially increased with the help of post-grouting. In this figure the theoretical skin friction (calculated from the borehole diameter and planned bond-to-ground length) is shown as a function of post-grouting pressure. Irrespective of borehole diameter, bore technique and amount of grouting, the theoretical skin friction (and therefore the carrying capacity of the anchor) seems to increase with increase in post-grouting pressure. Such results have also been reported by Jorge.[16] Before any conclusions are drawn it

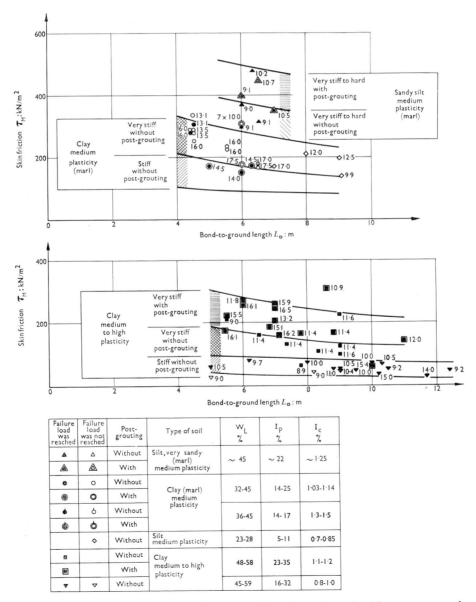

Failure load was reached	Failure load was not reached	Post-grouting	Type of soil	W_L %	I_p %	I_c %
▲	△	Without	Silt, very sandy (marl) medium plasticity	~ 45	~ 22	~ 1·25
⬛△	⬛△	With				
●	○	Without	Clay (marl) medium plasticity	32-45	14-25	1·03-1·14
◉	○	With				
◆	◇	Without		36-45	14-17	1·3-1·5
◈	◇	With				
	◇	Without	Silt medium plasticity	23-28	5-11	0·7-0·85
▪		Without	Clay medium to high plasticity	48-58	23-35	1·1-1·2
⬛		With				
▼	▽	Without		45-59	16-32	0·8-1·0

Fig. 12. Skin friction in cohesive soils for various bond-to-ground lengths, with and without post-grouting

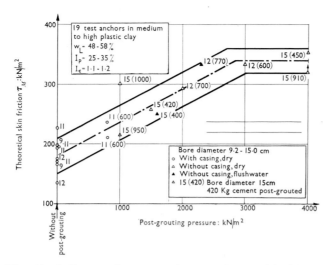

Fig. 13. *Influence of post-grouting pressure on skin friction in a cohesive soil*

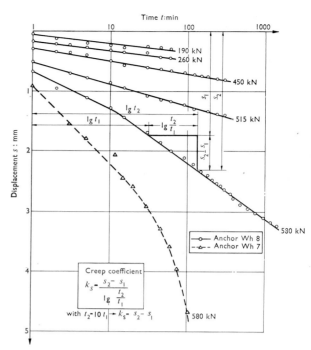

Fig. 14. *Time–displacement curves for various loading steps in a uniform sand*

is necessary to examine the relative importance of the pressure required to burst the grouted body and the steady end pressure.

34. Feddersen[17] has investigated the possibility of increasing the carrying capacity under unfavourable soil conditions by increasing the length of grouted body. With the help of strain gauges over a bond length of tendon of 18 m he was able to measure the skin friction in a stiff to very stiff, highly plastic clay. Except for peaks at the fore end of the bond length of tendon the skin friction can be taken as constant over the whole length. The values range between 40–80 kN/m² and agree favourably with those shown in Fig. 12. The interesting feature of these soil conditions is that the skin friction for type A anchors, in which the grouted shaft is under tension, is smaller than for type B anchors, in which the shaft is under compression.

CREEP BEHAVIOUR

Performance of single anchors

35. Before the failure load is reached, large creep displacements under constant load can take place in cohesive soils and also in uniform grained non-cohesive soils. For the design of permanent anchors it is essential to know the creep displacement as a function of time.

36. In general the relationship between creep displacement and time is an exponential function, i.e. a straight line is obtained when plotted to a semi-log scale. According to Fig. 14 the slope of such a line is defined as creep coefficient. The slope increases for each further loading step. When the ultimate failure load is reached the displacements do not decrease with time. For example, for loading of 580 kN the creep displacements of anchor Wh 7 did not decrease.

37. In Fig. 15 the creep coefficients are plotted with respect to the mobilized carrying capacity (ratio of test load to failure load). The values at the beginning are smaller but increase rapidly for medium to highly plastic clays of stiff consistency at 40% of the failure load, medium to highly plastic clays of stiff to very stiff consistency at 55% of the failure load, medium to highly plastic clays of very stiff to hard consistency at 80% of the

failure load, and uniform grained sands at 80% of the failure load. It is clear that this beginning of plastic flow of the soil around the grouted shaft in the case of permanent anchors should not take place. For this reason, in addition to the safety factor against failure, the safety factor against creep displacements must be introduced (see § 40).

38. That part of the creep displacement which is contributed by partial debonding in the steel/grout interface, creep of cement grouting and relaxation of tendon steel generally corresponds to a creep coefficient k_s of about 0·4 mm. Increase of k_s beyond 0·4 is a result of creep at the grout/soil interface. The straight lines in the semi-log plot of Fig. 14 are typical for straight shaft anchors only. In the case of multi-underreamed anchors, larger creep displacements are expected due to local stress concentrations causing consolidation and plastic deformation.

Approval and testing of permanent anchors

39. Because of creep under constant load and the behaviour, less well understood, under repeated loading, construction of permanent anchors at the present time is not approved[10] when the change in the prestress values of the steel tendon for repeated loading is more than 20% of the working load, or when the soil has a large organic content, or when the soil consists of loose sand, or when it is a cohesive soil with a consistency $I_c < 0·9$ or liquid limit $w_L > 50\%$.

40. The criterion for suitability tests and acceptance tests at the present time is that the creep coefficient k_s should be less than 2 mm under a load of 1·5 times the allowable working load (Fig. 15). It can be seen that although the creep coefficients for the working load lie under the critical break point of the curves (from then on the flow begins), the desired factor of safety of 1·75 against failure for permanent anchors in gravels and sands, and

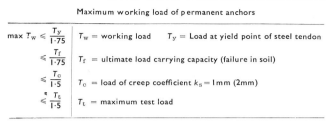

Maximum working load of permanent anchors	
$\max T_w \leqslant \dfrac{T_y}{1\cdot75}$	T_w = working load $\quad T_y$ = Load at yield point of steel tendon
$\leqslant \dfrac{T_f}{1\cdot75}$	T_f = ultimate load carrying capacity (failure in soil)
$\leqslant \dfrac{T_c}{1\cdot5}$	T_c = load of creep coefficient k_s = 1mm (2mm)
$\leqslant \dfrac{T_t}{1\cdot5}$	T_t = maximum test load

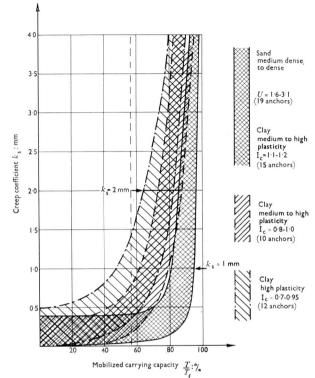

Fig. 15. Creep coefficients in relation to mobilized carrying capacity (results from 56 tests) and maximum working load of permanent anchors

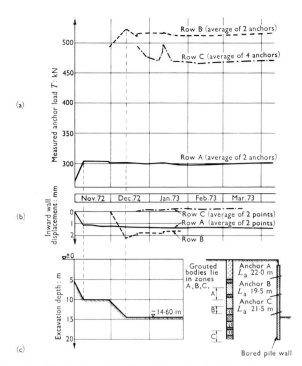

Fig. 16. Variation with time of anchor loads and movements of bored pile wall at Tivoli II, Munich: (a) anchor loads (average of rows A, B, C); (b) horizontal wall displacement at height of anchor rows A, B, C; (c) excavation depth

stiff to very stiff clays is not achieved. It is therefore recommended that the creep coefficient k_s should not be greater than 1 mm for 1·5 times the working load. This value of 1 mm for k_s corresponds to a displacement difference of 6 mm in a time interval between 30 minutes and 50 years.

41. To check their sensitivity to repeated loading the anchors should be tested for at least 20 loading and unloading cycles between half and full allowable working load.

42. When the installation and testing of anchors is done properly the chance that the safety of a construction site is endangered as a result of anchor failure is indeed very small. However, it is a good policy to examine the permanent anchors at regular intervals. For example, the deformations at site should be observed, and loads should be measured on anchors at random.

In situ creep measurements

43. Theoretically it is possible to calculate the creep displacement of the grouted body from the creep coefficient of the suitability test. From the creep displacement an idea can be formed about the prestress loss with time.

However, changes of anchor loads on site depend not only on the creep displacement of the grouted body but can also be due to wall movements and settlements. To separate these two factors, measurements of anchor head displacements should be very accurate. On the basis of such measurements it can be concluded that for anchor lengths of 20–25 m (bore diameters of 10–15 cm) a prestress loss of 6% in very stiff to hard clays and 12% in stiff to very stiff clays is due to creep. Data published so far[12,18] also lie in the above region. It is interesting to note that in general no further loss of prestress took place after a period of 2–4 months. The measured prestress loss up to this period was smaller than the value calculated from the creep coefficient. In other words, the calculated values lie on the safe side.

44. Figure 16 shows an example of a bored pile wall in Munich. The grouted bodies were situated in very stiff to hard silts and clays ($w_L = 35$–55%, $I_p = 10$–20%, $I_c = 1\cdot0$–$2\cdot0$) with sand layers in between. After the anchor row C was prestressed no further excavation took place. An outward wall movement of only 0·2 mm took place. The measured prestress loss of about 6% in two months was due primarily to creep displacement of the grouted body. For anchor rows A and B, on the other hand, the creep effect was superimposed on the increase in anchor loading due to further ground excavation and an inward wall movement of 1–2 mm.

45. It is pointed out here that no creep effects were observed for the anchors, numbering about 450 (working load 37 tons), used for the Olympic tent-type roof in Munich (Fig. 3). The grouted bodies lie in quaternary gravel or tertiary sands. The original prestress loading had a usual scatter of ±3 tons. Half to one year after prestressing a maximum of 3·5 tons prestress loss was

measured. These losses were traced back to foundation settlements only.

DISPLACEMENT OF ANCHORED WALLS

46. Displacement of anchored walls can be caused either as a result of displacement of each individual anchor or as a result of deformation of whole of the soil block (region ABCF in Fig. 17) and deformations of the underlying soil (region CDEF). The effective horizontal stresses acting on soil block ABDE after excavation are shown qualitatively in Fig. 17. For the equilibrium condition $\Sigma H = 0$, the resultant earth resistance of area C'D must be equal to the earth pressure acting on the area AE. This means that soil below the base of excavation experiences a reduction in vertical stress and also an increase in horizontal stress (compared to the pressure at rest). The soil therefore is pressed together in the horizontal direction. In addition shear strains are imposed on the slice ABDE which must be compatible with strains below the excavation base.

47. The main purpose of the model shown schematically in Fig. 17 (effect of moments has been ignored) is to show that the compressibility of the soil below the excavation base is responsible for wall displacement. For example, in the case of construction of Frankfurt Main Railway Station, where there were unfavourable soil conditions (stiff, highly plastic clay) and an excavation depth of 20 m, wall displacements of 14 cm were measured.[19] On the other hand, parallel displacements of only 0·05–0·11% of the wall height were observed in very stiff to hard clays and silts and dense non-cohesive soils, provided rigid construction (diaphragm walls, bored pile walls etc.) was adopted and the anchors were prestressed in accordance with earth pressure at rest.[20-22] The adjoining property suffered no damage under these conditions.

48. The main problem in the case of diaphragm wall construction in Munich (Fig. 18) was the arrangement of anchors so that they should not interfere with numerous wells and an underground network of various cables and tunnel regions. With great relief, here too no property damage was reported.

CONCLUDING REMARKS

49. Examples have been given of the many uses and possibilities of anchored walls, and the problem of dis-

placement of such walls has been described. From the results of numerous in situ and laboratory tests it can be said that, besides the extensive use of temporary anchors, permanent anchors can be put to use in a great number of instances. The chance that the safety of a construction is endangered due to failure of anchors has indeed become very small because the anchors undergo certain tests which have been duly described. Further research work is being carried out to trace the creep behaviour of anchors in various soils. With the help of long term measurements at site the possibility is to be explored of utilizing the short term tests for predicting the creep behaviour of anchors. Further points of interest are the change in the value and distribution of skin friction with time and the influence of pre-loading on long term behaviour. The ultimate aim is that the numerous post tests should be reduced to a minimum so that the installation of anchors will become more economical.

ACKNOWLEDGEMENTS

50. The tests were performed under the guidance of Professor Dr. -Ing. R. Jelinek at the Institut für Grundbau und Bodenmechanik, Technische Universität München. Research funds granted by the Bundesministerium für Städtebau und Wohnungswesen, the Hauptverband der Deutschen Bauindustrie and the firms Karl Bauer KG, Schrobenhausen and Philipp Holzmann AG, Frankfurt are gratefully acknowledged.

REFERENCES

1. BAUER K. Injektionszuganker in nichtbindigen Böden. *Bau und Bauindustrie*, 1960, **16**, 520–522.
2. JELINEK R. and OSTERMAYER H. Verankerung von Baugrubenumschließungen. *Vorträge der Baugrundtagung 1966 in München*, Deutsche Gesellschaft für Erd- und Grundbau, Essen, 1966, 271–310.
3. SCHWARZ H. Permanentverankerung einer 30 m hohen Stützwand im Stuttgarter Tonmergel durch korrosionsgeschützte Injektionsanker, System Duplex. *Bautechnik*, 1972, **49**, Sept., 305–312.
4. SOOS P. Anchors for carrying heavy tensile loads into the soil. *Proc. 5th Eur. Conf. Soil Mech., 1972*, Sociedad Espanola de Mec. del Suelo y Cimentaciones, Madrid, **1**, 555–563.
5. WERNER H.-U. Foundation and anchoring of the inclined ski-jumping-structure in Oberstdorf/Fed. Rep. Germany. *3rd Int. Conf. Rock Mech., Denver, Col., USA, Sept. 1974, paper submitted*.
6. OSTERMAYER H. Erdanker—Tragverhalten und konstruktive Durchbildung. *Vorträge der Baugrundtagung 1970 in Düsseldorf*. Deutsche Gesellschaft für Erd- und Grundbau, Essen, 1970, 5–35.
7. OSTERMAYER H. and WERNER H.-U. Neue Erkenntnisse und Entwicklungstendenzen in der Verankerungstechnik. *Vorträge der Baugrundtagung 1972 in Stuttgart*. Deutsche Gesellschaft für Erd- und Grundbau, Essen, 1972, 235–262.
8. DEUTSCHE INDUSTRIE-NORM. *Verpreßanker für vorübergehende Zwecke im Lockergestein: Bemessung, Ausführung und Prüfung*. DIN 4125, Blatt 1, 1972.
9. INSTITUT FÜR BAUTECHNIK. *Richtlinien über die Zulassungsprüfungen von Verpreßankern für vorübergehende Zwecke*. Institut für Bautechnik, Berlin, 1973.
10. INSTITUT FÜR BAUTECHNIK. *Zulassungsbedingungen für Verpreßanker für bleibende Verankerungen*. Institut für Bautechnik, Berlin, 1973.
11. OSTERMAYER H. Construction of ground anchors. Discussion, *Proc. 5th Eur. Conf. Soil Mech., 1972*, Sociedad Espanola de Mec. del Suelo y Cimentaciones, Madrid, **2**, 334–336.
12. LITTLEJOHN G. S. Soil anchors. *Ground Engineering*, Instn Civ. Engrs, London, 1970, 33–44.
13. LITTLEJOHN G. S. Ground anchors today—a foreword. *Ground Engng*, 1973, **6**, Nov., No. 6, 20–22.

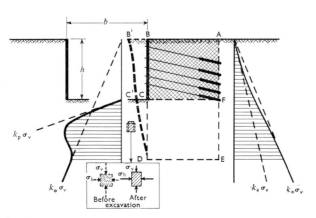

Fig. 17. Stresses and strains around an anchored soil block, shown schematically

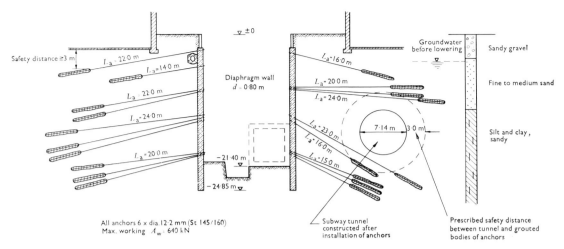

Fig. 18. Excavation site for subway station Implerstraße, Munich

14. WERNER H.-U. Die Tragkraft langzylindrischer Erdanker zur Verankerung von Stützwänden unter besonderer Berücksichtigung der Lagerungsdichte kohäsionsloser Böden. Dissertation, Reichsuniversität, Gent, Belgium, 1972.

15. WERNICK E. Mantelreibung von Verpreßankern und Verpreßpfählen im nichtbindigen Boden (Diskussion). *Vorträge der Baugrundtagung 1972 in Stuttgart*. Deutsche Gesellschaft für Erd- und Grundbau, Essen, 1972, 269–274.

16. JORGE G. R. Le tirant IRP réinjectable spécial pour terrains meubles, karstiques ou à faibles caracteristiques géotechniques. *Proc. 7th Int. Conf. Soil Mech., Mexico, 1969*, Spec. Session 15, 159–163.

17. FEDDERSEN I. Verpreßanker im Lockergestein. *Bauingenieur* 1974, **49**, Aug., 302–310.

18. ANDRÄ W. *et al*. Bohrpfahlwand für die Allianz-Neubauten in Stuttgart. *Bautechnik*, 1973, **50**, Aug., 258–264.

19. BRETH H. and RHOMBERG W. Messungen an einer verankerten Wand. *Vorträge der Baugrundtagung 1972 in Stuttgart*. Deutsche Gesellschaft für Erd- und Grundbau, Essen, 1972, 807–823.

20. NENDZA H. and KLEIN K. Bodenverformung beim Aushub tiefer Baugruben. *Vortragsveröffentlichungen, Haus der Technik, Essen*, 1973, **314**, 4–18.

21. JAMES E. L. and PHILLIPS S. H. E. Movement of a tied diaphragm retaining wall during excavation. *Ground Engng*, 1971, **4**, July, 14–16.

22. EGGER P. Influence of wall stiffness and anchor prestressing on earth pressure distribution. *Proc. 5th Eur. Conf. Soil Mech., 1972*, Sociedad Espanola de Mec. del Suelo y Cimentaciones, Madrid, **1**, 259–264.

PAPER 19

Behaviour of inclined groups of plate anchors in dry sand

W. J. LARNACH, MSc, PhD, FICE, FIStructE, *Senior Lecturer, University of Bristol*

D. J. McMULLAN, BSc, *Geotechnical Engineering Ltd, Gloucester*

An initial model investigation of the behaviour of plate anchors installed in sand in groups with inclined, parallel axes is described. The anchors were located in the practical deep range and were pulled out at constant rate of strain to determine post-peak behaviour. Group efficiencies and load distribution within groups of up to 16 anchors are reported.

INTRODUCTION

The work reported in this Paper forms part of a continuing experimental investigation into the behaviour of inclined ground anchors in dry sand. Previous papers[1,2] dealt with anchors arranged in lines. This Paper reports on initial investigations of groups of up to 16 anchors arranged in square or rectangular patterns. Figure 1 illustrates the terminology and layout. Inclinations of $\theta = 55 - 90°$ have been investigated, at a constant depth/anchor diameter ratio of $H/D = 25$, which establishes all of the anchors very clearly in the practical deep range[3] where anchor pull-out peak loads occur without visible surface effects. Previously spherical anchor models have been used but in this investigation these have been replaced by 30 mm dia. circular plates in order to facilitate easier computation in a theoretical prediction of their behaviour, which will be reported in a later paper.

2. This Paper deals with the load/displacement curves of a group, the efficiency of a given pattern of anchors (referred to a single anchor at the same inclination), and the distribution of load within a group and the way in which it varies with inclination and spacing. An important feature of the work is that loading of the groups takes place at a constant rate of pull-out. This enables post-peak behaviour to be studied, which is important since within a group individuals achieve their peaks at different group displacements.

EXPERIMENTAL DETAILS

3. The depth H (Fig. 1) was kept constant at 750 mm. The parameters varied were the number of anchors in the group, the group layout, the anchor spacing S (equal in both directions) and the inclination θ of the group. Spacings of $S = 2D, 4D, 8D, 12D$ ($D =$ anchor diameter) and $\theta = 90°, 70°,$ and $55°$ were investigated.

4. Figure 2 shows the general layout of the test apparatus, and the salient details of the sand tank are shown in Fig. 3. The inclusion of the 45° slope at the rear of the

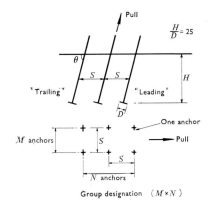

Fig. 1. *Geometry of anchor tests*

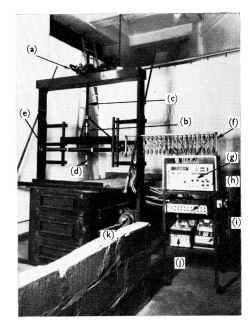

(a) Motor unit and worm drive
(b) Total load measuring gauge
(c) Pull rod
(d) Anchor tendon and load monitoring unit
(e) Pull rod head
(f) Remaining anchor load monitoring unit
(g) Digital voltmeter and data logger
(h) 20 channel strain bridge
(i) Printer
(j) Paper tape punch
(k) Vibrators

Fig. 2. *General test arrangements*

tank enabled the anchors to be pulled from the same location in the tank regardless of the inclination θ. It had the further advantage of reducing the volume of sand to be handled in each test. The motor unit (a), Fig. 2, turned a worm drive which operated on the screwed pull rod (c), producing a strain rate of 3·46 mm/min at the pull rod head (e).

5. Each anchor consisted of a 5 mm dia. steel tendon connected to a 30 mm dia. plate 5 mm thick, and was connected to the pull rod head via a load monitoring unit shown at (d) and (f), Fig. 2, and in detail in Fig. 4. The unit had ball joints at top and bottom, and these could be clamped whilst the anchor was positioned and the sand compacted. The clamps were loosened before the pull-out loading began and this gave sufficient degrees of freedom to ensure an axial pull on the anchor. The monitoring units were in turn connected to the pull rod head in the desired group pattern. The head was constrained to

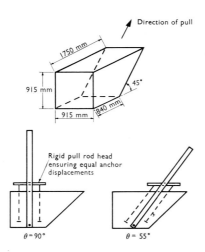

Fig. 3. *Testing tank: leading dimensions*

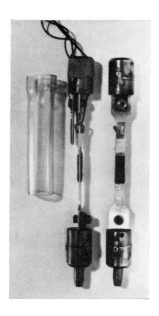

Fig. 4. Load monitoring unit

move parallel to itself in the direction of pull without any rotation.

6. The monitoring units used electrical resistance strain gauges attached to the gauge strip as shown in Fig. 4; corresponding dummy gauges were mounted on the units. A gauge system between the head and the pull rod ((b) in Fig. 2) was used to evaluate the total load applied to the anchor group. All of the strain gauges were connected via a 20 channel strain bridge to a data logger. This could sample the channels at rates from one channel every 12 seconds to 15 channels per second. The information in digital form could be produced by printer (slow rate of scan) or by paper tape punch (fast rate of scan). In the case of the paper tape the information was then processed by computer to produce a load/time plot for each anchor.

7. After careful location of the anchor group in the desired position, sand was carefully placed around the anchors using a standard technique until the tank was filled and surcharged for a few centimetres, determined from experience. The tank was then subjected to vibration for 15 seconds from a pair of eccentric mass vibrators, mounted as shown at (k) in Fig. 2. This technique was adopted after many trials. It is believed that the use of low vibratory accelerations and a short vibration duration avoided the condition of overconsolidation in which high horizontal stresses can develop. The effect of acceleration on porosity distribution has been investigated[4] and in the present tests no attempt was made to produce minimum porosity; the relative density after vibration was determined at 0·34. The density distribution was measured using embedded containers and values of 1570 ± 20 kg/m^3 were achieved throughout the tests. The gradings of the Leighton Buzzard sand before testing and after 200 tests show that some grain crushing or comminution occurred. The shear strength parameters for the as-delivered sand in the dry state and at the mean test density were determined in the shear box as $c' = 0$, $\phi' = 33°$ (peak). The effects just noted may have a slight effect on this ϕ' value.[5]

DISCUSSION OF RESULTS

8. In the majority of tests the individual anchor load/time (i.e. displacement) curves were produced automatically by computer. Simultaneously, individual anchor loads for a given sampling cycle were summed and compared with the total load (given by the monitoring gauge between the pull rod and pull rod head). An example of the pull-out curves is shown in Fig. 5. Because of the spiky nature of such curves (attributable to a slip–stick behaviour of the anchors) their interpretation was difficult, and it was therefore decided to smooth the curves by using a local least squares routine, again using the computer. The same curve after smoothing is also shown in the figure. The smoothing enables the peak of the curve to be more easily seen and facilitates comparisons between curves.

9. In the following discussion the individual anchor loads quoted are those existing at the time that the *total* load reached its peak. The group efficiencies discussed are defined as

Total load peak for a given grouping/
(Peak load of a single anchor at the same inclination
$\times$ Number of anchors in group)

Each of the tests in the programme was performed twice.

If the summed individual loads did not agree with the total load to within 5%, and the two total loads did not agree within 5%, then the test was repeated until the criteria were met. Greatest difficulty in meeting the criteria was experienced when large groups (nine or more anchors) were involved.

10. For a single anchor at the chosen inclination the test was performed five times (each time using a different channel and load monitor) and a mean was taken to provide the reference value for that inclination. This value was then used as a control in group behaviour. Figure 6 shows the variation of peak load for single anchors for various inclinations, indicating a maximum at about $\theta = 70°$. This agrees well with previous determinations.[1,2] It is worth noting that an effective doubling in this maximum load was achieved by increasing H/D from 16 (previous tests) to 25 (current tests), although there is some photographic evidence of displacement fields to suggest that the change in anchor shape may have contributed.

11. Figure 7 shows group efficiency related to spacing for the three inclinations investigated. It can be seen that despite the fact that an individual anchor achieves its maximum peak load at $\theta = 70°$ this is not so for groups. In fact groups at $\theta = 70°$ tend to produce relatively poor efficiencies, certainly as the number of anchors increases and the spacing decreases. These figures contain some evidence that the efficiency at all inclinations tends to increase at spacings greater than $S = 4D$. Space limitations within the tank (after making generous allowances to eliminate tank boundary interference effects) did not allow extremely large anchor spacings to be achieved, and hence the spacing required to eliminate mutual interference between anchors could not be established. There is evidence in Fig. 7 that interference persists up to and beyond spacings of the order of $14D$.

12. The distribution of loading amongst the individuals within a group is shown for typical groups in Fig. 8. The curves relate to individual loads achieved at the instant when the group load is at its peak. In Fig. 8 a distinct change in interference effects at about $S = 4D$ for a given inclination can often be seen, and the effect of a change in θ on the loads carried by two anchors is demonstrated. The interference caused by a trailing anchor can increase a leading anchor load above that of the corresponding single anchor, but this effect is reduced as the spacing is increased.

13. Figure 8(b) indicates what happens when two 1×2 groups merge to form a 2×2 group, and Fig. 8(c) shows that in a 1×3 group the load carried by individual anchors is very considerably affected by inclination and spacing.

14. When a 3×3 group is formed, Fig. 8(d), the situation becomes very much more complex, and the behaviour is more difficult to illustrate. It is clear that the leading outside anchors A_1 always carry more load than the trailing anchors A_2, and the central anchor A_5 plays a more important role as the spacing increases.

15. For the peak loads of individual anchors, the differences within a group are not as marked. For a 3×3 group at $\theta = 90°$ the loads follow the pattern $A_5 > A_1 > A_3$, but there is only a 10% variation within the group. For $\theta = 70°$ and $\theta = 55°$, A_3, A_5 and A_6 all carry slightly less load than their neighbours A_1, A_4 and A_2, though with much wider variation between the rows containing A_1, A_4 and A_2. If only the working load range

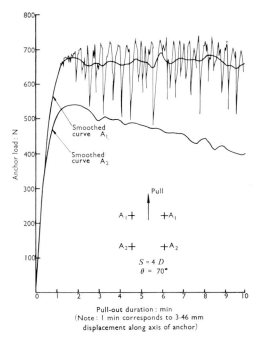

Fig. 5. *Typical load/pull-out curves*

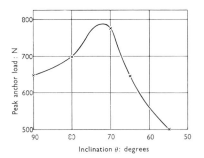

Fig. 6. *Single anchor peak loads*

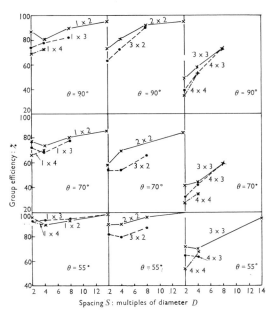

Fig. 7. *Group efficiency versus spacing for various groups*

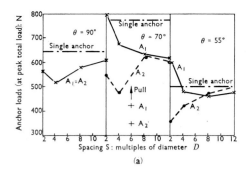

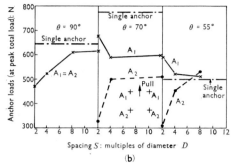

Fig. 8 (a) and (b). Load distribution within anchor groups

is considered, little or no difference can be seen in the above case. Each anchor takes up load at the same rate.

CONCLUSIONS

16. On the basis of the tests so far conducted on plate anchors installed in dry sand the following conclusions may be drawn.

17. Under constant rate of strain pull-out conditions, all anchors in a group carry a significant proportion of the group load, although the distribution amongst the anchors is complex and varies with group size and the other variables investigated. On this basis it would not be possible, for example, to exclude anchor A_5 from a 3×3 group without affecting group behaviour. Such a possibility does appear to exist under other circumstances.[6]

18. For maximum peak load in an individual anchor, installation at $\theta = 70°$ is required. For the inclinations investigated for groups, it appears that highest efficiencies are associated with $\theta = 55°$. Clearly investigation of other inclinations is necessary to establish the existence of a best group inclination.

19. In inclined groups the leading anchors establish their peak loads at relatively large displacements, whilst the trailing anchors reach their peaks more quickly and fall back to a residual value at larger displacements. The peak value of the group load is established at about the same displacement as that corresponding to the trailing anchor peaks.

20. Two phenomena of interference between anchors are suggested. At very close spacing ($S = 4D$) the interference is intense, and it seems that adjacent anchors effectively act as a single, larger anchor and efficiencies are

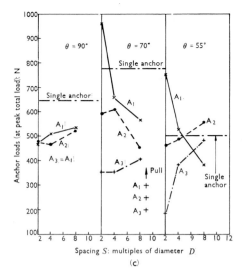

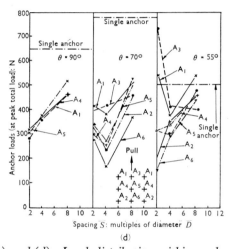

Fig. 8 (c) and (d). Load distribution within anchor groups

increased. At wider spacings this interlock effect is reduced, but the anchors tend more and more to act as individuals, and efficiencies again increase. The test evidence suggests that some interferences can still occur at spacings of the order of 14D.

REFERENCES

1. LARNACH W. J. The pull out resistance of inclined anchors installed singly and in groups in sand. *Ground Engng*, 1972, **5**, July, No. 4, 14–17.
2. LARNACH W. J. The behaviour of grouped inclined ground anchors in sand. *Ground Engng*, 1973, **6**, Nov., No. 6, 34–41.
3. HOWAT M. D. *The behaviour of earth anchorages in sand.* University of Bristol, MSc thesis, 1969.
4. ALYANAK I. Vibration of sands with special reference to the minimum porosity test. *Proc. Midland Soil Mech. Fdn Engng Soc.*, 1961, **4**, 37–72.
5. FEDA J. The effect of grain crushing on the peak angle of internal friction of sand. *Proc. 4th Budapest Conf. Soil Mechanics and Foundation Engineering, 1971*, Akademiai, Kiado, Budapest, 79–93.
6. YILMAZ M. *The behaviour of groups of anchors in sand.* University of Sheffield, PhD thesis, 1971.

Experimental excavation of length 50 m supported by strutted cast diaphragm walls: an analysis of stress distribution in the struts

Dr R. KASTNER, *Assistant, service géotechnique, Institut National des Sciences Appliquées de Lyon, France*

Dr P. LAREAL, *Chargé de cours de Mécaniqu edes Sols, service géotechnique, Institut National des Sciences Appliquées de Lyon, France*

As part of the design study for the Lyon underground railway scheme an experimental excavation, of depth 7 m and length 8 m, was carried out between two diaphragm walls of length 50 m, in order to test the working method to be used in the highly permeable alluvium of the Lyon subsoil. The walls were supported at seven levels by struts linked in pairs on the 6 central panels of the wall. In this Paper the Authors present detailed results of measurements with particular reference to variation of loads in the struts, and endeavour to analyse the influence of soil stiffness and wall rigidity and their effect on the pressure distribution on the wall.

INTRODUCTION

In 1970, the Société d'Exploitation du Métropolitain de l'Agglomération Lyonnaise (SEMALY) had an experimental section built in the borough of La Part Dieu for the study of the future underground railway of the city of Lyon. The method of construction used traditional cast diaphragm walling. Waldmann *et al*[1] have given a detailed description of this experimental section and of the measuring equipment. The essential points are summarized here.

2. The experimental enclosure was 56 m long and had a width of 8·30 m between the walls. Elevations are shown in Figs 1 and 2. Each wall was made of ten concrete cast diaphragm wall panels, 0·60 m thick, 5 m wide and 11 m deep. The two walls were connected by end panels. The rigidity *EI* of the wall was 6×10^5 kNm. Two side beams were constructed to carry a surcharge with the aid of rods anchored in the substratum. After a partial injection of the base, the excavation and strut laying were carried to a depth of 7 m. The six central panels (four measuring panels, two guarding panels) were equipped with special struts, arranged on a regular grid with depth intervals of 1·00 m and horizontal intervals of 2·50 m, giving two struts per panel at each level. On both sides the two end panels were supported for two layers (layer 2 and layer 6) by struts on Resplat springs (modulus of elasticity 10^8 N/ m²). Each strut was made of a steel tube (i.d. 17·5 cm, o.d. 22·5 cm) continued by a hydraulic jack. The hydraulic jacks were controlled by electric valves, one valve controlling two struts (both of the same panel and same layer) to apply equal forces on the wall. The distance between the two walls was measured with a differential transformer sensor. The information provided by the

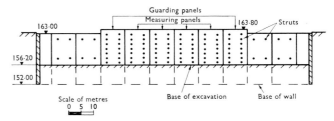

Fig. 1. *Side elevation of experimental site*

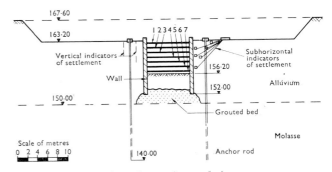

Fig. 2. *End elevation of experimental site*

displacement and pressure sensors was fed to a measure and control unit, allowing loading of each pair of struts to give the displacement ($\pm 0·1$ mm) or force (± 10 kN) chosen.

3. The soil was an alluvial gravelly sand ($D_{60} = 20$ mm, $D_{10} = 0·2$ mm) of high density (bulk density 2·2 Mg/m³) with friction angle (measured on site with a shear box 60 cm $\times$ 60 cm) in the range 27–35°. The water table was on average 70 cm below the test platform level.

4. The experiment was carried out in the following phases.

Phase A: excavation and laying of the struts

Phase B1: equalization of the forces in the struts

Phase B2: moving of the walls towards the excavation

Phase C: loading for simulation of adjacent buildings

Phase D: cancellation of all struts except for those at the top

Phase E: separation of the walls

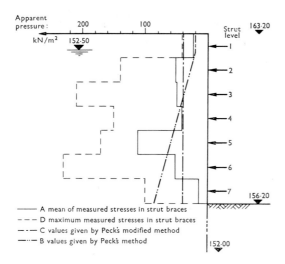

Fig. 3. *Apparent pressures for each layer in phase A*

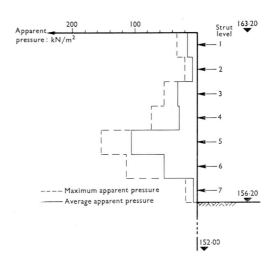

Fig. 5. *Apparent pressures for each layer after phase B1*

5. The aims of the work were to increase knowledge of the technology of the method in the alluvial soil of the Rhône, which is very permeable and has a high water table;[2,3] to study the possibility of dewatering the excavation;[1,3] to measure the side friction of the alluvial soil on the cast walls;[4] to determine the influence of a lateral surcharge on the walls;[5] and to study the effect of the soil on the walls by measurement of wall displacements. This last point was the object of a previous interpretation[5] concerned with research on the soil coefficients K_0, K_a and K_p.

6. This Paper provides an analysis of the distribution and variation of loads in the struts during phases A and B.

LOADS IN THE STRUTS DURING CONSTRUCTION

7. Phase A consisted of excavation to a depth of 7 m in such a way as to reduce to a minimum the movements of the wall (research on K_0, coefficient of earth pressure at rest). Each pair of struts was laid as soon as the corresponding digging was finished, and immediately brought under control lengthwise. From that moment each strut was very rigid ($\Delta 1_{max} = 0.1$ mm).

8. The results considered are those from the measuring

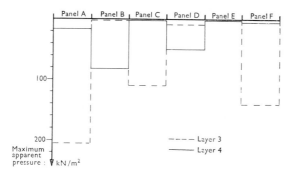

Fig. 4. *Apparent pressures in phase A for each pair of struts in layers 3 and 4*

panels only because the extreme panels, which had a different arrangement of struts, interfered with the stresses on the guarding panels.

9. Curve A of Fig. 3 represents the average estimated pressures per metre run on each layer, based on the strut load measurements. Except for layers 2, 5 and 7, values are near those given by Peck's method of calculation[6] for sheeted excavations, curve B. The divergence could be due to the fact that Peck's diagram of apparent soil pressure is superposed on the diagram of real water pressure. Substituting a rectangular diagram of apparent pressure (taking into account soil and water effects) gives the modified Peck curve C, which agrees well with the experimental values except at layer 5 which seems to be a hard area. Thus in the case of a rigid wall the distribution of forces in the struts is different from the real pressure distribution due to the soil and water.

10. Furthermore, in the system under study, consisting of a wall with little flexibility and very rigid struts, differences between struts within a layer can be very important. If, for each layer, one takes into account only the maximum value measured locally on one of the four pairs of struts, curve D of Fig. 3 is obtained for the whole of phase A. The values thus determined are much greater than those of the average stresses (curve A), and for layer 3 they are more than four times the pressures calculated by Peck's method. This great divergence is due to the rigidity of the strutted wall system. Examination of the forces taken up by the pairs of struts shows that some of them to a great extent take over the loads of their neighbours. Thus, in Fig. 4, the two struts in the third layer of panel C are heavily loaded, whereas the two adjoining ones and the pair immediately below (layer 4) are practically unloaded.

INFLUENCE OF THE RIGIDITY OF THE STRUTS

11. In order to decrease the variation of stresses within one layer of struts observed at the end of phase A, the following operations were carried out at the beginning of phase B. For layer 6 only, the struts were unloaded completely, the length obtained was taken as a reference and

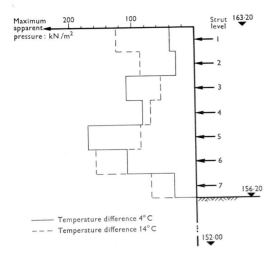

Fig. 6. *Influence of temperature on maximum apparent pressure*

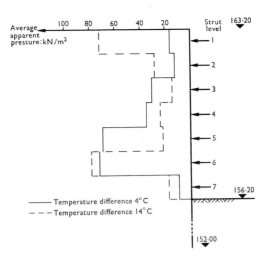

Fig. 7. *Influence of temperature on mean apparent pressure*

the new length control was imposed for the reloading of that layer. The procedure was then carried out in turn for layers 5, 4, 3, 2 and 1, and then again for layers 2, 3, 4, 5, 6 and 7. This cycle effected an approach towards equalization of the forces on each layer with very little wall displacement (0·15 mm). Figure 5 shows maximum stresses about double the average stresses at this stage, as opposed to a factor of four in phase A (Fig. 3).

12. The results indicate that for an excavation for which the struts are not lengthwise controlled—as is the case in practice—their natural deformation under load leads to a deduced pressure distribution which is more even than that observed in phase A.

13. Moreover, these operations demonstrate that the total force in all the struts is not directly related to the pressure of earth and water but rather to the geometry of the strut system. Although the walls did not move the total stress varied from $1·3 \times 10^4$ kN to $1·8 \times 10^4$ kN.

INFLUENCE OF TEMPERATURE

14. At the end of excavation, when layer 7 was temporarily out of order, there was an marked fall in temperature at Lyon (around 10°C). The air temperature fell 14°C below the temperature of the soil under groundwater, and caused a marked contraction of the external face of the walls. Deformation was prevented by the length control on the struts. The measured stresses changed appreciably, with a large increase in the first layer, a large reduction in layers 3, 4 and 5, and a small increase in layers 6 and 7 (Figs 6 and 7). These observations show that, in practice, provision must be made for eventual stress changes in the struts and changes of moments in the wall, which can occur when the wall is subjected to a large thermal gradient.

15. These changes may be still more important if the struts are not controlled lengthwise.

CONCLUSION

16. These observations show that the variation of stresses in the struts is greater when the strutted wall system is rigid. In the particular case of the experimental section, the empirical methods used to calculate the sheetings lead to significant underestimation of the apparent pressures of soil and water.

ACKNOWLEDGEMENTS

17. We wish to thank Mr Waldmann, Director of SEMALY, and also Mr Ferrand, for having allowed us to co-operate in the realization and exploitation of these tests. The experimental section was constructed by the SIF Bachy company under the direction of the SIMEC-SOL bureau, directed by Mr J. Kerisel, and the Direction Départementale de l'Equipement, directed by Mr M. Prunier.

REFERENCES

1. WALDMANN R. *et al.* Section experimentale du métro de Lyon: principe et technologie des mesures. *Revue Travaux*, 1971, Jan., 27–41.
2. WALDMANN R. and FERRAND J. Essais de parois moulées à Lyon. *Revue Travaux*, 1970, Nov., 73–80.
3. LAREAL P. *et al.* Utilisation de plusieurs techniques de parois moulées dans les alluvions de grande perméabilité et de granulométrie très variée du site de Lyon. *Int. Conf. Soil Mech. Moscow*, 1973, Speciality session no. 7.
4. KERISEL J. and LAREAL P. Essai d'arrachement d'un élément de paroi moulée dans des alluvions sablo-graveleuses. *Journées Nationales du Comité Français de Mécanique des Sols*, 1971, May.
5. KERISEL J. *et al.* Mesures de poussée et de butée faites avec 42 paires de butons asservis. *5th European Conf. Soil Mech.*, 1972, Sociedad Espanola de Mec. del Suelo y Cimentaciones, Madrid, Session III, **1**, 265–273.
6. TERZAGHI K. and PECK R. B. *Soil mechanics in engineering practice*. Wiley, New York, 1967, 402.

Permanent anchored wall in Caracas, Venezuela

Ing. S. PETRINI, *Vice-President, Aliva Stump CA, Caracas*

Ing. A. MILA de la ROCA, *Aliva Stump CA, Caracas*

The construction of large buildings in the congested and topographically complex city of Caracas has frequently required that techniques developed for anchored walls should be applied for such work as excavation of basements, cutting of slopes and reinforcement of foundations of nearby buildings. The Paper describes the design and construction of an anchored element wall in Caracas. The wall allowed excavation to a depth of 14 m, completely separating construction of the building from problems of high earth pressure, and also reinforcing two 10 storey residential buildings located on the adjacent lot. The Paper describes the characteristics of the locality, the calculation methods and their derivation, the system of construction, and practical experience gained during construction and in the following two years.

INTRODUCTION

The city of Caracas is situated in a narrow valley, approximately 15 km in length and 3 km in width in its central part. Its southern, eastern and western limits are formed by hills of varying height and of generally abrupt topography. The valley's northern limit is the Avila Sierra, a branch of the Cordillera de la Costa, which is unusable for construction because of its topography and its condition as a national park.

2. Rapid increase of population in Caracas has necessitated building on the hillsides and obtaining a maximum construction density on the flat areas. In the lower part of the valley, high cost of land necessitates use of as many stories and underground parking levels as possible: on the hills that surround the valley, the topography has to be substantially modified in order to allow a reasonable use of the area. The first situation brings up problems such as support of deep excavations, lowering of water level and underpinning of neighbouring foundations: in the second situation the principal problems are of slope stability.

3. Many of the retaining structures constructed recently in Caracas have been built with an anchored element wall technique, and several of them have been designed to serve as permanent systems. This Paper describes the design, construction and behaviour of the anchored wall (shown in Fig. 1) built to stabilize the northern face of the excavation for the Centro Comercial Eraso in the Las Mercedes district of Caracas, in the Authors' opinion one of the most interesting examples of a permanently anchored retaining structure.

DESIGN CRITERIA

4. The depth of excavation varies from 8 m to 14 m. On higher ground at a distance of 7 m is a 10 storey building which, although founded on piles, would be greatly affected if the movements originated by the wall should exceed certain limits. The subsoil is formed throughout most of its depth by alluvial plastic silty clays with varying sand and gravel contents. The plasticity index varies from 10% to 25%. Below this stratum is a layer of sandy gravel of very high static penetration resistance, and below that the micaceous schist, very weathered in the upper layers. The water level was found some distance beneath excavation level.

5. Soil parameters used in the calculations are given in Table 1. In consideration of the permanent character of the wall, the shear resistance parameters were obtained by a specialist firm using fully drained shear tests.

6. The magnitude of the earth pressure and the shape of its distribution diagram were thoroughly discussed between the consultants of the client and the designers of the wall.

7. It was the consultants' opinion that a K_0 condition (an at rest condition with the characteristic triangular distribution) should be assumed, in order to avoid lateral movements of the soil mass and consequent movements of the building behind the wall.

8. The designers of the element wall were of the opinion that the construction system for this type of wall, which is built from top to bottom by vertical and horizontal stages, cannot guarantee complete absence of movement in the soil mass because, although the prestressed anchor assures the immobility of the completed element, it is impossible to restrict the displacements of zones prior to anchoring during the unprotected excavation phase necessary for the construction of each element.

9. Placing the concrete wall against the already deformed soil causes a redistribution of horizontal stresses. German and Swiss experience indicates that the most adequate distribution in such cases is a rectangular one, and that the recommended magnitude of the total horizontal pressure should be 1·3 times the corresponding active pressure.

10. The magnitude of possible settlements was estimated on the basis of experience. The studies indicated

Table 1. Soil parameters

Soil	Friction angle ϕ, degrees	Cohesion, c	Density γ, t/m^3
Silty-clayey materials	25	0	1·8
Sandy gravel	35	0	2·1
Weathered schist	30	0	1·9

that settlements would be small and would not cause damage to nearby buildings.

11. As a result of the discussions the consulting firm, the client and the designers of the wall established the following design criteria.

(*a*) In view of the chosen construction method, a rectangular uniform pressure distribution with a total value of 1·3 active pressure was recommended.

(*b*) The shear strength parameters assumed for the static calculations were regarded as satisfactory even for the long term behaviour of the material.

(*c*) The settlement expected to occur behind the wall could be taken satisfactorily by the neighbouring structure without any damage.

(*d*) It was decided to study carefully the location and trajectory of the anchors in order to avoid interference with the piles of the nearby structure.

(*e*) It was decided to perform periodic revisions of anchor tension during the whole life of the wall, and the possibility of replacing some anchors was not disregarded.

(*f*) It was established that the limitations created by the permanent anchors in the neighbouring land would be clearly defined in the agreements to be made with the affected owners.

(*g*) As movements of the anchored wall could not be restricted by the structure of the future building, adequate space was to be left to allow for them and also to avoid the introduction of additional horizontal forces in the structure of the Centro Comercial. The space between wall and structure would be filled with a material which would flow under compression and could be used as an index of the magnitude of the movement.

(*h*) It was decided to install shallow and deep drains in sufficient quantity to guarantee an effective drainage of all the retained soil mass, thus avoiding the action of pore pressure on the wall and deterioration of the soil condition.

DESIGN

12. Taking into account these design criteria, the most convenient shape and intensity of the pressure diagram to be applied by the wall anchor system upon the soil was determined, together with the anchor length and anchor distribution.

13. Several types of anchor were studied. The one chosen was the 50 t working load Stump-Duplex, as this was considered best adaptable to local soil conditions and to offer the best corrosion protection. This type of anchor consists of two parts, the compression member and the tension bar. The compression member is a steel pipe, to the end of which the bar is fixed by a screw mechanism. It serves the double purpose of transmitting the anchor forces to the ground by compression of the grouting zone and of protecting the tension bar against corrosion. It is possible to use different lengths and diameters of compression member, depending on the strength characteristics of the soil in which the anchor is to be placed. After the application of the anchor force, the compression member and the surrounding grout are under compression, whereas the tension member is free and under tensile stresses only. Also, this type of anchor allows the control of deformation during application of the load, the periodic control of anchor loads and, if necessary, withdrawal and replacement of bars at any time during the life of the wall. The corrosion protection for the tension bar consists of sandblasting and application of several epoxy coats and special protection tapes. Finally, the annular space

Fig. 1

between the bar and either the exterior polyethylene pipe or the compression member is filled by flexible plastic products.

14. The magnitude of the total pressure was established in accordance with the criteria of § 11 and was uniformly distributed. The size of the wall elements was chosen in accordance with the particular site conditions, in such a way that no local failure would occur upon excavation of new elements. In this particular case the horizontal distribution of anchors was affected by the necessity of leaving slots 53–111 cm spaced every 6 m in the wall at the locations of future columns of the structure.

15. The wall was designed as a slab resting on the anchor points. This condition required a 35 cm thickness and a moderate amount of reinforcement on both sides and in both directions.

16. The anchor length was established as proposed by Dr H. Bendel[1] by analysing the possibility of a total failure. According to this method the anchors must reach a critical sliding surface (defined by limit analysis with logarithmic spirals) and reach a minimum length L_{min} beyond the active surface (L_{min} is a function of the boring diameter and the shearing resistance of the soil, and in this case was 1·2 m). The first condition is particularly important for the upper levels of anchors whereas the second one is important for the lower levels.

17. Eight anchor lines were used at the sections with higher wall height, and five lines at the lower sections. The anchor lengths varied from 17 m to 6 m in the former and from 14 m to 6 m in the latter sections. Anchors were placed with 15° inclinations except at the three lower rows where 30° inclinations were used.

CONSTRUCTION

18. Construction was completed in $5\frac{1}{2}$ months in accordance with the original working programme.

19. The construction stages of the element walls are shown in Table 2. All these stages were performed with no inconvenience. Vertical and horizontal movements were continuously checked by triangulation and were always below initial estimates.

20. The construction joints were carefully cleaned before the new concrete was poured. Dowels, which were later welded to the reinforcement of the element, were left.

21. During the final stages of wall construction, in order to reduce to a minimum the possibility of vertical

Table 2. Construction stages of element walls

1	Excavation of the soil to allow placement of the element
2	Placing steel reinforcement, with the vertical steel being driven into the ground below the element to act as dowels for the next row
3	Placing formwork
4	Pouring concrete
5	Drilling (with casing) holes where anchors will be installed
6	Installation of the assembled anchor and grouting
7	Application of the test load
8	Drilling drains and placing drain tubes

settlements, small vertical steel columns were placed during the excavation of the lower elements. These columns supported the overlying elements while construction was being carried out immediately below. These steel columns were left in place during pouring of the concrete of the new elements.

22. The borings for the anchors were drilled with rotative equipment, using 115 mm dia. casings. At the beginning of the job special implements were used to enlarge the diameter of the holes in the vicinity of zones where compression members were to he placed. After examination of the results from six test anchors constructed on the site, normal drilling procedures were adopted. The several tests performed on the test anchors showed that in this particular case the ultimate load did not depend on the possible enlargement of the boring diameter of the holes.

23. The six test anchors were loaded to failure and the minimum factor of safety obtained was 1·27 referred to the working load.

24. Drains were constructed with rotative machines and 2 in. slotted galvanized pipes. Their length varied between 12 m and 16 m.

LONG TERM BEHAVIOUR

25. The wall was completed two years ago and its behaviour has been totally satisfactory. Three verifications of anchor load have been carried out and small losses have been noted due to soil and steel creep. Anchor losses decrease rapidly with time and at present the typical losses are negligible.

REFERENCE
1. BENDEL H. Theorie und Versuche über die Berechnung der Tragfähigkeit und der Verankerungslänge mit Anwendungsbeispiel. *Schweiz. Bauztg*, 1966, Heft 6.

Discussion on Papers 18–21

Reported by C. P. WROTH

Papers 18–21 and all the contributions to this discussion except one were concerned with ground anchors formed in soil; further references to anchorages in rocks came in the following session.

2. The overall impression to be gained from the session was of the parallel between the development of ground anchors at the present time and that of bored piles eight years ago, when the timely conference on bored piles was held in London. The subject of ground anchors is developing rapidly and considerable progress can be expected in the next ten years or so. It is a branch of foundation engineering where current practice is ahead of theory; there is urgent need for a proper understanding of the behaviour under working conditions of the anchors and the surrounding ground.

3. Current practice was summarized in the valuable state-of-the-art Paper by Mr Ostermayer. In this Paper the collected field data are sufficiently comprehensive to give confidence in the results, presented in what are essentially design charts. The excellence of Mr Ostermayer's contribution to the subject was recognized both by the award of the special prize by the journal Ground Engineering and by the fact that it attracted by far the most discussion.

4. The eloquent plea made by Mr Mitchell in the previous session (Paper 17) for establishing a code of practice for the design, construction and acceptance of ground anchors was echoed by several speakers. Concern was expressed about the varying standards of workmanship in the construction of ground anchors, and the consequent need for an adequate procedure for testing a representative number of anchors after installation.

TESTING

5. Dr Bassett spoke about the variability of test results and showed some statistical data from pressure grouted anchors.[1] He felt that such wide variations must be borne in mind if use were to be made of the design charts presented in Mr Ostermayer's Paper (Figs 7 and 9).

6. Dr Littlejohn welcomed the points made by Mr Ostermayer in § 20 of his Paper, and commented that the German practice of testing permanent anchors to 1·5 times the working load agreed closely with the recommendations of the Bureau Securitas in France. Acceptance of this standard test was long overdue in Britain, where standards of testing had dropped in recent years. Permanent anchors were being installed with no overload test, and this meant that no measured factors of safety were obtained.

7. In reply to specific questions put by Dr Littlejohn, Mr Ostermayer reported that the test results quoted in Figs 7–10 of his Paper were for anchors which had been installed with inclinations of 10–30° to the horizontal,

with grouted bodies generally 10–15 cm in diameter and with overburden heights between 5 m and 8 m. The anchors had been stressed against a surface pad and the free anchor length (distance between grouted body and surface pad) was at least 5 m in order to avoid any influence of the reaction of the pad on the ultimate carrying capacity of the anchor.

8. Mr Vial discussed his firm's method of comparing the actual prestress load in each tendon with that required by the design, which allowed the friction losses to be isolated by means of a cyclic load test.

CONSTRUCTION TECHNIQUES IN COHESIONLESS SOILS

9. Mr Vadgama disagreed with Mr Ostermayer's statement (§ 22 of his Paper) that the boring techniques in cohesionless soils had no decisive influence on the quality and carrying capacity of anchors. He described the following method which gave superior results to ordinary boring methods with pressure grouting.

10. The free length of the anchor is drilled in the normal manner, with casing used to support the hole. The remainder of the hole is formed by the drill head being surged forwards and backwards without drilling fluid but with the hole being kept full of cement grout. This results in the grout being mixed into the soil, and the soil being compacted.

11. From laboratory tests he claimed that the spread of grout was 3–5 times the diameter of the drill head for sand and gravel ($k > 10^{-1}$ mm/s) and 2–3 times for fine sand.

12. Dr Littlejohn also doubted the validity of Mr Ostermayer's § 22 in relation to the growing use of vibratory driving methods in loose sands. Mr Ostermayer stood by his statement for medium dense and very dense sands, but had insufficient results to draw conclusions for loose sands.

UNDERREAMING IN CLAYS

13. Mr Ostermayer's brief dismissal of the effectiveness of underreaming in § 24 was strongly challenged. Dr Bassett referred to the success achieved by at least four companies in the UK, and outlined some findings from model underreamed anchors tested in remoulded London Clay at King's College. These tests showed the influence of an increasing number of underreams spaced at 1·5 times their (maximum) diameter. The ultimate loads increased in direct proportion to the number of underreams, but at working loads (or displacements) the effects were not so beneficial.

14. Mr Czechowski described the successful use in London Clay of underreamed anchors for the construction of retaining walls for an underpass.

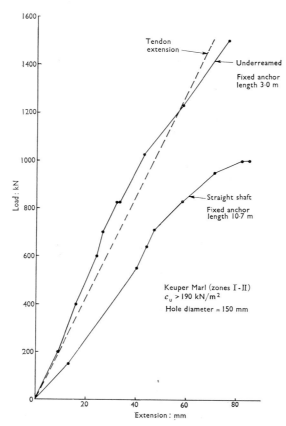

Fig. 1. Comparison of load–extension responses of an underreamed anchor and a straight-shafted anchor

illustrated by Fig. 1. Based on the same augered hole diameter of 150 mm, the straight shaft anchor with a fixed anchor length of 10.7 m failed at 1000 kN, whereas the underreamed fixed anchor of only 3 m withstood, without any signs of failure, a load of 1500 kN. As a result of tests of this type, it was now known that in stiff to hard clays safe working loads of 500–1000 kN could be obtained with underreamed anchors, compared with 300–400 kN using straight shaft anchors. These figures were based on load safety factors of 2.5–3.5 in order to keep prestress losses within acceptable limits, and in both systems grout injection was by an open hole tremie method, so that there was no danger of surface heave due to grout pressure.

17. Mr Ostermayer responded that there was no doubt that the carrying capacity of anchors in cohesive soils could be decisively increased by underreaming so long as the soil was not disturbed by the underreaming technique and the cavities were properly cleaned. This was possible in stiff to hard homogeneous clays for the case of vertical anchors. But there were other conditions where there had been only marginal increases in load carrying capacity with the penalty of a large increase in the plastic deformations of the soil. The point, however, was to compare the performance of underreamed anchors not with that of straight-shafted anchors without regrouting (as Dr Littlejohn and Mr Czechowski had done), but with regrouted anchors. He referred them to Figs 12 and 13 of his Paper, and believed that for any particular site only tests would show which method was the more economic.

MODEL TESTS

18. The clear conclusion to be drawn from the model tests on anchors is that a proper understanding of anchor behaviour (and indeed of all problems in foundation engineering) can only be reached by a careful study of the deformations that occur in the surrounding soil. Observation of the load–displacement response of an anchor (or group of anchors) alone is not sufficient.

19. Dr Larnach introduced Paper 19 by presenting some results from a photogrammetric study of the displacement fields associated with anchors. Figure 2 shows the limit of the zone of discernible displacements in the vertical direction associated with a group of two model anchors. Figure 3 indicates the growth of the zones as the

15. Dr Littlejohn quoted the case of the U Bahn in Hannover where in 1969 three underreamed test anchors were installed. The clay was hard and highly fissured (intact strength of 1000 kN/m², fissure strength of 200 kN/m²). The anchors were designed with 2, 3 and 5 underreams with lengths of 1.5–4.5 m, in the hope that anchor failure could be induced. All three anchors held 850 kN without sign of failure and with small fixed anchor movement; bearing in mind the low designed safety factors, prestress losses after two months were considered small, ranging from 3% to 6.5%.

16. The reason, according to Dr Littlejohn, for using multi-underreamed anchors in Britain could perhaps be

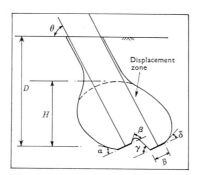

Fig. 2. Limit of displacement zone for two model anchors

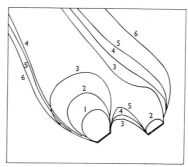

Fig. 3. Extent of displacement zone at successive displacements of about 2% of anchor diameter

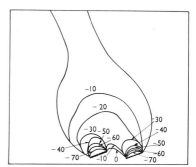

Fig. 4. Two anchors, $\theta = 70°$, $D/B = 15$ (see Fig. 2): contours within displacement zone indicating intensity of movement

anchors are pulled out. It should be noted how the zones associated with the individual anchors rapidly coalesce. Figure 4 shows the situation after a displacement of 18% of the anchor diameter, with the pattern of displacements within the sand given by the contours. The 'channel' at the top of the diagram is associated with movements of the sand grains contained between the two anchor tendons.

20. Mr Swain reported the preliminary findings of a model study at the University of Cambridge aimed at clarifying the fundamental mechanisms which govern the behaviour of underreamed anchors in clay. Tests were performed under both axisymmetric and plane strain conditions; but only the results relevant to plane strain were discussed.

21. The apparatus, shown in Fig. 5, consisted of a steel frame, with glass panels on the back and front, to contain the clay and the model anchors. The clay was consolidated from a slurry to the required consistency by the base piston, the final increment of load being applied using a flexible surcharge membrane. The sample size at the end of consolidation was 305 mm × 305 mm × 152 mm.

22. It is known that anchor placement techniques have a decisive influence on the behaviour of anchors. In an effort to overcome the influence of this effect, the anchors were placed in the clay prior to consolidation and provi-

sion was made to eliminate flow past the anchors during the consolidation process.

23. A regular grid of lead shot markers was placed at the mid plane of the clay so that the displacements within the clay could be observed by exposing radiographs at intervals during a test. The model anchor was pulled at a constant rate, except during the exposure of a radiograph when the anchor movement was stopped. The loads on individual underreams of the anchor were measured by load cells.

24. Three tests, each with clay samples at different values of the over-consolidation ratio (OCR), were performed on identical triple underreamed anchors. Figure 6 shows the relationship between the OCR and a function based on the ratio of peak load F to the undrained shear strength c_u of the clay. The shear strength c_u of the clay was measured in undrained triaxial tests. It appears that when the OCR is small the peak load depends only on c_u.

25. Several tests were carried out on single underream anchors in order to compare their behaviour with that observed for triple underream anchors at various values of OCR. The peak load observed for a triple underream anchor is compared with the peak load on a single underream anchor in Fig. 7. The fact that the peak load on the triple anchor is 1·5 times (not three times) the peak load on the single anchor suggests that at an underream spacing of 1·8 times the anchor diameter there is a complex inter-

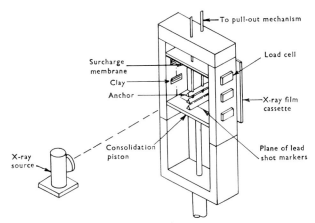

Fig. 5. Plane strain apparatus

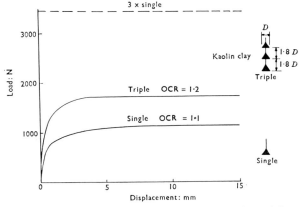

Fig. 7. Total load–displacement relationships for triple and single underream anchors

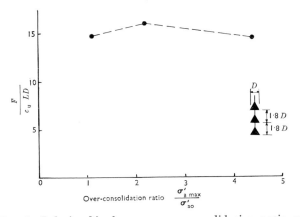

Fig. 6. Relationship between over-consolidation ratio and $F/(c_u LD)$, where F is the peak load and c_u the undrained shear strength of the clay. (The length L and diameter D of the anchor are introduced to provide a dimensionless variable.)

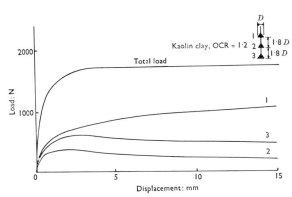

Fig. 8. Load–displacement relationships for individual underreams for the triple underream anchor

167

action between failure zones for each underream. However, the displacement to peak load was reduced from 8 mm for the single anchor to 3 mm for the triple anchor. Thus, although there is a relatively small increase in carrying capacity of the triple anchor over the single anchor, the triple anchor system is considerably stiffer.

26. Figure 8 shows, for the triple underream anchor, the load–displacement relationship for the individual underreams and their sum. The top underream continued to increase in load carrying capacity throughout the test as further consolidation took place in front of it, whereas the loads on the middle and bottom underreams reached a peak and then decreased. These two effects compensated so that the total load on the anchor was constant for a large displacement. It is notable that the load on the middle underream was the lowest.

27. The fact that the loads were different on the three underreams was probably due to two factors: the interaction between failure zones for each underream would mean that loads on the underreams would be different,

even in a rigid perfectly plastic material; and, in reality, the clay is not rigid perfectly plastic and so the mobilized shear stress depends on the shear strain. Both phenomena were being investigated by making observations of the deformations in the clay around the anchors. Figure 9 shows two radiographs superimposed so that the movements of the anchor and lead shot markers can be seen. There was considerable soil deformation in front of the top underream and behind the bottom underream; this observation goes some way towards explaining the relative sizes of the loads on the three underreams.

28. Mr Wernick commented that the skin friction of ground anchors is much greater than would be expected from the overburden pressure in the case of driven or bored piles. In cohesionless soils the large skin friction can be attributed to the suppression of the dilatancy of the soil in the thin shear zone, and the consequent interlocking of the particles. Skin friction[2] increases with decreasing diameter of anchor, with increasing stiffness of the soil, and in a highly non-linear manner with increasing density.

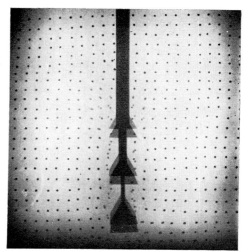

Fig. 9. *Two radiographs superimposed showing displacements of the anchor and of the lead shot markers at an anchor displacement of 7 mm and an over-consolidation ratio of 4·4*

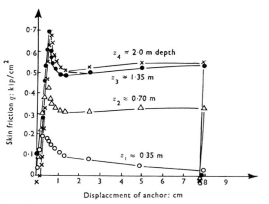

Fig. 11. *Development of skin friction as a function of displacement at several depths*

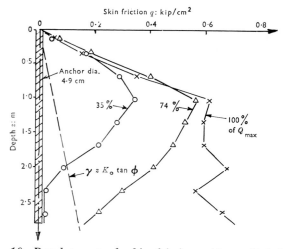

Fig. 10. *Development of skin friction with applied load (relative density ≈ 1·1)*

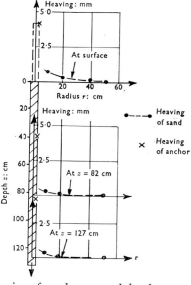

Fig. 12. *Heaving of sand at several depths at peak resistance for anchor of diameter 4·9 cm*

To study these effects large-scale laboratory tests were in progress at the University of Karlsruhe.

29. The pull-out resistance of vertical rough cylindrical anchors in a round test bin (dia. 250 cm, height 300 cm) has been investigated. The soil consisted of a medium sand poured so as to be of uniform density. The anchor consisted of pieces of steel pipe connected inside with threaded steel bolts. Strain gauges were fixed to the bolts in such a manner that the skin friction could be measured for each of the pipe sections.

30. The development of skin friction during various stages of loading (up to peak resistance Q_{max}) is shown in Fig. 10. At peak load the skin friction q increased only up to a depth of about 1 m below surface. Below this q remained constant and was (over the entire length of anchor) several times greater than would have been expected from the at rest pressure times the frictional coefficient (dotted line in Fig. 10, $K_0 = 0.4$, tan $\phi = 0.75$).

31. The development of skin friction at different depths as a function of the displacement of the anchor is illustrated in Fig. 11. The peak occurred at a displacement of only about 0.5 cm. From a displacement of 1 cm onwards, q remained constant at about 75% of maximum value. Figure 11 demonstrates the high dependence of the skin friction on the displacement of the anchor.

32. The vertical heave around the anchor at peak resistance, measured on the surface and at depths of 82 cm and 127 cm, is given in Fig. 12. The corresponding displacement of the anchor is also shown. These results reveal the limited width of the zone of large shear deformations.

33. To obtain an idea of the deformations near the anchor, breakable thin rods were placed radially near the anchor in the sand and dug out after the pull-out test. The zone of large shear deformations was limited to 5–8 mm around the anchor, which is about 5–10 times the diameter of soil grains.

34. These investigations have led to some important conclusions for the use of soil anchors in cohesionless soils.

(a) The bearing capacity of anchors is mainly due to the effect of dilatancy and its suppression; thus it cannot be expected that the frictional resistance on the surface of the anchor is a linear function of overburden pressure.

(b) The groundwater table may have only a limited influence on bearing capacity, and only then near the ground surface. The same applies to the inclination of the anchor.

(c) The skin friction is a function of the diameter of the grouted cylinder.

(d) During the placing of an anchor no loose soil should remain around the anchor surface, as otherwise the dilation effect is much less.

(e) Because of the high dependency of pulling resistance on density, it is preferable to use drive borings instead of wash borings or other boring methods where the soil is removed.

(f) The factor of safety has to take into account the dependence of skin friction on displacement. The peak of skin friction does not occur simultaneously over the entire length of the grouted cylinder because of its own deformation.

ANCHORS IN ROCK

35. Mr de Beer described the use of a unique type of rock-anchored abutment where the elasticity of the anchor cables is used to allow for ground movements due to expected mining subsidence. The Kazerne viaduct is part of the Johannesburg elevated motorway system and crosses ground undermined at the relatively shallow depth of 200 m.

36. Bearings and movement joints were provided to eliminate unacceptable deformations in the structure due to distortion of the ground. An essential requirement of the design was therefore to provide a joint between the abutments and the deck of the motorway.

37. A 15 m high abutment at the west end is inclined at 45° and retains fine mine tailings. The abutment is a beam and slab structure held against the tailings by rock anchors stressed initially partway between the least required force and the maximum capacity when full movement has caused elongation. Most forms of conventional rigid abutments would have been subject to unacceptably large forces generated by mining subsidence.

38. The permanent anchors, protected in a greased sheath, passed through the tailings and some overburden and were grouted into the rock beyond a distance of about 23 m from the abutment. The anchor heads were left exposed to allow for routine checking of the cable force, as well as checking if movement had occurred. With a free length of some 23 m the system could accept an elongation or shortening, due to mining subsidence, of 50 mm. This provides an acceptable and predictable increase or decrease in the anchor force.

39. The abutment was analysed by computer, taking into account the possible variations in anchor load. The design is complex in that the minimum factor of safety occurs when the ground is subjected to compressive strain which releases part of the load on the cables. However, should collapse be initiated the deformations in the ground would immediately reimpose the stress on the cables until restoration of equilibrium.

CONTRIBUTORS

Dr R. Bassett, King's College, University of London.
Mr M. K. Czechowski, Donovan H. Lee and Partners.
Mr J. H. de Beer, Ove Arup and Partners, Johannesburg.
Dr G. S. Littlejohn, University of Aberdeen.
Mr R. Parry-Davies, Ground Engineering Ltd, South Africa.
Mr A. Swain, University of Cambridge.
Mr N. Vadgama, Terresearch Ltd.
Mr C. Vial, Soletanche Co. (UK) Ltd.
Mr E. Wernick, Institute for Soil and Rock Mechanics, Karlsruhe.
Dr C. P. Wroth, University of Cambridge.

REFERENCES

1. BASSETT R. H. Contribution to discussion Session IV. *Proc. 5th European Conf. Soil Mechanics and Foundation Engineering, 1972,* Sociedad Espanola de Mec. del Suelo y Cimentaciones, Madrid, **2**, 330–334.
2. WERNICK E. Tragverhalten von Verpressankern in rolligem Boden. Contribution to conference Gründung unterirdischer Baukonstruktionen in gedrängten Bebauungsbedingungen, Nov. 1974, Hohe Tatra, CSSR.

Economics of basement construction

M. J. PULLER, DIC, MICE, MIStructE, *Contracts Manager, A. Waddington and Son Ltd*

The Paper describes those factors which influence the feasibility of basement construction on any one particular site. Factors discussed include the influence of subsoil and groundwater on choice of method of basement wall construction, the effect of plan area and shape of the basement, and the effect on overall construction time of the basement construction method chosen. Four typical basement excavations are examined, and those systems of basement wall construction which may be applied are compared in terms of UK construction price per unit area of basement wall. Whilst generalization is impossible from this small sample, the comparison indicates the procedure which is necessary at feasibility stage and the relatively large number of methods which remain available.

INTRODUCTION

Previous authors[1-5] have described in detail the various methods available for the temporary and permanent support of subsoil to the periphery of building basements. Zinn[1] in particular explained the evolution of a method of construction, similar to that described by Fenoux[3] as the downwards method, which incorporated contiguous pile walls supported by floor construction as excavation and floor construction were made successively with depth. Zinn concluded that three main criteria had the greatest effect on economy in design and construction as follows: the retaining wall which supported the side of the excavation had to be temporarily and permanently supported at every floor level, facilitating both economical design and construction; it was equally desirable that advantage should be taken of the ground as a means of supporting the soffit shutter to floors by casting directly on the ground before excavation; and the removal of excavation had to be rapid and continuous.

2. It is assumed that the choice of technique for a particular site follows from decisions regarding the necessity of a basement and the feasibility of construction. On the question of necessity, the requirements of a client owning land on which he wishes to speculate or to build for community use may be presumed wrongly. The presumption that high land values in city centres is an incentive to build deep as well as high may be false. Disincentives which may apply are suspected high cost risk of underground construction, particularly those cost risks associated with the frequent discovery of unforeseen subsoil conditions or existing constructions; and suspected high risk of demand on construction time for underground construction, causing delays to overall completion and commencement of return on capital. Building regulation requirements may require certain standards of natural lighting for habitable areas such as offices: this requirement would limit the location of such

areas to one storey below existing ground level. Also, the use of basements to office structures for car parking may be limited by planning requirements which exclude cars from city centres; and the use of modern air conditioning plant favours the siting of heating and ventilation plant at roof level rather than below ground level.

3. The remainder of the Paper deals with factors influencing the feasibility of basement construction.

PERMITTED PRICE OF BASEMENT FLOORSPACE TO CLIENT

4. Distinction must be drawn between cost and price. The total price to the Client, ignoring finance charges, will be constituted in a conventional bid situation as follows:

> Final contract price to Client
> = Contractor's estimate of cost of site labour, plant and materials and site overheads
> + Contractor's margin to allow for Company overheads and profit or loss
> + Cost of variations and claims
> + Cost of professional fees including supervision fees

The difficulty in assessing the true merits of different methods of basement construction is increased by the large fluctuations which occur in each of the first three items of the right hand side of this equation. In the first two items, the Contractor's estimate of site costs and his margin applied to these costs, experience in the UK has frequently shown a wide spread in the prices for specialized foundation engineering subcontracts. Typically, with diaphragm wall subcontracts, the price of which may constitute more than 30% of the total basement price, the subcontractors' prices themselves commonly vary from the lowest to a high figure equal to half as much again. Such spread of price may be caused by commercial pressure on margins rather than inaccurate estimating of the cost of resources to carry out specialized work and this in turn may lead to the risk of high cost of variations and claims.

5. In these circumstances, specialized foundation techniques which appear at first sight technically efficient and economically attractive may well produce a very wide spread of prices at bid stage, the lowest of which are unacceptable because of the opportunity of high contractural claims in underground works or the high financial risk to the Client of the specialized works being done badly at a very low price. It should also be noted that the total price of any construction includes its ultimate demolition price. This definition of total price is applicable to the basement construction also, yet such consideration is rarely given in a total feasibility study for a basement or in a comparison of the merits of different forms of basement

construction. Perhaps this is due in part to the fact that nowadays a building's first owner may not pay for its demolition.

6. The viability of building deep will depend on a comparison between expenditure and income, the expenditure represented by the total price to the Client and the income represented by an equivalent commercial rent which he receives for the floor space. In these circumstances, on a given site a single-storey basement in timbered excavation may prove the most economical choice of structure to store canned food, whilst a three-storey structure below ground constructed within an anchored diaphragm wall may prove feasible, economically, for the storage of rare books or works of art.

PERMITTED COST OF SUPPORT TO SUPERSTRUCTURE

7. Whilst the decision to build a basement solely as a means of supporting the superstructure must be rare, excepting the case of a city site with an existing basement construction, both a technical and financial advantage will accrue from basement accommodation when considering the cost of transference of load from the superstructure to the subsoil. Firstly, it allows transference of load to a bearing stratum at a lower level, partly perhaps through the periphery basement wall, and of course it also allows a reduction of net bearing pressure by the displacement of overburden.

8. The use of contiguous bored pile walls, secant pile walls and diaphragm walls allows a threefold purpose to be obtained from the periphery basement wall: both temporary and permanent subsoil support together with means of transference of vertical load from the superstructure (or vertical upward loads caused by buoyancy). Fenoux[3] refers to the transfer of load by a diaphragm wall on stilts. This form of construction uses individual bites of each panel taken to a lower and improved bearing stratum but reinforced without discontinuity with the remainder of each panel. Whilst the lack of published observations of load tests on diaphragm wall panels still continues, Fenoux reports the use of a 600 mm wide diaphragm with adequate subsoil conditions to transfer loads up to 300 t/m, and loads of the order of 150 t/m have been transferred to Thanet Sand (a dense fine sand) from 800 mm walls in Croydon, south London.

9. At working load state it is possible that much of the transfer of load occurs from the diaphragm wall by friction at the back of the wall rather than by direct bearing at the base of the wall.

10. The structural form of the building's superstructure will predetermine the distribution of dead and live loads through the basement substructure. Conventional column and beam construction will allow the dead and live loads from the periphery of the superstructure to be transferred into load bearing walls of the substructure, whilst high rise blocks supported from a central core will not derive this advantage unless the basement is locally deepened at the core.

SUBSOIL AND GROUNDWATER CONDITIONS

11. The final price of the basement structure will depend to a large extent on subsoil and groundwater conditions. The choice of overall construction and excavation method will be directly dependent on these conditions. Methods of construction downwards whilst superstructure proceeds simultaneously upwards minimize construction time, but higher excavation costs may deter adoption of such methods on sites other than in areas of high land values.

12. The system of retaining the subsoil at the periphery of the excavation is only part of the overall construction method. The principal development in recent years has been the adoption of systems which support the ground both in the construction period and in the permanent condition after completion of construction. The use of battered excavations in urban areas is very much limited because of the low utilization of the land area. The influence of ground conditions upon the chosen method of ground support is summarized in §§13–31.

Berlin method

13. The use of vertical soldiers at spacings of 2–4 m together with timber, concrete or steel laggings has been stimulated by the availability of efficient drilling plant and the development of tie-backs allowing the use of the system beyond the limits of the trench method of wall construction. The system is limited to subsoil with little or no groundwater and to such soils that may be economically augered and may be excavated without support for sufficient depth to allow the placing of the horizontal lagging boards.

14. The method is particularly economical in stiff clays and in such clays overlain by limited thicknesses of cohesionless strata, providing groundwater is low or can be made so. Compared with other temporary support methods for deep excavations, the Berlin method may appear so economical that, even with subsoil not ideal for its application (such as fills, open gravels, shales and gritstones), inclusion of a contingency sum for chiselling, dewatering or grouting may still show an overall saving.

15. White[2] reports that in California it is common practice to fasten a mesh to the soldier piles and use gunite where subsoil will stand for a height of at least one metre. Elsewhere reinforced plastic sheeting secured to the soldiers has been used successfully.

16. In terms of initial price the system using timber laggings, such as secondhand railway sleepers, may be 80% or less of the cost of sheet piling. However, total price of the system may not prove so attractive.

17. The system provides only temporary support. The verticality of its installation may be critical if it is used as a back shutter to the permanent wall to be built inside it. Drilling tolerances may require an additional thickness of sacrificial cover to the permanent wall reinforcement. If the temporary wall is not used as a back shutter, recovery of part of the temporary wall materials may be possible. However, sufficient working space must be allowed to fix the back shutter and this loss of permanent basement area and the cost of such additional excavation and subsequent backfill must be considered as cost penalties against the system. It should be noted that the use of a projecting toe at the rear of a permanent concrete retaining wall will preclude the use of any temporary sheeting as a back shutter. Equally so, variations in the vertical line of the back of the wall will also make the use of the sheeting as a back shutter impractical.

18. Figure 1 shows a typical installation, at an excavation for a telephone exchange at Wolverhampton.

Underpinning

19. The use of underpinning by casting small sections of reinforced concrete wall working from ground level downwards and supported by tie-backs is limited to sub-soil without groundwater and of sufficient strength to be self-supporting over small areas. The method has the virtue of maximum site utilization. Figure 2 shows application of the method in dense dry sand in Madrid.

Steel sheet piling

20. The use of steel sheet piling may be limited in city centres for environmental reasons. Although current methods of 'silent' driving may diminish this objection, undue vibration during driving may still limit its application. Economical driving will be achieved in loose to medium dense sands and gravels and cohesive soils; conditions giving poor driving will include these soils when rock strata are interspersed, also very dense fine sands,

and clays which include boulders or thick claystone strata. The possibility of increasing the designed pile section to allow hard and deep driving conditions should be remembered.

21. The viability in financial terms of the use of sheet piles as a means of ground support to basements is dependent upon whether they may be extracted for re-use. Extraction of piles may not be possible in a large basement occupying a large proportion of the site area because of lack of access for a suitable crane.

22. The previous remarks (§17) referring to the use of temporary sheeting as a back shutter equally apply to steel sheet piling used in this way. An expendable lining would be necessary, however, between the permanent wall and the piles to allow their easy extraction, and some wastage of concrete into the pile pans must be allowed for. The use of tie-backs more recently has allowed the economical use of steel sheet piles as a means of temporary soil support.

23. Figures 3 and 4 show the excavation support in weathered London Clay for the basement to a telephone exchange. Figure 5 shows the construction proceeding

Fig. 1. Berlin method

Fig. 3. Steel sheet piling

Fig. 2. Underpinning

Fig. 4. Steel sheet piling

within the basement and emphasizes the advantage provided by an excavation unimpeded by struts or rakers. This saving may not accrue to the Client's benefit, and indeed the true costs of an excavation impeded by timbering may not have been known to the Contractor.

Contiguous bored piles

24. The use of highly efficient augers has allowed the economical construction of piles cast in line around the basement periphery as both temporary and permanent soil support. Stiff homogeneous clay without the presence of groundwater is the ideal subsoil for such construction, and in these soil conditions contiguous bored piles probably form the cheapest wall type. Even so, a gap of some 50 mm is likely between adjacent piles (and perhaps much more in deep basements with a typical verticality tolerance for augered piles of 1%). This gap would need to be sealed by gunite or the use of a false wall either cast against the piles or built with a cavity between the wall

Fig. 5. Steel sheet piling

Fig. 6. Secant interlocking piles

and the piles. In either case this additional construction must be considered a cost penalty to the system.

Secant interlocking piles

25. The Benoto piling rig is used to provide a structurally interlocked pile wall, and since the piles are cased throughout their length during construction the method has application in ground conditions which preclude other methods.

26. Such conditions include running sands and boulder clay and, since excavation is made by hammer grab, economical excavation can also be made through varying rock strata both sedimentary and igneous. In conditions of groundwater entering the pile bore concreting is undertaken by tremie methods. The system provides a waterproof wall even in poor subsoils, but suffers the cost penalty of contiguous pile walls associated with the construction of an internal lining where aesthetics require this. The secant pile wall may be simply cantilevered or propped by rakers, struts or tie-backs in the temporary construction condition. Figure 6 shows a typical installation.

Diaphragm walls

27. The means of excavation for slurry trench construction vary from rope operated grabs to hydraulic grabs mounted on kelly bars to sophisticated reverse circulation equipment. The subsoils economically excavated by grabs include soft to stiff clays and loose to medium dense sands and gravels, whilst the Japanese Tone reverse circulation equipment is particularly efficient in cohesionless soils. Other reverse circulation equipment using rotary–percussive means of excavation are best suited for excavation through dense strata and rock. The extent of overbreak of excavation influences considerably the site costs of the operation. Excavation to depths in excess of 30 m is not uncommon. At Dartford, Kent, excavation by kelly equipment to 32 m depth was carried out in 1972, the lower 15 m of this excavation being through rock chalk containing flints.

28. Obstructions and rock strata are broken up by chiselling which may well detract from the advantages of the method if extensive. Very soft estuarine clays have caused instability during excavation in slurry trenches and mud losses have occurred in open gravels. Additionally, the compressibility of peats has caused local severe overbreaks at construction joints.

29. The working platform from which slurry trenches are excavated is necessarily 1 m in height above maximum groundwater level. The price in the UK of construction of the lightly reinforced concrete guide walls at this level is of the order of £30/m with an additional price for demolition and removal from site. A diaphragm wall of 15 m depth thus carries a cost penalty of £2/m² (or approximately 5% of total cost) for the minimum construction cost of temporary guide walls. On city sites it is not unusual to find existing basement construction obstructing guide trench construction. Whilst the removal of such obstructions would be needed for any support system, the cost of removal and the construction of guide walls through the increased depth adds a further cost penalty to the slurry trench system.

Comparisons

30. In some instances the soil conditions will them-

selves predetermine the system of soil support to be used as described in §§ 11–29. In less difficult conditions, alternative methods are available and a cost estimate comparison is necessary for each scheme. Tables 1 and 2 show the results of such a comparison (using prices for plant, labour and material current in London in January 1974) for soil conditions typical in London: fill overlying sand and gravel in turn overlying a stiff fissured silty clay. Tables 3 and 4 show the results of a similar cost estimate comparison in more varied subsoil conditions. The basement in each case was taken to be rectangular in plan, 50 m × 70 m; ground conditions and cross-sections are shown in Fig. 7. The calculations for Tables 1–4 are given in Appendices 1–4.

31. Whilst generalization is not possible from either these or a much larger statistical sample, some indications do emerge for construction in the UK. The indications are that contiguous bored pile walls with an internal lining wall are an inexpensive form of construction in subsoil conducive to piling, despite the inefficiency of the circular cross-section in bending; secant piling is an economical form of walling in poor ground conditions and is competitive in easily bored subsoil for deeper basements; if sheet piling is left in place a very expensive wall results; and anchored diaphragm walling is most competitive in deeper basements in good soil conditions and particularly where the intended basement use does not require a false lining to the diaphragm. Furthermore, the apparent high cost of anchors even in shallow basements is offset by unimpaired space for subsequent construction operations, but this advantage may not be shown in the Contractor's price to the Client; conversely, the use of rakers and temporary bunds may indicate inexpensive support systems in shallow basements but hidden costs caused by obstruction to construction may not be passed to the Client.

PLAN AREA AND SHAPE OF THE BASEMENT

32. The need to utilize all or only a proportion of a city site varies considerably even within the centre of one city. The use of basements circular in plan does not indicate a prime need to obtain high site area utilization, whereas at the other extreme the need to do so will limit to a minimum the allowed working space between the basement wall construction and an adjacent boundary or building. A summary of soil support systems and the minimum working space needed to construct them is shown in Table 5. Additionally, when maximum site area utilization is important, care is required to minimize the total thickness of wall construction, and in this respect the inefficiency of the circular cross-section in bending may detract from the use of piled walls.

33. Mention should also be made of a possible compromise between the cost of a battered excavation and the cost of a vertically supported excavation. Schnabel[6] reported the effectiveness of using sloped sheeting to the sides of deep excavations and contended that, where sheeting had sloped at an angle of about 10° from the vertical, the measured brace loads were consistently less than two thirds of the computed brace loads for vertical sheeting in the same soil.

34. The plan shape of a basement will itself affect the price of its support. Walls with re-entrant angles in plan cause greater movement to plant used in their installation. The support of walls by struts, rakers or tie-backs often becomes more complicated and expensive in such an excavation than one either rectangular or circular in plan.

35. Whilst excavations of irregular plan shape may sometimes be supported by using both cross strutting and tie-backs in separate areas, care must be taken in the support of adjacent walls at a re-entrant angle to avoid tie-backs either clashing with each other or being founded

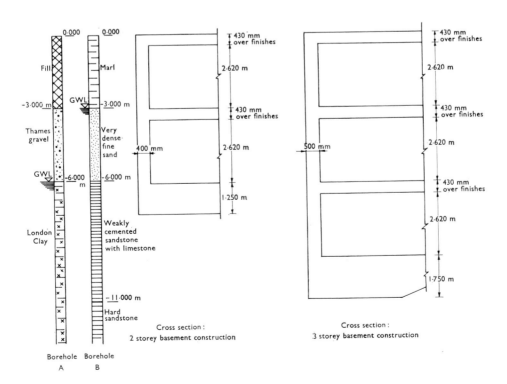

Fig. 7. Typical soil conditions and cross-sections for cost estimate comparison

Table 1. Estimate of price of support system for two-storey basement wall construction, soil conditions shown in borehole A of Fig. 7

Type of construction	£/m² above formation level						
	Permanent works			Temporary works			Total for temporary and permanent works
	Concrete and steel	Shutter	Total	Sheeting	Support	Total	
1. 400 mm r.c. basement wall, permanent sheet piles and temporary rakers	11·98	4·50	16·48	25·65	3·06	28·71	45·19
2. 400 mm r.c. basement wall, temporary sheet piles and temporary rakers	11·08	5·50	16·58	16·97	3·06	20·05	36·61
3. 400 mm r.c. basement wall, temporary sheet piles and temporary tie-backs	11·08	5·50	16·58	16·97	8·27	25·24	41·82
4. 400 mm r.c. basement wall, king posts, horizontal lagging timbers and rakers	11·98	5·00	16·98	19·29	1·07	20·36	37·34
5. 400 mm r.c. basement wall, king posts, horizontal lagging timbers and tie-backs	11·98	5·00	16·98	19·29	9·45	28·74	45·76
6. 600 mm dia. contiguous bored piles and rakers			25·23		3·06		28·29
7. ditto + 100 mm r.c. internal skin			32·58		3·06		35·64
8. 600 mm dia. contiguous bored piles and tie-backs			25·23		8·27		33·50
9. ditto + 100 mm r.c. internal skin			32·58		8·27		40·85
10. 880 mm secant piles at 750 mm centres and rakers			40·70		3·06		43·76
11. ditto + internal skin 112 mm brickwork			43·44		3·06		46·50
12. 600 mm thick diaphragm wall and tie-backs			40·00		4·90		44·90
13. ditto + internal skin 112 mm brickwork			42·74		4·90		47·64

Table 2. Estimate of price of support system for three-storey basement wall construction, soil conditions shown in borehole A of Fig. 7

Type of construction	£/m² above formation level						
	Permanent works			Temporary works			Total for temporary and permanent works
	Concrete and steel	Shutter	Total	Sheeting	Support	Total	
1. 500 mm r.c. basement wall, permanent sheet piles and temporary tie-backs	14·75	4·50	19·25	34·91	17·61	52·52	71·77
2. 500 mm r.c. basement wall, temporary sheet piles and temporary tie-backs	13·85	5·50	19·35	23·10	17·61	40·71	63·81
3. 500 mm wall, king posts, timber horizontal laggings and tie-backs	14·75	5·00	19·75	23·41	19·13	42·54	62·29
4. 750 mm contiguous bored pile wall and tie-backs			25·20		11·75		36·95
5. ditto + internal skin 100 mm r.c. wall			32·55		11·75		44·30
6. 880 mm dia. secant piles at 750 mm centres and tie-backs			42·50		10·11		52·61
7. ditto + internal skin 112 mm brickwork			45·24		10·11		55·35
8. 800 mm diaphragm wall and tie-backs			46·72		6·61		53·33
9. ditto + internal skin 112 mm brickwork			49·16		6·61		55·77

within a potential failure wedge of the adjacent wall. Advantage can sometimes be taken at a re-entrant angle of a wall to tie horizontally and diagonally across the soil within the re-entrant angle rather than use tie-backs to each wall.

TIME FOR COMPLETION

36. The following principles may be listed if programme completion time is to be minimized. Basement walls should be built in one operation for both temporary and permanent support, and permanent and temporary works design should proceed simultaneously; construction methods which allow simultaneous construction upwards and downwards should be adopted if possible; and for basements constructed in open excavation the working space for floor and wall construction should be unimpaired as far as possible, for instance by the use of tie-backs.

37. The construction time and financial penalties of basement construction impeded by either natural or unnatural obstructions has been referred to previously. Whilst complications such as adjacent structures with shallow foundations, multi-level basements, awkward plan shapes and the presence of existing services obviously add to time and cost, mention should also be made of other less

Table 3. Estimate of price of support system for two-storey basement wall construction, soil conditions shown in borehole B of Fig. 7

Type of construction	£/m² above formation level						
	Permanent works			Temporary works			Total for temporary and permanent works
	Concrete and steel	Shutter	Total	Sheeting	Support	Total	
1. 400 mm r.c. basement wall, permanent sheet piles and temporary rakers	11·98	4·50	16·48	28·18	3·06	31·24	47·72
2. 400 mm r.c. basement wall, temporary sheet piles and temporary rakers	11·08	5·50	16·58	19·15	3·06	22·21	38·79
3. 400 mm r.c. basement wall, temporary sheet piles and tie-backs	11·08	5·50	16·58	19·15	8·27	27·42	44·00
4. 400 mm r.c. basement wall, king posts, horizontal laggings and rakers	11·98	5·00	16·98	20·08	1·07	21·15	38·13
5. 400 mm r.c. basement wall, king posts, horizontal laggings and tie-backs	11·98	5·00	16·98	20·08	9·45	29·53	46·51
6. 880 mm secant piles at 750 mm centres and rakers			42·00		3·06		45·06
7. ditto + internal skin 112 mm brickwork			44·74		3·06		47·80
8. 600 mm thick diaphragm wall and tie-backs			39·73		4·90		44·63
9. ditto + internal skin 112 mm brickwork			42·47		4·90		47·37

Table 4. Estimate of price of support system for three-storey basement wall construction, soil conditions shown in borehole B of Fig. 7

Type of construction	£/m² above formation level						
	Permanent works			Temporary works			Total for temporary and permanent works
	Concrete and steel	Shutter	Total	Sheeting	Support	Total	
1. 500 mm r.c. basement wall, king posts, timber horizontal laggings and tie-backs	14·75	5·00	19·75	24·19	19·13	43·32	63·07
2. 880 mm dia. secant piles at 750 mm centres and tie-backs			37·90		10·11		48·01
3. ditto + internal skin 112 mm brickwork			40·64		10·11		50·75
4. 800 mm diaphragm wall and tie-backs			44·76		6·61		51·37
5. ditto + internal skin 112 mm brickwork			47·50		6·61		54·11

obvious restraints which are likely to cause lengthened completion times, such as embargoes on noise of construction plant, headroom limitations caused by flying shores between propped adjacent structures, restrictions on working caused by the presence of underground railways beneath the site, and difficult site access for heavy excavation plant in city areas.

CONCLUSIONS

38. The choice to build deep may not be as attractive for the building Owner as the Engineer may believe.

39. The overall method of basement construction will define construction time and price to the Client. Reductions in both time and price are achieved by methods of simultaneous construction of superstructure and substructure and by systems of shared superstructure support and subsoil horizontal support.

40. Although site location, subsoil conditions, basement shape and extent of site utilization will each influence the choice of soil support system on a particular basement excavation, frequently several possibilities are still available after an initial procedure of rejection of

Table 5. Minimum distance between soil support system and site boundary

Support system	Installation plant	Distance,* mm
Underpinning	Conventional bulk excavation plant e.g. hydraulic excavator with hydraulic grab	0
Steel sheet piling	Crane and piling hammer	500
Contiguous bored piles	Rotary piling equipment 600 mm pile	600
	Tripod piling equipment 500 mm pile	150
Secant piles 880 mm thick	Benoto rig	1060
Diaphragm walls	Rope grab, kelly mounted grab or Tone reverse circulation drill	150
Soldier and horizontal lagging— Berlin method	Rotary piling equipment 600 mm dia. bore	600

* Minimum distance between outer face of support system and site boundary. Distances quoted are those at ground level; consideration must be given to verticality tolerances of support system.

unsuitable systems has been made. A cost estimate comparison exercise is then required, at permanent works design stage, in order to choose the most economical system and to include in the permanent works design and detail any features required by that system.

41. Advantages and disadvantages in cost are indicated by a relatively small sample of cost estimate comparisons of soil support systems.

REFERENCES

1. ZINN W. V. Economical construction of deep basements. *Civ. Engng Publ. Wks Rev.*, 1968, **63**, Mar., 275–280.
2. WHITE E. E. Sheeting and bracing systems for deep foundations. *Wld Constr.*, 1973, **26**, Jan., Part 1, 20–26.
3. FENOUX Y. Deep excavations in built-up areas. *Travaux*, 1971, Aug.–Sept., Parts 437 & 438, 18–37.
4. PECK R. B. Deep excavations and tunnelling in soft ground. *7th Int. Conf. Soil Mech., Mexico, 1969*, State of the Arts Paper, 225–290.
5. TOMLINSON M. J. Lateral support of deep excavations. *Ground Engineering*. Instn Civ. Engrs, London, 1970, 55–64.
6. SCHNABEL H. Sloped sheeting. *Civ. Engng, Easton, Pa*, 1971, **41**, Feb., Part 2, 48–50.

APPENDIX 1. TABLE 1 COST ESTIMATE CALCULATION

400 mm r.c. wall
Basic prices, gross

Excavation, in the solid to site stockpile	£1·00/m³
Concrete, placed in 400 mm wall, including waste	£18·00/m³
Steel: cut, bent, fixed h.t. reinforcement	£174/t
Shutter: fair faced front shutter	£4·50/m²
Lining shutter to sheet piles, fixed	£1·00/m²
Lining to horizontal laggings	£0·50/m²

Gross price of wall

Concrete + excavation: 0·4(£18 + £1)	£7·60/m²
Steel @ 20 kg/m²: 0·02 × £174	£3·48/m²
Shutter, single face	£4·50/m²
	£15·58/m²

Sheet pile scheme
Allow for waste into pile pans @ £0·90/m²
Basic prices

Transport and handle piles	£10/t
Material cost piles	£90/t
Pitch and drive	£20/t
Net price	£120/t
Oncost	£22/t
Gross price	£142/t

Allowing Larssen 3 section @ 155 kg/m² with 1·5 m cut-off
Gross price per m² above formation = £142 × 0·155 × 8·85/7·35 = £25·65/m²

Temporary sheet piles

Extraction, handling and transport off site	£25/t
Allow for piling use, based on 3 uses	£33·3/t
Pitch and drive	£20/t
	£78·3/t
Oncost	£15·6/t
Gross price	£93·9/t

Allowing Larssen 3 section @ 155 kg/m² with 1·5 m cut-off
Gross price per m² above formation = £93·9 × 0·155 × 8·85/7·35 = £16·97/m²

Cost of support: raker
Allow 300 mm sq. timber at 3 m centres
Allow supply of timber @ £1·64 per lin. m net, allowing 3 uses
Supply timber per raker = 5·18 × 1·64 = £8·5/raker
Allow for fixing £8·5/raker

	£17/raker
Oncost	£3·40/raker
	£20·40/raker

Gross price per m² above formation = £20·40/(7·35 × 3·00) = £0·91/m²

Waling

Allow net cost of fabricated steel waling		£180/t
Place at required level		£20/t
		£200/t
Credit for return less handling and transport		£20/t
		£180/t
	Oncost	£36/t
	Gross price	£216/t

Allow waling @ 75 kg/m
Therefore cost per m² of wall above formation = 216 × 0·075/7·35 = £2·15/m²

Tie-backs
Gross price of tie-back £180 @ 4 m centres
Gross price per m² of wall above formation = 180/(7·35 × 4) = £6·12/m²

Add waling as before	£2·15/m²
	£8·27/m²

King post scheme
Basic prices, gross

Bore @ £8·20/m, depth of bore 9·2 m	£75/king post
Concrete at base of bore—allow 1 m³	£18/king post
Steel 680 kg × £216	£144·7/king post
Sleepers at £2·40 each on site	£86·4/king post
Place sleepers at £1·20 each	£43·2/king post
	£367·3/king post

Gross price per m² of wall above formation level = 367·3/(7·35 × 2·60) = £19·29/m²

Tie-backs
Allow 1 frame of tie-backs @ 2·60 m centres
Gross price per m² of wall above formation = 180/(7·35 × 2·60) = £9·45/m²

Rakers
As before gross price per raker = £20·40
Gross price per m² of wall above formation = 20·40/(7·35 × 2·60) = £1·07/m²

Contiguous bored pile wall scheme
Allow 600 mm dia. piles, gross price £13·12 per lin. m
Allowing 50 mm between piles and a cut-off of 1·8 m
Gross price per m² of wall above formation = 13·12 × 9·15/(0·65 × 7·35) = £25·23/m²

Skin wall construction, 100 mm min. thickness, gross price

Shutter	£4·50/m²
Mesh and concrete av. 150 mm thick: 0·15 × £19	£2·85/m²
Gross price per m² of wall above formation	£7·35/m²

Rakers + walings: as before at 3 m centres, gross price = £3·06/m² of wall above formation
Tie-backs: as before at 3 m centres, gross price = £8·27/m² of wall above formation

Secant pile scheme
Gross price of piles @ 750 mm centres = £26 per lin. m
Price per m² = £34·60
(If piles @ 770 mm centres, price per m² = £33·80)
Allowing for cut-off 1·3 m, price per m² above formation level = £34·60 × 8·65/7·35 = £40·70 /m²

Diaphragm scheme
Gross price £32/m²
Allowing for cut-off 1·8 m
Gross price per m² of wall above formation = £32 × 9·15/7·35 = £40/m²

Anchors @ 5 m centres, one frame.
Gross price per m² of wall above formation level = 180/(7·35 × 5) = £4·90/m²

112 mm brick skin

Material price	£12/1000
Lay	£25/1000
Mortar	£1/1000
	£38/1000
Oncost	£ 7·60/1000
	£45·60/1000

Therefore gross price per m² = £45·60 × 60/1000 = £2·74/m²

APPENDIX 2. TABLE 2 COST ESTIMATE CALCULATION

500 mm r.c. wall
From previous pricing
Gross price, excavation and concrete: $0·500 \times £19·00 = £9·50/m²$
Steel: Allow 25 kg/m², therefore gross price = $0·025 \times £174$
$= £4·35/m²$

Sheet piling scheme
Permanent sheeting
Gross price from before £142/t
Allowing section of weight 200 kg/m²
Gross price allowing for cut-off of 2·5 m = $142 \times 0·2 \times 13·4/10·9$
$= £34·91/m²$
Temporary sheeting
Allowing section of weight 200 kg/m²
Gross price allowing cut-off of 2·5 m = $£93·96 \times 0·2 \times 13·4/10·9$
$= £23·10/m²$
Walings
Allowing 3 frames, total weight of waling steel per lin. m = 270 kg
Gross price per m² of wall above formation = $£216 \times 0·270/10·9$
$= £5·22/m²$
Tie-backs @ 4 m centres
Cost per m² above formation = $540/(10·9 \times 4) = £12·39/m²$
Total for waling and tie-backs = £17·61/m²

King post scheme
Basic prices, gross

Bore @ £8·20/m, depth of bore 13·4 m	£110/king post
Concrete at base, allow 1 m³	£18/king post
Steel 1·59 t per king post × £216	£339/king post
Sleepers 54 × £3·6	£194/king post
	£661/king post

Gross price per m² of wall above formation level
$= 661/(10·9 \times 2·60) = £23·41/m²$
Tie-backs
Gross price per m² above formation = $540/(10·9 \times 2·60) = £19·13/m²$

Contiguous bored pile wall
Allow 750 mm piles @ 800 mm centres
Gross price £16·40/m
Gross price per m² above formation allowing 2·5 m cut-off
$= (16·4/0·8) \times (13·4/10·9) = £25·20/m²$
Tie-backs: allowing 2 frames, ties @ 4 m centres
Total weight of walings = 180 kg/m
Gross price per m² above formation = $0·18 \times 216 \times 10·9$
$= £3·49/m²$
Gross price of tie-backs = $360/(4 \times 10·9) = £8·26/m²$
Total gross price per m² of wall above formation = £11·75/m²

Secant pile scheme
Gross price of piles, 880 mm @ 750 mm centres = £26 per lin. m
Price for m² = £34·60
Allowing for 2·5 m cut-off, price per m² above formation
$= £42·50$
Tie-backs @ 5 m centres
Allowing 2 frames
Walings: $5·22 \times 2/3 = £3·50/m²$
Tie-backs: gross price per m² = $£360/(5 \times 10·9) = £6·61/m²$
Total gross price per m² of wall above formation = £10·11/m²

Diaphragm scheme
Gross price of diaphragm wall, 800 mm thick = £40/m²
Gross price per m² above formation = $£40 \times 12·7/10·9 = £46·72/m²$

Allow 2 frames, tie-backs @ 5 m centres
Gross price per m² of wall above formation = $360/(5 \times 10·9)$
$= £6·61/m²$

APPENDIX 3. TABLE 3 COST ESTIMATE CALCULATION

Sheet piling scheme
Basic price, allowing increased price for driving in difficult ground

Transport and handle piles	£10/t
Material cost of piles	£90/t
Pitch and drive	£30/t
Net price	£130/t
Oncost	£26/t
Gross price	£156/t

Allowing section @ 155 kg/m² and allowing 1·5 m cut-off
Gross price per m² of wall above formation
$= 0·155 \times £156 \times 8·85/7·35 = £28·18/m²$
Temporary sheet piles
Basic prices

Extraction, handling and transport	£25/t
Piling use allowing 3 uses	£33·3/t
Pitch and drive	£30/t
	£88·3/t
Oncost	£17·66/t
	£106/t

Allowing pile section @ 155 kg/m²
Gross price per m² of wall above formation
$= 0·155 \times £106 \times 8·85/7·35 = £19·15/m²$

King post scheme
Basic prices, gross, allow bore @ £9·84/m
(depth of bore 9·2 m)

	£90/king post
Concrete at base of bore, allow 1 m³	£18/king post
Steel, as before	£144·7/king post
Sleepers, as before	£129·6/king post
Gross price	£382·3/king post

Gross price per m² of wall above formation = $382·3/(7·35 \times 2·89)$
$= £20·08/m²$

Secant pile scheme
Gross price per lin. m of 880 mm pile: £26
Gross price, allowing cut-off of 1·5 m
$= 26 \times (1000/750) \times (8·85/7·35) = £42·00/m²$

Diaphragm wall scheme
Allowing gross price of £33/m²
Gross price, allowing 1·5 m cut-off = $33 \times 8·85/7·35 = £39·73/m²$

APPENDIX 4. TABLE 4 COST ESTIMATE CALCULATION

King post scheme
Gross prices

Allow bore at £9·84/m (depth 13·4 m)	£132/king post
Concrete at base of bore	£18/king post
Steel from before	£339/king post
Sleepers from before	£194/king post
	£683/king post

Gross price per m² above formation level = $683/(10·9 \times 2·60) = £24·19/m²$
Tie-backs as before, £19·13/m²

Secant pile scheme
Basic gross price = £26 per lin. m
Gross price per m² above formation
$= 26 \times (1000/750) \times (11·9/10·9) = £37·90/m²$
Allowing 1 m cut-off

Diaphragm wall scheme
Basic gross price = £41 per m² for 800 mm wall
Gross price per m² above formation = $41 \times 11·9/10·9 = £44·76/m²$
Allow 1 m cut-off

Anchored walls adjacent to vertical rock cuts

R. E. WHITE, FASCE, PE, *President, Spencer, White and Prentis, Inc., New York*

A frequent problem in the construction of earth-retaining walls arises when the wall is tied back to rock with inclined anchors which put a heavy downward thrust through the wall onto the rock. Dead weight of earth overburden and rock plus live loads from traffic or construction equipment also add vertical load, and consequently there is danger of rock shearing off beneath the retaining wall, leading to collapse. Tendency of rock to shear is aggravated by weaknesses within the rock: clay-filled seams, fissures, joints, faults, etc. Various expedients can be used to guard against rock failures. These are rock bolting, installing vertical supports in holes drilled to below subgrade, underpinning of rock by means of concrete piers installed in pits, inclined shoring to subgrade of rock excavation alongside retaining wall with subsequent removal of rock berm, pinning of tips of soldier piles, flattening the angle of installation of anchors, and using a permeable sheeting system. Specific examples are given of failures and near failures of retaining walls and the underlying rock. The remedial measures that were taken are described. Also, examples are given of wall constructions designed from the start to avoid the possibility of failure. Actual examples of projects are used for illustration and these encompass different types of earth-retaining structures, such as soldier piles and horizontal wood lagging (Berlin method), interlocking steel sheet piling and diaphragm wall (soldier pile system). The Paper concludes with a discussion of contractual aspects.

INTRODUCTION

A frequent problem in urban construction is that of building an earth-retaining wall, tied back by rock anchors, which after an excavation is made will be adjacent to and above the vertical rock wall of the excavation. This Paper points out the dangers of this type of construction and the remedies available to the constructor.

2. Because of the downward component of the tendon of the ground anchor, which when grouted into a rock formation is usually sloping at a 45° angle, a high, downward, vertical load from the anchor is imposed on top of the rock surface within a metre or two of the vertical face (see Fig. 1). In addition to the vertical load from this source, there is also the dead load from the earth overburden and very likely there is a live load from traffic and possibly construction cranes and other equipment.

FACTORS AFFECTING THE STRENGTH OF ROCK

3. The ability of the bedrock to withstand the shearing stresses from the retaining wall and weight of overburden depends not only on the strength of the rock itself, which is generally sufficient, but even more on how free the rock is from fissures, clay-filled seams, joints, cracks and the

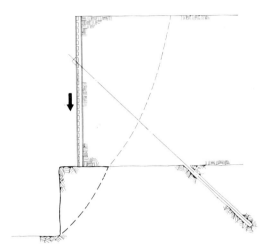

Fig. 1. Vertical loads from tie tension plus hard driving of soldier beam tend to cause failure in rock

like. A clay-filled seam—even if only a few millimetres thick—when oriented in an unfavourable way can almost completely rob the bedrock mass of its strength, and in such cases disastrous collapses of the rock, the retaining wall and the earth overburden behind the wall can occur and have occurred. (Such collapses have also happened where, instead of anchors, more conventional horizontal struts in compression have been employed, as in subway or other trench work; and, of course, they happen all the time in nature as rock slides.)

4. Besides the weakening of the rock from naturally occurring seams and fissures, the excavation of the rock itself may aggravate weaknesses in the remaining rock. This comes about from the drilling and blasting of rock to enable its removal. In the case of heavy blasting and widely spaced drilled blast holes the effects can be pronounced: they can be considerably reduced by close spacing of line holes, pre-splitting, use of light blasting charges and other such precautions.

5. Careless operation of excavating equipment may also damage rock supporting the sheeting system. In one case, where the rock consisted of various strata of horizontally bedded soft shales and sandstones, a heavy power shovel bucket actually ripped out the supporting rock back of the line holes, and caused a foot or two settlement of a stretch of the tied-back interlocking steel sheet piling holding back the earth overburden.

6. Hard driving of piling onto the surface of the rock and even well into it can also weaken the rock. Examples of this are given below.

7. There is also a time factor at work as rock tunnel men well know. They give the name 'stand up time' to indicate how a rock will stand unsupported for a period of

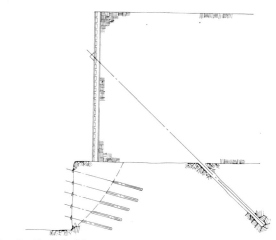

Fig. 2. Rock bolting

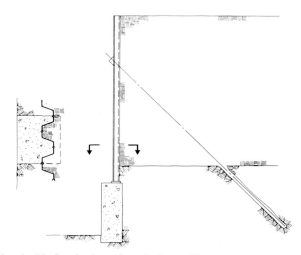

Fig. 4. Underpinning of steel sheet piling

time and then collapse. This period can be very short—a matter of minutes—or it can be as long as a year, as in the case of one tie-back job in Texas which stood up very well for a year and then collapsed, causing great damage to the adjoining street.

8. Another cause of vertical movement can be the assumption of excessive lateral earth pressures, which can be self-defeating in a policy of conservatism perhaps otherwise commendable. This will cause an extra unnecessary downward thrust to be put on the vertical elements of the earth-retaining system (i.e. soldier beams, interlocking sheet piling, diaphragm wall) when the tie-back is proof-tested with hydraulic jacks to the tension called for by the conservative design plus an overload. The obvious deduction to be made here is that the design engineer should try to be realistic in computing lateral earth pressures, as well as in specifying the test load of the tie-back. In the Author's opinion, for the ratio of test load to design load values of 125% to 150% should be sufficient.

PREVENTIVE AND REMEDIAL MEASURES

9. To guard against rock failures a variety of measures have been used.

10. Rock bolting, illustrated in Fig. 2, is the use of nearly horizontal tension members (they can be strands, wires or rods) which start at the vertical rock face and go back into the rock mass a sufficient distance to be anchored behind a possible sliding plane in the rock. Design of rock bolts is based on rock mechanics and is beyond the scope of this paper. Rock bolting may also be used in conjunction with a sheeting system to prevent falls of pieces of rock between bolts. Sheeting may consist of boards held by steel channels, gunite and mesh, or even as little as a spray of bituminous emulsion to prevent drying out and spalling of certain types of shale.

11. A second method (Fig. 3) is the use of vertical supports installed in holes drilled to below subgrade. If the vertical component of the anchor load is transferred directly to rock at or below adjacent subgrade, this obviously relieves the rock of any responsibility for carrying this portion of the superimposed load. In soft rocks, 18–30 inch dia. holes may be drilled to elevation of adjacent subgrade or below using large diameter kelly-bar drills equipped with, say, spiral flight augers fitted with tungsten carbide teeth. Soldier beams are then set in these holes and the holes are backfilled with concrete. In hard rocks the holes to below subgrade may have to be drilled with cable tools (percussion bits, chisels) handled by churn

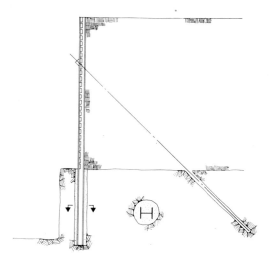

Fig. 3. Soldier beams set into holes drilled below subgrade

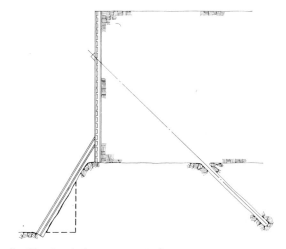

Fig. 5. Shoring before removal of rock berm

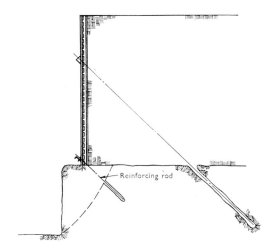

Fig. 6. Pinning tips of soldier beams

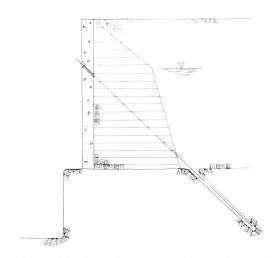

Fig. 8(a). Full hydrostatic pressure on impervious wall

drills (spudding machines). Such drilling may be slow and expensive, but nevertheless unavoidable.

12. A case history is given (§ 20) in which interlocking steel sheet piling was underpinned by means of concrete piers installed in pits dug in soft rock (Fig. 4). This can also be slow and expensive. To the Author's knowledge it has been used only once, as a remedial measure when the rock on which the sheeting was driven very hard began to fail.

13. Another method (Fig. 5) is the use of inclined shoring to the subgrade of the rock excavation alongside the sheeting, with subsequent removal of the rock berm. In the case of diaphragm walls, a similar function has been performed by concrete buttresses.

14. Pinning of tips of soldier piles (Fig. 6) is similar to rock bolting. Its purposes are to take lateral load and to reinforce the rock at the point where the concentrated load of the soldier piles comes onto it. As with rock bolts, the pins may be grouted, or used with expansion shields; with the latter, the advantage is a saving in time which also may be achieved with epoxy (resin) grouts or high alumina cements.

15. Another method (Fig. 7) involves flattening the angle of installation of the anchors. Throughout this Paper the anchors referred to have been rock tie-backs,

which are grouted into bedrock and steeply inclined, usually at an angle of 45°. By going to earth anchors (the quality of the overburden permitting) the designer may have to use more tie-backs at lesser loads, but as the earth anchors are much flatter—at angles of about 10–30° to the horizontal—the vertical component of load on the sheeting system is greatly diminished, thereby eliminating much of the problem.

16. Use of a permeable sheeting system, as opposed to an impervious wall, reduces lateral pressure due to hydrostatic forces. The less the lateral pressure, the lower will be the downward load on the underlying rock mass. As an example, a diaphragm wall would be subject to hydrostatic pressure whereas a soldier beam and horizontal sheeting system would not (Figs 8(a) and 8(b)).

SOME SHORT CASE HISTORIES

17. In excavating the rock for an office building at 18th & K Street, NW in Washington, DC (1968), it was found that tetrahedron-shaped pieces of the rock (weathered gneiss with clay seams) tended to fall out from the vertical rock faces below the tip elevations of the soldier beams, endangering not only these members but also the men working below. The entire rock face was protected

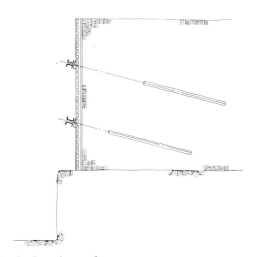

Fig. 7. Anchors in earth

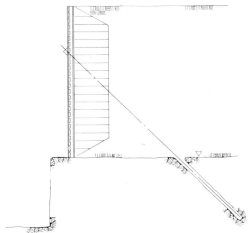

Fig. 8(b). Soldier beams and horizontal sheeting giving lower groundwater level and reduced lateral pressure

Fig. 9 (left). A rock tie-back project at 18th & K Street NW, Washington, DC. Upper quarter of photograph shows earth overburden retained by pairs of channels set in holes augered to top of decomposed rock and backfilled with lean concrete; tie-backs are placed between webs of channels. Lower part of photo shows exposed blocky and decomposed rock at right, and the same bedrock protected by boards and steel channels held by rock bolts

Fig. 10 (below). Telephone building, White Plains, NY. Rock protection by means of timber panels with steel battens anchored by long 3-strand prestressed tendons grouted into rock. Where rock is of sufficient soundness, as at left, panels are omitted. Note at top left of photo a rock-anchored system for soldier beams and horizontal sheeting in overburden. Top right shows tied-back concrete underpinning of existing six-storey telephone building. Total depth of excavation is 75 ft

(Fig. 9) by a system of boards and battens held by rock bolts, which were single half inch 7 wire strands grouted into holes of length 20–30 ft and diameter 2 inches drilled just below the horizontal in the rock. Boards were 2 in. × 6 in. × 8 ft long and were held by two 4 inch steel channels, each channel being secured in turn by two bolts. Boards were placed horizontally and channels were placed vertically in 4 ft lifts as the rock was excavated.

18. A more elaborate system (Fig. 10) was used for the very deep basement of a telephone building in White Plains, NY (1972) to protect a 54 ft deep vertical face of unstable rock, over which was a 21 ft high system of soldier beams and horizontal sheeting tied back by 45° anchors into a gneiss rock in a fault zone. In this case the boards and channels were prefabricated into 5 ft × 10 ft panels, which were set into place and bolted back by half inch 7 wire strands, and then the irregular void between the boards and the rock face was solidly backfilled with concrete. Tendons were unusually long—up to 105 ft— as rock mechanics studies showed that sliding planes within the rock existed at an angle of 33° to the horizontal.

19. On another building excavation in Washington, DC, the Bureau of National Affairs at 25th and P Streets,

NW (1966), the rock was very blocky, but not particularly hard. A kelly-bar drill turning flight augers equipped with tungsten carbide teeth was able, despite slow progress and high bit costs, to predrill soldier beam holes a couple of feet below subgrade. Horizontal boards between the flanges of the soldier beams, and steel channels welded to the soldier beams, served to protect the workmen from falling pieces of rock, as well as to prevent loss of over-burden and undermining of streets (Fig. 11).

20. The US Steel Corporation headquarters building in Pittsburgh (1968) provided some interesting examples of what can happen on rock tie-back contracts. Because of a desire to prevent lowering of groundwater level, possible settlement of a church across the street and most likely a resulting lawsuit, it was specified that interlocking steel sheet piling be used rather than the more economical soldier beams and horizontal lagging. It was specified that the sheet piling be driven through weathered rock and onto a sounder rock. As it turned out the rock, which was soft shale and sandstone, was actually split vertically by the hard driving so that during excavation—in spite of careful line drilling, blasting and shovel operation—the sheet piling was found to be settling about a half foot or

Fig. 12. US Steel Corporation headquarters, Pittsburgh, Pa.
(a) (below left) Interlocking Z piling in centre is settling; emergency operation is under way. Note careful line drilling of vertical rock face 4 ft from face of sheet piling
(b) (below) Two months later; sheet piling underpinned

so (Fig. 12(a)). Inclined shores were hurriedly placed against the wale brackets, relieving the downward load of the sheeting on the weak rock. Next, for a few feet of each interlock, the sheets were welded together; this was so that they could span over 3 ft × 4 ft pits to be hand dug at 15 ft centres. The rock being soft, no dynamite was necessary and pneumatic clay spades and picks sufficed to loosen it. Upon reaching subgrade, the pit excavations were filled with structural concrete as high as necessary to support the tips of the sheet piles; thus the piling was underpinned. After this, the emergency inclined shoring was removed (Fig. 12(b)).

21. It was observed that considerable groundwater was emerging through seams and fissures into the vertical rock face, and occasionally through the interlocks of the sheeting. This made it hardly likely that the original groundwater level was being maintained behind the interlocking steel sheet piling as originally anticipated. Inasmuch as a second phase of the work had yet to be done, it was proposed by the contractor that soldier beams and horizontal wood lagging be employed, the soldier beams to be placed in pre-augered holes. Every other hole would be augered to a few feet below subgrade; these

were for soldier beams adjacent to which the anchors were to be placed. The alternate soldier beams, not being close to anchors, would have no downward load and thus were required only to reach to the surface of the rock. The design did not have to cope with hydrostatic pressure from groundwater. This scheme was adopted and successfully performed without movement. A similar operation is illustrated in Fig. 13.

22. This principle of carrying soldier beams to beneath adjacent subgrade in predrilled holes was also used for the cofferdam for the pump house of the Hudson River anti-pollution project off Manhattan Island (1973). In lieu of a sheet pile cofferdam, 225 ft × 143 ft supported by four tiers of horizontal wales and struts, the contractor proposed that three sides of the cofferdam be of the gravity type, consisting of circular steel sheet pile cells filled with sand, and the remaining landside be a diaphragm wall tied back to rock. This landside wall would protect the elevated West Side Highway carrying very heavy automobile traffic. The depth to rock of 30–50 ft and hydraulic pressure caused the tie-backs to be highly loaded to 420 kips each. Inasmuch as it would have been very expensive to excavate in the hard mica schist bedrock a

continuous slot 3 ft wide and 10–25 ft deep for a conventional reinforced concrete diaphragm wall, it was decided to build a diaphragm wall of the drilled-in caisson type (Figs 14 and 15). By this design, the only drilling in rock would be 3 ft dia. holes at 6 ft centres to receive the core beams. The beams were 27 inch rolled steel sections weighing 177 lb/ft. Between core beams the diaphragm wall was unreinforced, arching action from beam to beam being relied on to support the lateral earth and water pressures.

23. For the diaphragm wall at the World Trade Center (1968) in lower Manhattan, a reinforced concrete wall 50–60 ft deep and 36 inches thick, the entire bottom of the wall of 3000 ft perimeter was concrete tremied into a 5 ft deep slot laboriously chiselled into the same hard mica schist. At 10 ft intervals, corresponding to the horizontal spacing of the tie-backs above, concrete buttresses 3 ft 6 in. wide were placed to safeguard the toe of the wall from moving into the excavation (Fig. 16(a)). Another design for accomplishing this was the pouring of a continuous reinforced concrete inclined slab against the bottom of the wall (Fig 16(b)).

FEDERAL RESERVE BANK, WASHINGTON, DC

24. A striking example of the cumulative effect of vertical loads on soldier piles was given during the installation of the tied-back excavation protection for the Federal Reserve Bank in Washington, DC at C and 21st Street, NW (1971). The excavation at this site was irregular in plan, and had a perimeter of 1770 ft, a depth to rock of about 50 ft and a rock excavation depth of about 11 ft. The rock was typical of gneiss to be found in the northwest of the city in that the mass was criss-crossed by many fissures and joint planes which contained thin seams of decomposed rock resembling clay. The top few feet of the bedrock was also decomposed.

25. Running the length of C Street was a very large sewer which, with other utilities, prevented the top row of rock ties from being installed. Accordingly, the uppermost wales had to be supported by compression members (braces) sloping downward to subgrade where rock was the reaction. To maximize the unobstructed working area the braces were made of $\frac{1}{2}$ in. thick 24 in. o.d. steel pipe which, because of their large l/r, required no secondary bracing. The braces were 24 ft on centres. Instead of pulling down on the soldier piles, as did the rock anchors, they tended to lift up the soldier piles. Thus an average soldier pile on C Street had a computed downward load at the tip of 172 kips, whereas just around the corner on 21st Street—where anchors only were used as there was no big sewer—an average soldier pile had a tip load of 372 kips.

26. The excavation started at one end of C Street and progressed to 21st Street, where it turned the corner and went away from C Street. When excavation was about 100 ft away from the corner the shelf beneath the 21st Street soldier piles collapsed, with a resultant maximum settlement of these soldier piles of about 2 ft. No anchors broke—indeed, they were all subsequently tested and found to be secure—but excessive bowing of the wales occurred and they had to be supplemented by another set of wales with inclined braces, which were later removed before the foundation wall was built at this location. A hundred feet of half of 21st Street had to be repaved after

replacement of a 30 in. dia. sewer which ran through this area.

27. Meanwhile, much of the perimeter at the base of the protection system had to be excavated in rock. This was safely done by installation of inclined shores.

CONTRACTUAL ASPECTS

28. It is most common for separate contracting organizations to perform different roles in the foundation work. These subcontractors are most frequently engaged by the General Contractor: they have no relationship with each other, only with the General Contractor who coordinates their different operations for the Owner.

29. Because the sheeting and tie-back subcontractor does none of the excavating and the excavating subcontractor is responsible only to the General Contractor, the tie-back subcontractor's work can easily be put into jeopardy by a careless excavator. The tie-back subcontractor therefore should protect himself as best he can by the use of exculpatory language in his subcontract agreement with the General Contractor. For example, the following language has been used.

Excavation of earth and rock is not part of this Subcontract. Excavation of rock shall be done with special care and may include: presplitting, close line-drilling and broaching so as to prevent the breaking out of the rock beyond the sheeting lines and the undermining of the soldier piles. Any expense in connexion with

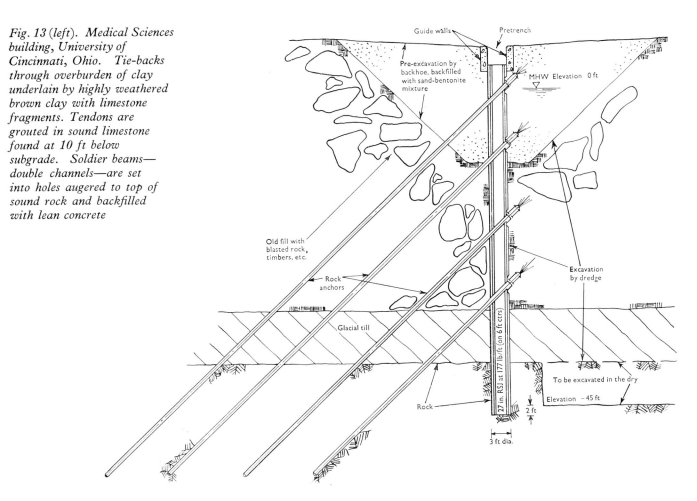

Fig. 13 (left). Medical Sciences building, University of Cincinnati, Ohio. Tie-backs through overburden of clay underlain by highly weathered brown clay with limestone fragments. Tendons are grouted in sound limestone found at 10 ft below subgrade. Soldier beams—double channels—are set into holes augered to top of sound rock and backfilled with lean concrete

Fig. 14 (above). Rock anchored diaphragm wall, North River Pollution Control Plant, New York City

Fig. 15. North River Pollution Control Plant, New York City: diaphragm wall with drilled-in soldier piles; rock excavation under way

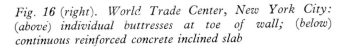

Fig. 16 (right). World Trade Center, New York City: (above) individual buttresses at toe of wall; (below) continuous reinforced concrete inclined slab

filling of voids, rock face protection such as shotcrete, rock bolting, shoring or lagging of rock faces, maintenance of proper rock shelves to support the vertical loads in the soldier piles, inclined shores to soldier piles, wales and brackets or additional shoring required by reason of 'break back' of rock shall be borne by others at no expense to this Subcontractor.

30. So that there may be no loose ends, the responsibility for rock bolting, line drilling, etc. described above should be clearly included in the excavator's subcontract agreement.

31. Another way of contracting for the work is to make the excavator a subcontractor of the tie-back and sheeting subcontractor. This will give the latter more control over the way the excavator does the rock excavation work so that the sheeting system is not damaged. General Contractors are not always in favour of this approach because the sheeting and tie-back contractor is certainly entitled to a mark-up on his subcontractor's work (i.e. the excavation) and the General Contractor would prefer to pocket this mark-up himself.

32. Another problem arises in connexion with the anchors. Since the Owner benefits through the opening up of the work and the elimination of interior obstructions (braces) for almost all of the many subcontractors, he should obtain and pay for permits and fees required by the owners of adjoining lands where the tie-backs are installed. Also, the Owner has much more time to arrange for these matters than does the anchor subcontractor who must often start work on extremely short notice.

Contiguous bored piles

J. P. NORTH-LEWIS, FICE, *Director, Soil Mechanics Ltd*

G. H. A. LYONS, BSc(Eng), MICE, *Senior Engineer, Soil Mechanics Ltd*

This Paper defines contiguous bored piles and deals generally with the subject. Some aspects of the design of contiguous pile retaining walls are mentioned, including current reinforcing practice. Construction and installation techniques are discussed in the light of available piling plant, and reference is made to working tolerances. The application of this technique as a form of retaining wall for various subsoil conditions is described with supporting examples. Material content and relative construction costs are compared with those for diaphragm and sheet pile walls. Future developments involving post-tensioning pile shafts to save reinforcement and new grouting techniques are mentioned.

INTRODUCTION

The word contiguous is defined in the Shorter Oxford English Dictionary as touching, in contact; adjoining, continuous, (loosely) neighbouring.

2. Contiguous bored piles are therefore regarded for the purposes of this Paper as those which are installed close together or touching and not interlocking as is the case with the secant pile method.

3. Contiguous bored pile walls do not appear to have been in general use in this country until the early 1950s, some 30 years or so after the introduction of the bored piling system. Probably the reason for this long gap in the adaptation of individual bored piles into a wall of touching piles lies in the rebuilding activities in our towns and cities after the last war and the need to find a system that could be used up against existing buildings and structures, or in confined spaces, without risk of damage and interference.

4. Contiguous bored pile walls are therefore frequently associated with deep and shallow excavations for basements to buildings and for cofferdam work; particularly in built-up areas where noise or the effect on adjacent property are of importance and in industrial complexes where access, headroom or a restriction on vibration may make other methods such as steel sheet piling or diaphragm walling less suitable or more expensive.

5. On small sites in open areas and on relatively short runs, such as retaining walls for the sides of cuttings and embankments for roads and bridge works, bored pile walls can often show a distinct advantage with the use of light equipment quickly and cheaply erected and able to operate on slopes or stagings.

6. As well as the practical advantages under these conditions, contiguous bored pile walls constructed in suitable ground such as clay can be more economical than diaphragm walls and give an added bonus in the absence of mess and problems in the disposal of large quantities of drilling mud associated with the latter.

7. With the increasing use of auger rigs for both large and small diameter piles, the speed of construction and economy of constructional cost of contiguous bored pile walls has become even more apparent in recent years.

8. In ground conditions such as predominantly water-bearing sands and gravels, silts etc., which are in any event difficult for deep bored piling, the contiguous bored pile wall is at an obvious disadvantage to sheet piling or diaphragm walling. In these circumstances the secant bored pile wall constructed by oscillating casings fitted with cutting teeth and grabbing out the spoil can be an alternative.

9. The purpose of this Paper is to show some of the uses of contiguous bored pile walls in typical situations and to discuss the design and construction technique currently in use. At the same time, the Authors have attempted to show comparisons where appropriate with the other well known techniques of diaphragm walling and sheet piling, to assist engineers in selecting the most suitable method for a given situation.

DESIGN ASPECTS

10. Contiguous bored piles when acting as retaining walls are reinforced to resist the applied bending moments and shear forces. The reinforcement requirement can be calculated by conventional load factor design techniques after the applied loading has been determined.

11. There is great flexibility in the choice of pile diameter, but the circular pile section usually necessitates the use of more main reinforcement per metre run of wall than the constant section of a diaphragm wall to produce an equivalent moment of resistance. Figure 1 compares the lever arms of conventionally reinforced piles and a diaphragm wall where the diameter of the piles is equal to the thickness of the wall.

12. It has become accepted practice to position the steel reinforcement in a pile shaft to achieve the maximum resistance to bending from a given number of bars, as indicated in Fig. 1.

13. Where the loading dictates so great a number of reinforcing bars that the congestion might hinder the free flow of concrete between the bars, it is possible to use steel sections such as universal beams or columns as an alternative.

14. Contiguous bored piles, although usually installed vertically, can be installed at an inclination to the vertical in suitable conditions, thus alleviating the required moment of resistance.

15. Although it is possible to design free-standing contiguous pile walls to retain vertical faces in excess of

Diaphragm walls and anchorages. Institution of Civil Engineers, London, 1975, 189–194

10 m, ground anchors are often incorporated in the retaining scheme.

16. The use of ground anchors makes it possible to limit deflexions and consequential movement of the retained ground. It is also possible by this means to reduce the amount of reinforcement required to withstand the large bending moments and shear forces developed during the temporary stage between excavation and construction of propping slabs.

17. In order to transfer the load uniformly from the piles to the anchors, reinforced concrete waling beams are normally cast in situ as the excavation is taken down in stages, as shown in Fig. 2.

CONSTRUCTION AND INSTALLATION TECHNIQUES

18. Contiguous bored pile walls are usually constructed with mechanical auger rigs or tripods, the piles being bored and concreted independently and nearly touching or in very close proximity to each other.

19. Normally it is necessary to use temporary casings during construction not only to support the sides of the borehole through non-cohesive ground but also, by pre-setting a number of casings in advance of the piling, to ensure the correct spacing of the piles in the line. The depth of casing employed depends on the ground conditions and where, for example, the piles are in clay only a short casing is usually necessary to seal into the top of the clay and the boring is continued in an uncased hole.

20. The use of temporary casing does mean that gaps are formed between piles. With casings of 12–15 mm wall thickness the spacing between piles need not normally exceed 75 mm at the head of the pile and the verticality of the pile shafts should also remain within the normally accepted tolerances for bored piles. The

Federation of Piling Specialists specification for bored piles gives the tolerance in verticality as 1 in 75.

21. Where auger machines are used the diameter of the auger for boring the hole, which is equal to the diameter of the pile shaft, has to be smaller than the internal diameter of the casing to allow for the passage of the auger up and down through the casing, the difference being normally 50–75 mm. This leaves the cased portion of the pile oversize. Alternatively, casing of the required pile diameter can be used, and the shaft bored undersize with the auger then reamed out to the correct diameter using a special drilling tool or heavy drop tool.

22. With auger boring machines the accuracy of installation, both in plan position and verticality, can be improved where the kelly bar is guided rather than allowed to hang suspended from the jib head.

23. A critical function in bored pile wall construction by auger or tripod rig is in the installation and extraction of the temporary casings. In difficult ground conditions these can become very tight, even where moderate lengths of casing are used, and particular attention needs to be given to keeping the casing as free as practicable and to having adequate means for extraction. It is of equal importance, therefore, that a proper working surface, sufficiently firm and adequate for the extraction loads, be provided for the piling work.

24. In certain circumstances, such as very dense gravels and sands, preboring with drilling mud or use of the 'mudding in' technique can greatly facilitate the installation and extraction of temporary casings. An example of this occurred when constructing a bored pile wall very close to an existing building, where it was found that the London Clay was overlain by some 8 m of dry dense sand and gravel. This was too compact to allow casings to be driven without risk of damage to the building and would have made extraction of the casings very difficult. Pre-

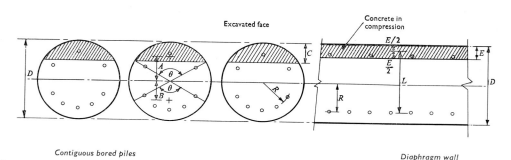

Lever arm pile = A + B   *Contiguous bored piles*

Lever arm wall = $L = \frac{D}{2} + R - \frac{E}{2}$ *Diaphragm wall*

$A = \dfrac{(D \sin \theta/2)^3}{12 \times \text{hatched area}}$

B = centroidal distance of tensile steel from centre

Example: comparison of lever arms, for applied bending moment equal to maximum moment of resistance of circular section due to concrete in compression, assuming piles are touching, D = 600 mm, R = 200 mm, P_{cb} = 10 N/mm², P_{st} = 210 N/mm² and ignoring steel in compression.

Circular section

First approximation, for maximum moment due to concrete in compression

M_r (pile) = 0·58 D^3 = 125·28 × 10⁶ Nmm

Lever arm = A + B = 0·46 D = 267 mm

Diaphragm wall length D

Applying M_r (pile) to determine L

$D \times E \times \frac{2}{3} \times 10 \times L = 125 \cdot 28 \times 10^6$

Substituting $L = \frac{D}{2} + R - \frac{E}{2}$ and solving for E

E = 134·3 mm

Lever arm = L = 432·8 mm

Note. For maximum moment of resistance due to concrete in compression, L = 375 mm

Fig. 1. Plan section through contiguous bored pile and diaphragm wall with comparison of lever arms for circular and constant section

Fig. 2. Detail of cast in situ reinforced concrete waling and anchor head

boring was therefore adopted, using drilling mud, down to the top of the clay, and the temporary casings were then installed to seal into the clay. After this, the drilling mud could be cleaned out and the borings continued unlined in the clay.

25. Contiguous bored piles are concreted using the accepted practice for cast in situ bored piles: by placing the concrete directly into the pile shaft through a short trunking and hopper for dry conditions and by the tremie technique under water. Concrete should be of high slump in accordance with the specification recommended by the Federation of Piling Specialists.

26. It is common practice in the experience of the Authors to construct contiguous bored pile walls on the hit and miss method when using auger rigs; that is, by constructing alternate or primary piles in a row to allow the concrete time to harden before boring the intermediate or secondary piles. This part of the technique further illustrates the function performed by the temporary casings in setting the correct spacings and line for the prim-

ary piles; the practice being, where piles are very close together, to set a number of casings in a line before boring out a pile to depth.

27. It is sometimes preferable with an auger rig to work more than one section of a row or line of piles at a time, in order to give room for the rig to continue working whilst concreting of the previously bored pile shaft is taking place. Figure 3 illustrates a typical installation sequence when using high production auger machines.

28. Tripod rigs, with their much slower output, often confined to one pile a day, can usually be accommodated in a relatively small section of the working area at a time and the installation procedure commonly adopted allows for consecutive piles to be constructed. Figure 4 gives the sequence of construction for contiguous bored piles by the tripod rig.

APPLICATION

29. From the preceding remarks on the method of construction, it follows that contiguous bored pile walls are generally best suited to dry cohesive soils and where the depth of non-cohesive or water-bearing ground is limited.

30. In consequence, they are to be found very frequently in areas such as London where the depth of made ground and gravels overlying the London Clay may only be a few metres. This allows for rapid and economical boring and construction particularly where auger rigs can be employed. In these situations and conditions, the system can provide a most viable method of construction for both shallow and deep retaining walls taking substantial superimposed loadings from adjacent structures.

31. It is important to mention the flexibility of this method, particularly regarding underground obstructions and services. In built-up areas and on confined sites retaining walls often need to be constructed close to adjacent structures or buildings, and problems can arise through the presence of old foundations or existing services at depth. With bored piling it is often possible to chisel

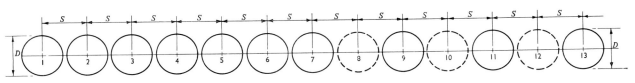

When $S \geq D + 150$ mm odd numbered piles constructed first then, when concrete in adjacent piles has acquired sufficient strength, even numbered piles are constructed.

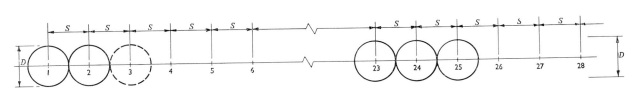

$S \simeq D +$ twice casing thickness

At end of line

(a) Casings set at positions 1 and 2.
(b) Pile 1 bored and concreted.
(c) Casing from position 1 set at 3.
(d) Pile 2 constructed when concrete in pile 1 is sufficiently strong.
(e) Casing from 2 set at position 4 and process repeated.

At intermediate position

Proceed as if at end of line from 24 leaving casing at position 23 for continuity.

Fig. 3. Installation sequence for contiguous bored piles using auger boring machines

through an obstruction or to bore close to services with the minimum risk of damage, but with sheet piling or diaphragm walling such conditions can present a much more difficult problem, and here a serious risk inherent with the diaphragm method is the loss of drilling mud with consequential danger to the stability of the excavation.

32. There are times when lack of headroom, access or the risk of damage to adjacent property or structures preclude the use of steel sheet piling and also diaphragm walling even though the ground conditions may be more suited to one of these methods and rather less to bored piling. In circumstances such as these, and particularly where headroom or space is a governing factor, the use of the ubiquitous tripod rig may be the only practical solution. It is here, where ground conditions are less favourable for bored piles because of unstable ground at depth or a high water head, that due cognisance must be given to the limitations of the method, particularly in regard to seepage between the piles and also the exposed face and line of the piles which is likely to be less uniform in weak or unstable water-bearing ground. A factor that can arise in this connexion is the wastage or flow of concrete into weak ground or cavities as the casings are withdrawn. It will be appreciated that in non-cohesive water-bearing soils, particularly where the boring is hard and slow due to boulders or difficult ground, the amount of overbreak resulting from driving and boring out casings at very close centres in a line is likely to be greater than that arising from single piles at wide spacings.

33. Account needs therefore to be taken and provision made for such factors in the subsequent excavation and

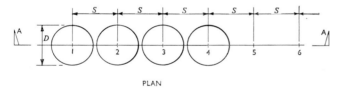

PLAN

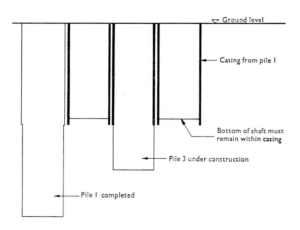

SECTION AA

$S \geqslant D +$ twice casing thickness

(a) Set casings at position 1, 2 and 3.
(b) Bore and concrete pile 1.
(c) Casing from 1 set at position 4.
(d) Bore and concrete pile 3.
(e) Casing from 3 set at position 5.
(f) Bore and concrete pile 2.
(g) Casing from 2 set at position 6.
(h) Repeat cycle for piles 4, 5 and 6.

Fig. 4. Installation sequence for contiguous bored piles using a tripod rig and drop tools

construction work where contiguous bored piles are used in these soil conditions.

SEALING JOINTS

34. Where contiguous piles pass through water-bearing strata, various grouting techniques can be employed to seal the joints. It should be emphasized, however, that grouting is not certain to provide an absolutely watertight seal, but is generally adequate in low water head conditions such as Thames gravel over London Clay.

35. Grouting of joints is carried out behind the piles and must be done before any excavation is carried out. It must be borne in mind that grouting will not be effective close to ground level without additional cover or surcharge of soil.

36. As a supplement to grouting, the watertightness of the structure can be improved by guniting or raking and sealing the joints on the exposed face.

EXAMPLES

37. There are numerous examples today of contiguous bored pile walls and some of these are illustrated in this Paper to show differing applications. In particular, the Authors would like to mention the contiguous bored pile walls constructed by auger machines and tripods for the deep basement excavation at the new National Westminster Bank building in Bishopsgate, London, where the 600 mm piles were constructed in part only 300 mm clear of the existing bank building (Fig. 5). The piles had to withstand the surcharge from these buildings and retain an 11 m excavation. In this section of the wall the piles are reinforced with steel H beams to withstand the shear and bending moments imposed by the ground and the superimposed loads. The resultant deflexion of the piles was monitored and after excavation was found to be at maximum 15 mm. A total of 350 piles was installed, varying in length between 7·30 m and 17·70 m.

38. A special application of the tripod rig working in restricted headroom is shown at the GLC Archway Road widening scheme (Fig. 6) where 27 contiguous bored piles of 600 mm dia. were installed at high level in front of each abutment of the Archway Bridge in limited headroom, prior to breaking down the existing brick retaining walls and building new reinforced concrete retaining walls to give increased clear road width. Because of the high surcharge loads, the bored piles were each reinforced with 12 in. × 12 in. × 190 lb universal column in three lengths, connexions being by splice plates and 64 high strength friction grip bolts in each joint.

39. Two further examples are shown in Figs 7 and 8.

COST COMPARISONS

40. For practical reasons it is not possible to give accurate comparisons in terms of material content and constructional costs between sheet piling, diaphragm walling and contiguous bored piles. These cannot be generalized and will depend on a number of factors particular to a given case.

41. An indication of the likely order of comparison, taking purely hypothetical examples based on two differ-

Fig. 5. National Westminster Bank redevelopment, Bishopsgate: 600 mm contiguous bored piles constructed with cased auger rig for 11 m depth excavation to basement: the top of the London Clay is at approximately the level of the second waling above formation

Fig. 7. New depot and housing at Warwick Road, London for the Royal Borough of Kensington and Chelsea: 450 mm contiguous bored pile wall for basement; curved section with capping beam was unpropped during excavation

Fig. 6. GLC Archway Road widening scheme: tripod rigs constructing 600 mm contiguous bored piles with plate jointed universal column reinforcement

Fig. 8. New underline bridge for LTE, Chigwell: 600 mm contiguous piles constructed by tripods at rake of 1:5

ent soil and groundwater conditions, has been attempted in Table 1.

42. It must be emphasized that the contiguous bored piling under example B—sand and gravel into a high water table—cannot be expected to give the same degree of watertightness (even though grouting has been allowed) as the sheet piling or diaphragm walling.

FUTURE DEVELOPMENTS

43. The current steel shortage has prompted research into ways and means of reducing the reinforcement content of contiguous bored pile walls. A method under investigation is the incorporation of tendons which can be post-tensioned upon completion of the curing of the concrete pile shaft.

44. Work carried out to date indicates savings in steel content, although problems with profiling and maintenance of position of the tendon during construction remain

to be solved. A promising application of this technique would appear to be the use of concentric tendons in conjunction with multi-anchored or multi-propped retaining schemes.

45. Post-tensioning the concrete will also increase the stiffness of the section and consequently decrease the deflexions.

46. Research is also being carried out into methods of forming an effective seal between piles by pre-setting grout tubes in the pile shafts at the construction stage.

CONCLUSION

47. A contiguous bored pile wall has application in many situations:

(a) as an alternative to sheet piling or diaphragm walling which may be precluded for environmental or practical reasons of construction;

(b) in the right ground conditions such as clay and

Table 1

Subsoil profile	Type of construction	Diameter, thickness or section	Depth to dredge level, *m*	Overall depth, *m*	Approximate material content per metre run of wall			Relative construction cost per metre run of wall (approx.)	Remarks
					Net concrete volume, m^3	Horizontal steel reinforcement, *kg*	Vertical steel reinforcement, *kg*		
A Stiff fissured clay	Contiguous piles	0·450 m	5	6	2·12	18·6	81·3	1·00	1 row of props
	Diaphragm wall	0·500 m	5	6	3·00	93	57	1·58	1 row of props
	Sheet piles	Frodingham 1A	5	7–8	—	—	—	1·37	1 row of props; alternate pairs of piles 7·0 and 8·0 m long
A Stiff fissured clay	Contiguous piles	0·600 m	10	13	6·6	198	827	1·00	2 rows of props
	Diaphragm wall	0·800 m	10	13	10·4	427	685	1·60	2 rows of props
	Sheet piles	Frodingham 2N	10	11·5–12	—	—	—	1·08	2 rows of props; alternate pairs of piles 11·5 and 12·0m long
B Gravel and sand with water at 1 m below ground level	Contiguous piles	0·550 m	5	8	3·3	43	264	1·00	1 row of props; grout sealing treatment included
	Diaphragm wall	0·500 m	5	8	4·0	112	82	0·71	1 row of props
	Sheet piles	Frodingham 2N	5	8–9	—	—	—	0·62	1 row of props; alternate pairs of piles 8·0 and 9·0 m long

Notes

(*a*) No restrictions due to access, working space, etc., have been taken into account, and the differences in establishment costs have been ignored on the basis that the volume of work is sufficiently large.

(*b*) A factor has been included for walings to the sheet piling and bored piles. For the diaphragm wall extra reinforcement in place of walings has been included.

(*c*) Cost factor for grouting to the bored pile walls included under B.

(*d*) Cost of ground anchors/strutting not taken into account, being common to all systems.

where the problems of groundwater seepage are limited it can provide the most economical solution.

48. The flexibility of the system in terms of diameter, construction at a rake, and ability to be carried out in restricted conditions close to buildings and services with the minimal risk of damage, gives it great versatility of use. The system has further particular attraction in that it is relatively free from noise, vibration and mess.

49. However, like all methods, it has its disadvantages in certain situations:

(*a*) in water-bearing ground, seepage of groundwater through the joints will occur unless special measures are taken to alleviate this problem, and if the ground is particularly unstable excessive overbreak can occur;

(*b*) generally, the circular section of the bored piles requires heavier reinforcement than a constant section;

(*c*) where anchors are employed with bored pile walls, cast in situ reinforced concrete or steel walings usually have to be added.

50. The Authors hope that this short Paper will have in some measure highlighted the pros and cons of the contiguous system and be of assistance to engineers when making their appraisal of a particular problem.

ACKNOWLEDGEMENTS

51. The Authors would like to thank the following for their kind permission to publish photographs.

Figure 5: The National Westminster Bank Ltd
R. Seifert & Partners
Pell Frishman & Partners
Mowlem—Main Contractor.

Figure 6: Mr. T. L. G. Deuce
Chief Engineer Construction, Department of Planning and Transportation, GLC
Costain Civil Engineering Ltd—Main Contractor.

Figure 7: Arup Associates
Architects and Engineers and Quantity Surveyors
Mowlem—Main Contractor.

Figure 8: Mr E. W. Cuthbert
Chief Civil Engineer
London Transport Executive
Costain Civil Engineering Ltd—Main Contractor.

REFERENCE

1. FEDERATION OF PILING SPECIALISTS. *Specification for cast in place piling.* FPS, London, 1969, amended 1971.

Ground anchored sea walls for Thames tidal defences

J. K. PICKNETT, BSc, DIC, FICE, MIStructE, MIWE, MASCE, *Senior Engineer, Binnie & Partners*

D. L. GUDGEON, BSc, DIC, FICE, MIWE, MASCE, *Senior Engineer, Binnie & Partners*

E. P. EVANS, MA, MICE, *Assistant to Chief Engineer, British Waterways Board (formerly Engineer, Binnie & Partners)*

Problems in raising the existing flood defences in industrialized areas along the north bank of the Thames estuary are discussed. The development of a design consisting of a concrete wall which resists the floods and waves by the combined action of the soil resistance and permanent ground anchors installed below the river is described and the results of a full-scale test on a prototype section of the wall are presented.

INTRODUCTION

In 1971 Binnie & Partners were retained by the Essex River Authority to investigate and prepare designs for the raising of the flood defences along the north bank of the Thames estuary, between Barking and Leigh-on-Sea. This Paper describes the development of a novel form of flood defence for certain lengths, and its subsequent full-scale testing.

2. The lengths of existing defence involved consist of a bank formed of clay excavated from ditches behind the bank, surmounted by a short crest wall of sheet piling. Industrial development has taken place behind the wall, with installations built right up to the toe of the bank. Figure 1 illustrates a typical length. The existing banks are founded on soft organic clays and silts and often have very low factors of safety against seaward slip (1·2–1·4).

Fig. 1. Typical length of existing defence

3. The primary design requirement of the new defences is to defend the hinterland against high water levels resulting from northerly winds funnelling water down the North Sea and causing surges. Examination of records of such events had led to a surge of probability 1:1000 per year in the year 2030 being set as the design standard. Severe wave action would be unlikely under these conditions because of the general east–west orientation of the estuary. There is, however, the possibility of a surge of lesser magnitude, caused by an easterly wind, but associated with bigger waves. The defences were therefore also designed against these secondary conditions.

DEVELOPMENT OF THE DESIGN

4. The restricted space available and the low strength of the clay foundation made a straightforward raising of the banks by further earthworks a practical impossibility. Conventional designs for sheet pile walls, and retaining walls on bearing piles, would be difficult to construct due to the extremely narrow working area and restricted access available. A plain retaining wall with integral concrete cut-off founded on the trimmed-off top of the existing bank was also examined, but while this added no extra weight to the existing bank and hence did not decrease its already low stability against seaward slip, it did not possess an adequate factor of safety against a slip to landward under flood conditions.

5. The effect of adding a ground anchor, installed at a low angle to relieve the clay bank of a part of the horizontal loading, was then studied and shown to give a design which had adequate stability. It could be constructed using comparatively light equipment. The cost of this solution was estimated and its feasibility discussed with contractors. It was found to cost no more than raising with further earthworks, had space been available for this solution. A retaining wall on bearing piles at a 1:3 rake would cost 50% more. The use of steel H piles at a 1:1 rake could possibly halve this cost difference but would render the construction even more difficult.

6. The low stability of the banks against seaward slip at low tide meant that the tendons of the ground-anchored wall could not be prestressed. The ground anchor load would therefore depend on the total movement of the wall and the bank itself under water load. The proportion of the applied load carried by the anchor and the soil could therefore not be calculated with certainty. Preliminary designs were based on both the soil and anchor being cap-

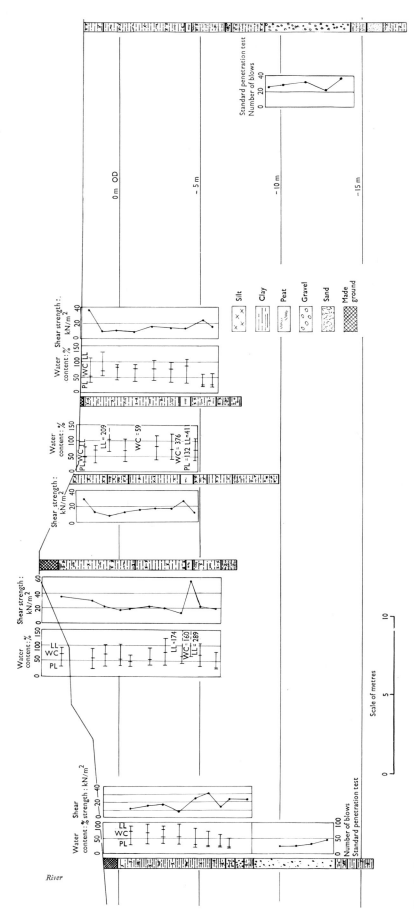

Fig. 2. Cross-section through existing bank showing results of site investigation

able of carrying the total loading, albeit with a low factor of safety.

7. The defence was required to be in full working order, with its design factor of safety, in 60 years' time, which meant that the anchor tendon would have to be extremely well protected against corrosion. The specification will require the tendons to be totally enclosed in a watertight insulating sheathing, which will be subjected to electrical tests for integrity before and possibly after installation.

TESTING THE DESIGN

8. The design is, as far as the Authors are aware, a new and untried concept. A full-scale test to observe the behaviour of the wall, foundation and anchors under working load, and ultimately at failure, was necessary to reveal any unexpected mode of failure. Secondly, since it is proposed to build considerable lengths of this wall costing several million pounds, it was hoped that investigation of the load–deformation characteristics of the soil–anchor system could lead to a less conservative design.

9. A suitable test site was chosen, near Mucking Creek, Stanford-le-Hope, Essex. A number of boreholes were put down at the site, taking continuous undisturbed samples. The cross-section of the existing bank, and the results of the site investigation, are shown in Fig. 2.

Test arrangement

10. The layout and details of the test facility are shown in Figs 3 and 4.

11. It was proposed to test the wall by subjecting it to a raised water level, giving the static effects of a surge, and to apply the design wave loads by means of jacks connected to the wall by bar or strand. Failure would ultimately be induced by increasing the jack load. Wave loads applied by jack were represented by the static value of the 1% wave and also by cyclic loading representing repeated impacts of the average wave.

12. Three bays of the wall were built: the central test section, and two independent wing bays to provide reasonably realistic end conditions.

13. The single ground anchors installed in the centre-line of each bay were of the TML type of Entreprise Bachy, Paris. At each position a cased hole was bored down into the gravel, and a tube à manchettes was inserted into the borehole. The casing was then withdrawn and the tube grouted into the gravel. The inflatable sac at the top of the anchor lengths was grouted first to seal this length off. Finally, the tendon, consisting of seven 18 mm Dyform strands, was inserted into the tube à manchettes and grouted in. The design called for anchors of nominal working load 400 kN, with a factor of safety of 2 on pull-out, but the cross-sectional area of the tendon was greater than necessary to encourage a

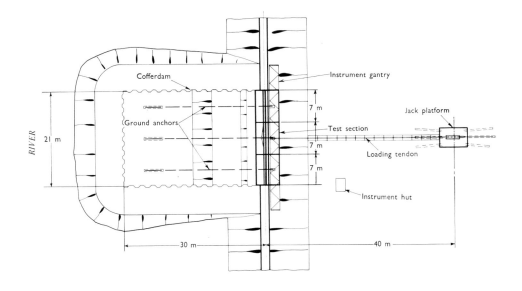

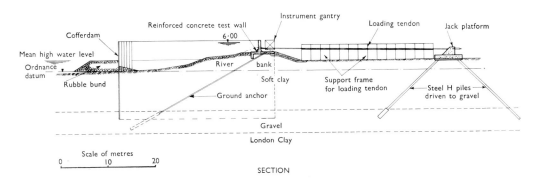

Fig. 3. General arrangement of test facility

larger share of the load to be taken by the anchor at small strains as the wall deflected.

14. The jacking platform, supported on steel H piles, provided support and reaction for the jacks. For the static working load tests a 500 kN hollow centre jack was used, connected to the wall by a 60 mm high yield steel bar. For the dynamic working load tests an electronically controlled hydraulic actuator was used, with the same bar. For the failure test 2000 kN pulling capacity was needed, and this was supplied by a large prestressing jack, with a 16 strand cable to the wall, bifurcated to clear the anchor head. Loading points were provided at appropriate levels on the wall for the various loadings.

Instrumentation

15. The central test bay was instrumented with inclinometers, piezometers, earth pressure cells, displacement transducers, and load cells on the anchor head and the loading bar. The earth pressure cells were as described by Wong.[1] The instrumentation is shown in Figs 5–7. All instrument terminals were housed in the instrument house. The transducers, earth pressure cells and load cells were energized at 10 V d.c. and their outputs were continuously observed and recorded during tests on a 25 channel ultra violet recorder.

Tests

16. The working load tests were carried out with water levels and loads as shown in Table 1. The results of the north wind surge static test are shown in Fig. 8. A sample of the recorder trace for the east wind surge dynamic test is also shown, in Fig. 9.

17. The working load tests showed up a number of points.

18. The behaviour of the wall under test was satisfactory.

19. Deflexions were well within acceptable limits. For the north wind surge static test, which gave the biggest movement, the displacement transducers indicated a horizontal displacement of about 6 mm, and settlements of 6 mm at the centre and 3 mm at the rear of the base. As can be seen from (c) and (d) of Fig. 8, the horizontal movement was less than that indicated by the inclinometers, probably due to movement of the abutments of the instrument gantry. Trouble was also experienced with vertical thermal movement of the gantry; it was difficult to make absolute measurements of small movements in these circumstances. Horizontal recovery was rapid and virtually complete, aided by the ground anchor, but vertical recovery was not complete.

20. The anchor load, as indicated by the load cell on the anchor head, remained low, due to the small displacement of the wall.

21. The earth pressure cell results were difficult to interpret, but showed that most of the load was transferred to the bank through the base slab rather than by the cut-off.

22. The piezometer results reflected the water load

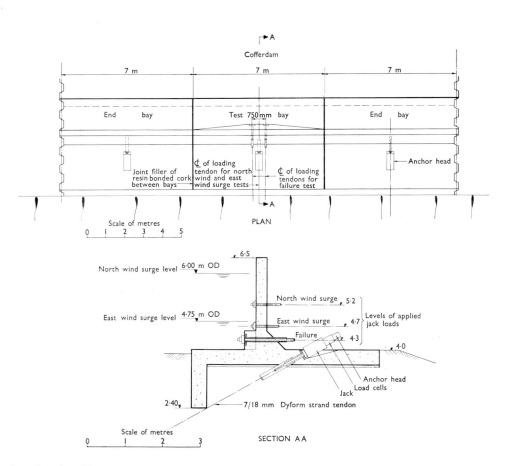

Fig. 4. Details of anchored wall

Table 1

Test	Water level, *m* OD	Jack load, *kN*
East wind surge, static	4·750	370
North wind surge, static	6·000	80
East wind surge, dynamic	4·750	0–140 cyclic
North wind surge, dynamic	6·000	0–30 cyclic

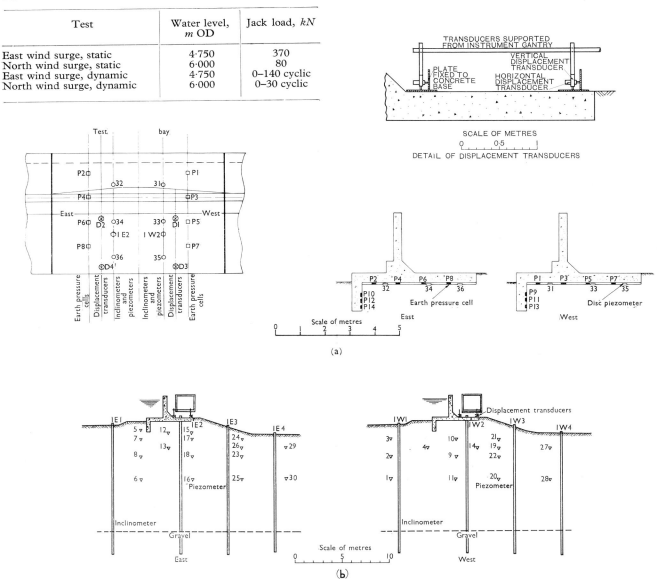

Fig. 5. Instrumentation: (a) plan showing instrumented sections, and sections showing earth pressure cells and disc piezometers; (b) sections showing inclinometers and piezometers; (inset) detail of displacement transducers

Fig. 6. Anchor head, load cell and jack

Fig. 7. Displacement transducers

in a straightforward manner and recovery was complete after the tests. As anticipated, there was no evidence of an increase in pore pressure due to seepage.

23. The dynamic load had little effect. The earth pressure cells and anchor showed little response, and there was no progressive movement of the wall.

24. After considering the results of the working load tests it was decided that, since the anchor load had been comparatively small throughout, the wall should be tested without the anchor connected before proceeding to the failure test.

25. With the anchor connected, the water in the

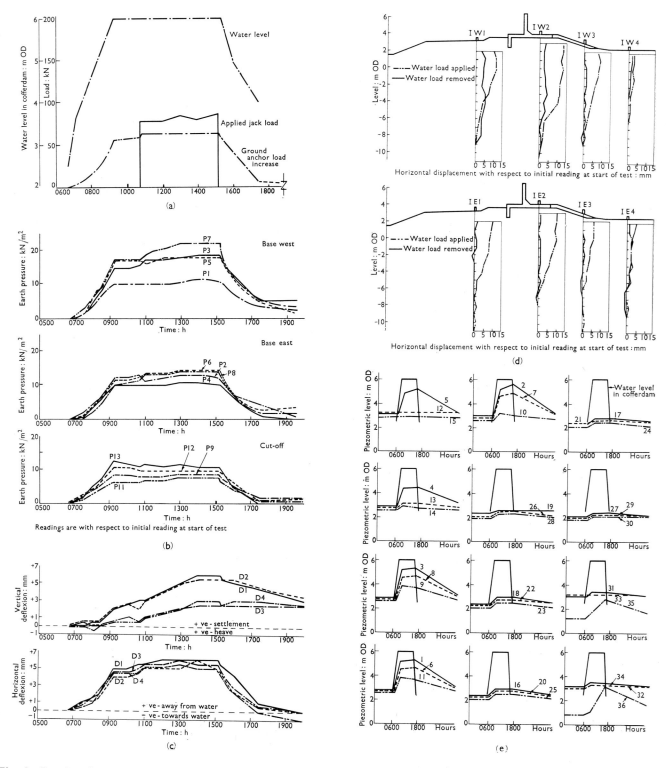

Fig. 8. Results of north wind surge static test (6 Dec. 73): (a) loading; (b) earth pressure cells; (c) displacement transducers; (d) inclinometers; (e) piezometers

cofferdam was raised to +6·00 m OD, with a jack load such that twice the horizontal working load was applied. Deflexion and anchor load increased linearly and remained small. A further increment of jack load was then applied, equal to the horizontal component of anchor load, and since there was still no significant change in behaviour, the load was relaxed, the anchor was disconnected, and the loading up to twice horizontal working load repeated. Horizontal deflexion increased by a further 7 mm.

26. Following this test, the test to failure was carried out. The water level was raised again to the north wind surge level, and a jack load applied in 100 kN increments, pausing 10 min at each increment, up to a load of 600 kN, which was held for 1 h. Loading was then resumed as before up to a maximum of 1900 kN. Jack load, anchor load and horizontal deflexions are shown in Fig. 10.

27. Only a small amount of creep was observed while the load of 600 kN was held for 1 h. When loading was resumed behaviour was almost entirely linear up to a load of about 800 kN. Thereafter the horizontal deflexion became progressively non-linear and at 1300–1400 kN the load in the anchor increased sharply. By the time a load of 1900 kN was reached, horizontal deflexion was approaching 90 mm and the anchor load was over 700 kN. This load was held for 30 min during which time the anchor load increased to 800 kN. As displacement was by then substantial, and the anchor had reached twice its working load, the test was terminated.

28. Vertical deflexion at the middle of the base was practically nil, and at the rear edge there was 33 mm settlement. Horizontal displacement of the top of the wall was 110 mm.

CONCLUSIONS

29. The tests showed that at working loads the behaviour of the wall was very satisfactory. No unexpected mechanism of failure was observed and the margin of safety as demonstrated by the failure test was more than adequate.

ACKNOWLEDGEMENTS

30. The Authors are indebted to the Essex River Authority, Chief Executive and Engineer Mr E. L. Snell, OBE, FICE, FIWE, and to their consulting engineers Binnie & Partners for permission to publish this Paper.

31. The use of land at Mucking for the test was by permission of the Greater London Council who own the site.

32. The main contractor for the construction and testing of the wall was Edmund Nuttall Ltd. Entreprise Bachy was the subcontractor for the ground anchors and Testwell Ltd for the dynamic test equipment. The main contractor for the instrumentation was Soil Instruments Ltd. The site investigation and soil testing was carried out by Foundation Engineering Ltd.

33. The resident engineer for Binnie & Partners was Mr R. I. White, BSc, MICE, and the agent for Edmund Nuttall Ltd was Dr D. W. Allenby, BSc, PhD.

34. The Authors gratefully acknowledge the helpful discussions with Professor A. W. Bishop, MA, DSc(Eng), PhD, FICE and his colleagues at Imperial College.

35. The photographs for Figs 1, 6 and 7 were kindly provided by the Essex River Authority and reproduction of Fig. 1 is by permission of Shell UK Ltd.

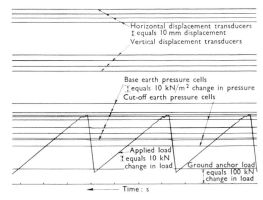

Fig. 9. Recorder trace for east wind surge dynamic test

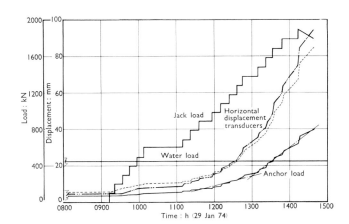

Fig. 10. Results of failure load test

REFERENCE

1. WONG H. Y. Field instrumentation of vibroflotation foundation. *Field Instrumentation in Geotechnical Engineering.* Butterworths, London, 1974, 475–487.

Discussion on Papers 22–25

Reported by I. K. NIXON

Although the four Papers submitted to this session covered rather diverse topics, it was possible to identify two quite separate main themes. One theme was the commercial and practical choice available of supports for deep excavations in urban environments, as discussed in Papers 22 and 24; the other was the stability of the base of retaining walls, as discussed in papers 23 and 25.

CONTRIBUTIONS ON PAPERS 22 AND 24

2. Mr Puller's opening remark was that, in the final event, basement design was still an art—even crystal ball gazing—notwithstanding the detailed and rational analysis he had presented in Paper 22. He then returned to systematic consideration and drew attention to other Papers in the Conference that described particular reasons for the selection of an excavation support method: Paper 7, referring to the considerable time saved on the contract by using a diaphragm; Paper 12, describing the economic advantages of precast diaphragms; and Paper 24, on contiguous bored piling.

3. The process, he emphasized, should be one of elimination, firstly on technical grounds and then on cost, rather than one of highlighting the virtues of any particular method.

4. Conditions of contract and methods of contract measurement could also have some influence. Whereas the Engineer's approval was necessary for temporary works under the ICE contract, the Building Method of Measurement often referred to a 'planking and strutting' item to be priced by the Contractor, and an estimator often had only a short period in which to tender. In other instances the Architect, Quantity Surveyor or Engineer responsible for the particular contract would cast discretion (and the Contractor's experience) aside, and specify the method.

5. Mr Puller quoted examples of diaphragms being chosen to protect precious trees, only to find them cut down after contract completion; or to avoid noise and vibration with unyielding stop-end casings, extracted by pile extractor without any complaints. The reasons for elimination of a possible soil support method must be very carefully considered before its final rejection.

6. Dr Greenwood remarked that, with rapidly changing costs, the validity of a choice could be affected by a shift in the relative cost of labour, plant or materials in the course of the long period before completion of a project. He also referred to the wide variation in subcontract prices, which Mr Puller had attributed to commercial pressures on margins, and said he believed that a major part of such variation was due to widely differing interpretations of the site investigation data available at tender stage. For success in terms of effective completion within original budget cost, geotechnical processes needed to be relatively insensitive to small local variations in soil conditions. The important point was that site investigation should reveal detail on a scale which was relevant to the scale of the single basic unit operation of the process, e.g. a single panel of a diaphragm wall, a single anchor or a single injection. Since diaphragm wall panels were relatively large they could cope with some reasonable degree of soil variation. However, anchors and grout holes were sensitive to the soil in their immediate vicinity—within three or four diameters—and lensing or layering on this scale must be appreciated if proper judgements were to be made. Often a too literal interpretation was made of having records without regard for the geological inferences. Moreover, boreholes should be located where the work was to be done; for example, Dr Mascardi (in relation to Paper 16) had described the use of angled boreholes to reach the vicinity of a proposed fixed anchor zone.

7. Mr Hodgson spoke of the importance of the time taken for construction, including clearing up afterwards, and its influence on the return on investment. As an example, he described a recent large commercial job for which the diaphragm wall would have taken 8–10 weeks to construct, whereas the sheet piling and reinforced concrete retaining wall would have taken 10 weeks and 8 weeks respectively. The time saving was therefore 8–10 weeks, and although the tender price for the diaphragm wall was 20–25% higher than for the sheet piling and reinforced concrete wall, the very large saving in time more than outweighed this from an overall financial viewpoint. Mr Puller replied that often there was difficulty in obtaining agreed output times, suggesting that a decisive advantage was difficult to predict except possibly in the case of a large project where the support works predominated.

8. The use of permanent sheet piling is indicated in Paper 22 to be probably disadvantageous in cost for deep basements, but Mr Little noted that the figures suggested a marked improvement at shallower depths. Anchored walls were often necessary in congested urban conditions where trunk or motor roads were constructed in cutting; these were usually of modest height. The case was described of such a wall for the Sharston bypass near Manchester where it was essential not to disturb the free flow of traffic some 7–8 m above the level of the projected motorway. The cutting, some 900 m in length, was mostly in marl and the systems of support investigated were contiguous bored piles, diaphragm walls and permanent sheet piling. The visible finish to the wall had to be to a high standard; taking this into account, the overall comparative estimates of cost weighed significantly in favour of permanently anchored sheet piling faced with in situ concrete (Fig. 1). The piles were raked at 1/24 to alleviate the 'canyon' effect on drivers and the reinforced facing concrete was held by its grip on the walings and by

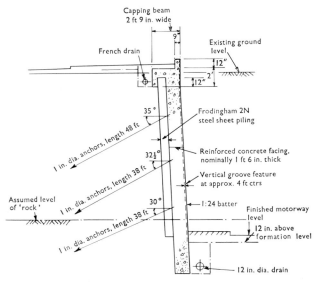

Fig. 1. Sharston bypass, Manchester: section through anchored sheet pile wall

Fig. 2. Northern approaches, Second Mersey Tunnel: exposed surface of piles

Fig. 3. Northern approaches, Second Mersey Tunnel: tops of prefabricated cages showing starter bars for arch

hooking over the top of the sheet pilings, in this way also forming a collector channel for surface water. The facing concrete was always extended to formation level even if the sheet piling stayed higher due to ground hardness. A low level of noise during construction was achieved by the use of suitable specialist equipment for driving the piles. There were generally three rows of anchors and the horizontal spacing was limited, first to keep down the size of the walings and, hence, the thickness of the facing concrete, and second, to limit the size of panel so that even in the unlikely event of deterioration of the sheet piles over the years, by corrosion from the back face, the support system, formed by the concrete-encased walings, the anchors and the reinforced facing concrete itself, would still be able to resist failure of the wall.

9. Amplifying information given in Paper 24 on the National Westminster Bank site, Bishopsgate, Mr North-Lewis said that the geology consisted generally of approximately 6 m of Thames gravel and sand overlying London Clay. Movements of the wall were critical due to the nature and proximity of the old multi-storey buildings. The anchors, in three rows with cast in situ walings, were designed generally to a working load of 500 kN. Measurements showed that maximum deflexion on the deep sections had been of the order of 15 mm (somewhat less than predicted) and there had been some loss of stress in anchors with time. Referring to the limitations of the system, Mr North-Lewis said that problems could arise when bored piles were constructed through non-cohesive water-bearing ground, particularly where long lengths of casing were involved with rigs of limited capacity, and the problems could be enhanced when the piles were contiguous.

10. The use of contiguous bored pile walls on the northern approaches of the Second Mersey Tunnel in Liverpool was described by Mr Holland. The consulting engineers had designed the arched approach as an in situ reinforced concrete arch of some 30 m carried on 2·6 m nominal diameter bored piles. The method of construction was to bore the piles from arch springing level, which was approximately 13 m below general ground level, through approximately 10 m of boulder clay and into Bunter Sandstone to an average bored length of 13 m. The excavation to springing level was generally between vertical Berlinoise walls. The piles were very heavily reinforced with prefabricated cages each weighing 7 t. The arch was constructed on specially made travelling formwork supported on the ground at arch springing level. Subsequent excavation was done under the completed arch to road formation level (Fig. 2). The exposed surface of the piles was faced with in situ concrete using connector bars cast into the face of the piles.

11. Particular care was required to ensure that the tops of the cages remained accurately in the correct direction and level so that the starter bars for the arch were properly aligned (Fig. 3). Similar difficulties were mentioned by Mr Puller in maintaining the alignment of unidirectional cages when extracting with reciprocating casing devices, because of the tendency for the cages to become twisted.

12. Referring to the considerable amount of reinforcement that was evident in the piles at Liverpool (Fig. 3), Mr Lyons emphasized the potential for development of more economical designs, such as may result by means of post-tensioning the pile shafts.

13. In a written contribution, Mr Rowlands pointed out that contiguous bored piles have a decided advantage over diaphragm walls in sloping ground, particularly when used transversely across a slope where there is any risk of landslip.

CONTRIBUTIONS ON PAPERS 23 AND 25

14. A wealth of practical experience and a very frank exposition of possible problems were contained in the introduction by Mr White on Paper 23, *Anchored walls adjacent to vertical rock cuts*. Although the situation does not occur very often in the UK, several contributors to the Conference discussions made reference to it: one showed a soldier pile precariously balanced on the edge of vertically fissured rock; another remarked that every case of failure of anchored king piles that he had collected pointed towards bearing failure of the toe.

15. In a written contribution, Mr Compton described a site in downtown Toronto, where a 10 m deep excavation in clay overburden was supported by soldier piles concreted 1 m into shale bedrock and tied back with two rows of anchors. Close boarding, wedged at the soldier piles, was used to transfer the load. On completion of the support works a further 10 m of excavation was carried out in the shale directly up to the face of the soldier piles, and during a late afternoon check it was found that the bottom 1 m of concreted pile had moved into the site sufficient to enable an arm to be placed around the circular concrete toe. Checks on the residual loads in individual anchors revealed that changes had taken place and it was decided to undertake immediate propping of the whole face (Fig. 4) which was completed by noon the next day. Subsequently, the toe of each pile was underpinned and tied back using rock bolts. He suggested that a three-tier system of anchors introduced with one level near to the bottom of the piles would have obviated such action.

16. A review of the position concerning rock anchor design was contributed by Dr Littlejohn. He reported that, in 1972, the Geotechnics Research Group of Aberdeen University had sent out a questionnaire to engineers in some 20 countries, which showed that although rock anchors had been installed successfully for 40 years there were many important design and construction problems which required clarification.

17. Invariably an inverted cone of rock was considered to fail, but the angle and position of the apex of the cone with respect to the fixed anchor length were chosen differently by various engineers in different countries. The uplift capacity was normally equated to the weight of the specified cone, employing safety factors in the range 1·0–3·0, and often no account was taken of the effect of overburden pressure on rock strength (tensile or shear) at the failure plane. In order to calculate the anchorage depth accurately with a known safety factor it was necessary to utilize all the tools of rock mechanics, e.g., detailed mapping of joints, together with an assessment of their shear strength characteristics. This approach was currently the rare exception. Discussion was required on the type of site investigation and field data needed to facilitate rock anchor design. With regard to the rock/grout interface, the straight shaft anchor relied mainly on the development of bond or shear in order to transfer load, but estimation of the magnitude and distribution of the bond along the fixed anchor was difficult. Nevertheless, for all rock

Fig. 4. Commercial Court, Toronto, Canada: propped soldier piles

types it was common practice to assume a uniform bond stress in order to design the fixed anchor. Typical bond stresses for three categories of rock are shown in Table 1, but care must be taken before applying the values, since degree of weathering (seldom quantified) affects not only the value of bond stress at failure but also the load–deflexion relationship during service.

18. The problem was similar for the tendon/grout interface. The Australian Anchor Code stipulated a maximum bond stress of 2·1 N/mm² for clean strand, which was close to the 2·2 N/mm² recommended by CP 110 for deformed bars embedded in concrete of characteristic strength 30 N/mm². No guidance was given on reduction factors for groups of strands, or the possible decoupling effects of spacers and centralizers, or how much steel was allowed to be packed into the hole. On the construction side, water tests were seldom specified on completion of drilling; their value, however, was questionable, as the problem was loss of cement.

19. Introducing Paper 25, *Ground anchored sea walls for Thames tidal defences*, Mr Picknett gave further information on the failure load test. Figure 5 shows the total applied load (i.e. water plus jacking force) and horizontal displacement of the wall. The lowest line shows the soil reaction against the cut-off, measured by the earth pressure cells. The ground anchor reaction measurements, which agreed well with calculations, are at 30° to the horizontal. The horizontal load transmitted through the base slab by shear to the bank below was obtained by deducting the horizontal component of the ground anchor reaction, and that of the cut-off, from the total applied load. The large difference between the proportion of the

Table 1. Guide to bond stresses for three categories of rock (after Koch[1])

Rock type	Bond stress	
	lb/sq. in.	N/mm²
Weak rock	50–100	0·35–0·70
Medium rock	100–150	0·70–1·05
Strong rock	150–200	1·05–1·40

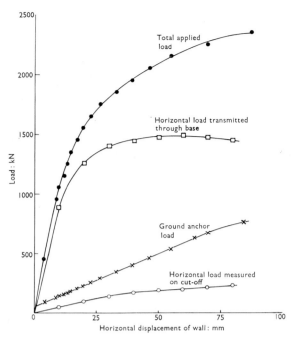

Fig. 5. Anchored sea wall trial, Thames tidal defences: load in individual components versus horizontal displacement

load taken by the cut-off and that taken by the base slab was partly due to the fact that the base slab reaction included the end effect, i.e. the shear force developed between the soil under the test panel and the soil under the adjacent panels. By calculation, it added roughly 20% to the soil resistance. Another feature was the small proportion of the total load mobilized in the ground anchor.

20. One of the primary objects of the test was to examine the compatibility of the wall displacements needed to mobilize soil resistance and anchor resistance respectively, in order to take both into account. There was a fairly wide range of movements where high soil resistance was available and (for instance by increasing the tendon cross section) the full ground anchor resistance could be mobilized, so that the resistance of the combined system might be taken as the sum of the maximum soil strength and the anchor ultimate resistance.

21. Dr Greenwood raised the question of the effect on anchor deflexions of variations in the particle size distribution within the gravelly sand. Mr Picknett confirmed that the gravel layer was variable in strength and that some differential movement of adjacent anchors could therefore be expected. However, subsequent testing of the anchors had shown that anchor creep was likely to be small and, since for practical reasons it would be impossible to pre-stress the anchors, movement of the anchor head (and hence the wall) would depend almost entirely on tendon extension rather than anchor creep. Large differential movements of adjacent wall panels were therefore not expected.

22. A description was given by Mr Guillaud of a new type of load cell, designed for insertion between the bearing plate and the clamping head of the tie-back. Two high carbon steel rings ensured the centring and the homogeneous distribution of forces on to an intermediate ring made of special steel (highly sensitive to variations of magnetic field), in which was housed an electric coil. The magnetic permeability of the special steel was a linear reverse function of the compression stress into it; when the coil was fed with an alternating electric current, the variation of magnetic permeability generated a variation of impedance —in this case inductance—which was measured on a classical impedance bridge. The cell was claimed to be insensitive to eccentricity of the resultant force.

CONTRIBUTORS

Mr A. J. Compton, Cementation Ground Engineering Ltd.
Dr D. A. Greenwood, Cementation Specialist Holdings Ltd.
Mr M. A. Guillaud, Sol-Expert International.
Mr F. T. Hodgson, Clarke, Nicholls and Marcel.
Mr G. R. Holland, Lind Piling Ltd.
Mr G. Little, Howard Humphreys and Sons.
Dr G. S. Littlejohn, University of Aberdeen.
Mr I. K. Nixon, Soil Mechanics Ltd.
Mr G. O. Rowlands, Glamorgan Polytechnic.

REFERENCE

1. KOCH J. L. (BBR Australia). Private communication, 1972.

Diaphragm wall specification

W. G. K. FLEMING, BSc, PhD, MICE, *Project Manager (Technical), Cementation Piling and Foundations Ltd*

M. FUCHSBERGER, Dipl Ing, MSc, *General Manager, ICOS (Great Britain) Ltd*

O. KIPPS, BSc, FICE, MASCE, FGS, *Director, ICOS (Great Britain) Ltd*

Z. SLIWINSKI, BSc, MICE, *Consultant, Cementation Piling and Foundations Ltd*

This Paper owes its origin to the preparation of a specification for cast in place concrete diaphragm walling by the Federation of Piling Specialists in 1973. It takes stock of the present state of the art in regard to specification writing for this type of construction and contrasts the position with that existing in the piling field where an excessively large number of specifications are at present in use. The hope is expressed that a common specification document can be accepted before a large number of individual specifications come into existence. The principles of specification writing are briefly discussed, and attention is drawn to some of the apparent differences of opinion between the writers of specifications for diaphragm walling now in existence.

INTRODUCTION

In the civil engineering industry, as in other fields, new developments at first pose the problem of establishing technical and practical feasibility. If in initial cases these aspects are found to be satisfactory, and if economic considerations look favourable, then a period of gradual development and growth takes place until the new process becomes generally accepted as a solution within its field. Increasing demand brings with it new practitioners, variations of the plant used and a flow of secondary techniques and developments, and creates a need for proper training of operatives and supervisors.

2. At this stage many prospective users are introduced to the process and would welcome a set of standards and site controls enjoying a wide consensus of support, which they could apply to the work knowing that they would lead to a sound and satisfactory finished product. It is unfortunate that, communication being as it is, each new user has to learn by his own experience, often with only the most general information for background.

3. The art of diaphragm walling dates back to about 1948, and it is true to say that it is still a developing subject. Nevertheless, the contracting specialists in the UK felt in 1973 that a sufficient fund of common experience was available for the preparation of a specification for cast in place diaphragm walling, which would be reasonably comprehensive, would generally make clear the standards which could normally be accepted, and would be of benefit to the industry at large. A specification[1] was therefore prepared by a special committee of the Federation of Piling Specialists and it is from the preparation of this document that the present Paper originates. The committee was representative of the major specialist companies operating in the UK, and the Authors' companies were closely involved in the drafting work and discussion prior to publication.

OBJECTIVES AND GENERAL PRINCIPLES OF SPECIFICATION WRITING

4. Before any specification is prepared it is necessary to establish several basic principles upon which it is to rest. This can be done in various ways, but the Authors believe that the following considerations are relevant to the establishment of a satisfactory basis.

5. The specification is a technical document. Contentious items occurring in specifications in general are usually found to be those which combine a technical matter with a condition of contract. It is far better to clearly separate the two matters and to treat them in separate documents, the technical items alone being considered in the specification.

6. The specification should set out a course of action to perform the work which is reasonable from the contractor's point of view and gives a reasonable expectation of achieving the desired end product. A specification quickly loses credibility on site if its requirements are outside the range of practical achievement, and such a specification can only be detrimental to the whole job.

7. The specification needs to set out a course of action which will lead to an end product with which the client can be satisfied. If dissatisfied he will not wish to repeat the experience and will abandon either the specification or the use of the new development in its entirety.

8. The specification should be direct in its approach. Whilst the document should set out to be neither a text book nor a code of practice, it should nevertheless relate cause and effect as closely as possible. Obscure requirements will naturally be the first to be forgotten. Thus, for example, it is better to say 'after mixing, concrete shall be placed before it loses the specified workability' than to say 'concrete shall be placed within x minutes from mixing'.

9. Having established basic principles it is then necessary to define the particular items which should be included. At this stage a decision has to be made to cover one or more of the following subjects:

(*a*) design
(*b*) materials
(*c*) workmanship
(*d*) performance requirements of the finished structure.

10. Item (*a*) may or may not be included, depending on whether the contractor or the engineer is intended to do the design. Item (*d*) may stand on its own as a particular form of specification, but if it is used in combination with any or all of the preceding items then these other items need to be compatible with it. Items (*b*) and (*c*) normally stand together with or without the others.

EXISTING SPECIFICATIONS

11. There are apparently available at the present time only two general published specifications for diaphragm walling, one prepared in Austria[2] and the FPS Specification from the UK. The Austrian specification is still open to comment before final issue. A preliminary document in the form of a Code of Practice has also been prepared in East Germany, but details of this are not yet available to us.

12. In addition to these, various consulting engineers have prepared their own specification documents for particular contracts, and these documents are varied from job to job to take into account particular features of the work and new experience as it arises, so that in the end specifications of fairly wide general application are developed. However, given the same starting point no two engineers would ever arrive at precisely the same end product. Two consulting engineers' specifications have been used for comparison purposes in this Paper, but it would be unfair to identify their sources since they are clearly on-going documents, not prepared for general publication.

13. In the closely related field of piling, the industry has long suffered from an over-proliferation of such specifications. The result is that much time is spent by the staff of consulting engineers and contractors in preparing and digesting a variety of individual documents when it would be much simpler and more satisfactory if reference could be made to a single nationally accepted document which could be quickly identified and understood. Specific amendments, suited to the particular work in question, could be added if necessary.

14. In the opinion of the Authors a suitable universal specification could be prepared for both piling and diaphragm walling, and it is a real hope that at the present stage in diaphragm walling such a specification may be adopted, so that a confusing collection of specifications as afflicts the piling industry may be avoided.

COMMENTS ON EXISTING SPECIFICATIONS

15. The FPS Specification differs from the other specifications available in that it presents a series of notes for guidance interspersed with the text. This was done in order to clarify the items, but the notes are not intended in themselves to be a part of the specification. Basically the notes are directed to the user, whether contractor or engineer, in order to help him to appreciate the background of the work. Perhaps, when the technique has been more widely used, and when the details of the special construction method are more fully appreciated, these notes could be separated from the main document or discarded.

16. All of the specifications studied deal with cast in place diaphragm walls. None makes mention of other types of wall, such as plastic diaphragms which are some-

times used for dam and reservoir cut-offs, or precast walls, fixed within a bentonite-filled trench. These particular forms are less common than the cast in place concrete wall. However, the design of these forms of diaphragm is a specialist job, particularly with regard to the infilling materials, and there would not seem to be any major problem in modifying a basic specification to accommodate such variations. Neither do any of the specifications make any reference to noise limitation during construction, though one of the more significant advantages of the method over, for example, conventional sheet piling in urban areas must surely be in the reduction of both noise and vibration which might affect adjoining properties.

17. A comparative study of the existing specifications has been made (see Appendix) and the following comments may be of value.

18. In many specifications there does not seem to be a clear demarcation between contract conditions and specification. This is particularly so with documents prepared by consulting engineers.

19. The FPS Specification is probably the most detailed document available. The specifications from the consulting engineers are also on the whole very comprehensive documents. The other documents considered are in general much less detailed.

20. There are clearly differences of opinion between the specification writers in matters of detail, but on the whole there is much less disagreement concerning the sort of items which should be included.

21. The remainder of the Paper covers the main points of divergence.

Design

22. *Bond stress on steel reinforcement.* The FPS Specification allows the use of ordinary round bars, but allows only a 10% increase of bond on deformed bars over the allowable value for the round bars. The Austrian specification rules out the use of anything but deformed bars.

23. Factual evidence on this item does not appear to be available from many sources, and in preparing its specification the FPS committee made reference mainly to the CIRIA Interim Report No. 9.[3] It seems probable that rising concrete sweeps cleanly away most of the bentonite suspension from the bar surface. Two points of view might be taken: that a deformed bar gives a stronger and more reliable mechanical grip, even though strain may be relatively large to bring it into action; or that the strain necessary to mobilize stress in a deformed bar, if any bentonite is trapped by the surface irregularities of the bar, may be sufficient to give rise to significant cracks in the concrete, and if the amount of bentonite trapped at each rib is not consistent, transfer of bar tension to the concrete by bond stress may be erratic. More evidence on this aspect of design would be useful.

24. *Cover to reinforcement.* The FPS Specification gives a value of 75 mm, the Austrian gives 50 mm for internal walls and 60 mm for visible outside walls, and the consultants seem to be united in requiring 75 mm. The British Code of Practice CP 2004[4] recommends a cover of 75 mm for this class of work and there would appear to be no valid reason for changing this.

25. *Spacing between bars.* The FPS Specification gives a minimum clear spacing of 100 mm, whereas the

Austrian document gives 50 mm clear between bars. This item does not seem to be included in other specifications, although it is of vital importance to the successful placing of the concrete. The Codes of Practice CP 114[5] and CP 110[6] would permit the distance between bars to be as close as the Austrian document suggests, but the Authors feel that this spacing would give rise to risk, possibly preventing full displacement of bentonite suspension and therefore leading to inclusions of mud in the wall.

26. *Factors of safety.* These are not generally specified in any specification except for one from a consultant. It appears to be the case that the factors of safety are derived from the Codes of Practice and the interpretation of the codes is left to the skill of the individual design engineer.

27. The Austrian document suggests that when calculating active pressure the angle of friction between the wall and soil may be taken as between $\frac{1}{3}$ and $\frac{2}{3}$ of the angle of internal friction of the soil, whereas when calculating passive pressure a wall friction equivalent to only $\frac{1}{3}$ of the angle of internal friction of the soil may be considered effective.

28. In clays, for design purposes a choice has to be made as to whether to use effective stress design parameters or to use a method based on the results of undrained triaxial tests, possibly amended by an arbitrary strength reduction or fissure factor. Selection of the method depends on the circumstances and the duration of full load conditions, but there are many circumstances in which the correct design procedure can be debated, and any statement of factors of safety on its own, without consideration of the most realistic design approach, may be misleading.

Materials

29. *Concrete workability.* The FPS Specification sets limits on concrete workability through the conventional slump test, and also gives a maximum water/cement ratio. Minimum slump as specified by FPS is 150 mm, and no maximum is suggested. One of the consultant's specifications suggests 125 mm with a maximum slump of 200 mm. It is perhaps unfortunate that the slump test is stretched to the limit of credibility in the range required, and that other tests, such as the compacting factor test, appear not to have much relevance. A new type of test needs to be developed for fluid mixes of the type used in diaphragm walling and piling.

30. *Cement content in concrete.* The FPS Specification gives a minimum cement content in concrete of 400 kg/m³ unless otherwise approved. Other specifications do not make this a requirement, but this amount is the same as recommended in CP 2004 for conditions where concrete has to be placed under water.

31. *Bentonite control.* All the specifications require control of the properties of the bentonite suspension at the stages of mixing, supply to trench, and immediately before concrete is placed. Table 1 indicates the tests recommended in each of the specifications.

32. It is noticeable that the consulting engineers have asked for all the tests which could be thought to have any bearing on the subject, though at this stage it is probable that this is a research exercise rather than an attempt at control. The principle involves the view, it would seem, that once one has established a site laboratory, then one

Table 1. Recommended tests for bentonite control

Parameter	Specification			
	FPS	Austrian	Consultant A	Consultant B
Density	•	•	•	•
Viscosity (Marsh cone)	•	•	•	•
Plastic viscosity	—	—	•	•
Apparent viscosity	—	—	•	•
Bingham yield strength	—	—	•	•
10 min gel strength	•	—	•	•
Fluid loss	—	•	•	•
pH	•	—	•	•
Sand content	—	—	•	•

should collect maximum information. However, relationships between the various properties of bentonite suspensions which are conventionally measured are obscure, and it does not appear likely that these relationships will easily be clarified by random site observations without collection of additional details not now requested.

33. There seems to be some difference of opinion concerning the limiting values to be attached to each parameter. The FPS Specification sets out a range for each included item; the Austrian document suggests that standard values should be obtained from a control suspension at the commencement of work, but is less definite on how to proceed with contaminated fluid. The consulting engineers differ in approach, from requiring the contractor to state limits before beginning work to specifying limit values for each parameter.

34. While the FPS specifies properties at 20°C, the other specifications do not appear to set a similar standard. Suitable compensation factors for other temperatures are not yet available from any source.

35. The frequency of control tests is also a matter upon which opinions diverge, but this is clearly a point on which a reasonable approach needs to be adopted depending on the circumstances and the rate of contamination of the fluid under each particular ground condition.

Workmanship

36. *Bentonite suspension level.* The level at which the bentonite suspension should be maintained in the excavation is specified variously as 'within 300 mm of the top of the guide wall', 'within the depth of the guide walls and not less than 1·00 m above the level of external standing water' or simply at a 'sufficient level to maintain stability'.

37. The 300 mm value would appear to be over stringent for general use. Depending on grab volume and panel length it could be very difficult to attain, especially if the ground has an open structure and fluid loss also occurs. In general terms, the higher the bentonite suspension level, the better is the overall stability of the trench walls.

38. *Loss of suspension.* When sudden loss of suspension occurs the FPS Specification requires the trench to be backfilled. One continental specification requires a reserve of suitable filling material to be kept in stock, and suggests about 30 sacks of cement. It also suggests that fluid sufficient to fill twice each panel in course of excavation should be kept in reserve.

39. These latter suggestions may appear to be prudent precautions, but they are probably too academic in concept. The important thing is to act without delay to seal off the loss and subsequently to look for an answer which can allow the work to proceed safely.

40. *Tolerances.* Insofar as tolerances are concerned, verticality appears to be the main consideration, and on this there is fairly general agreement on a ratio of between 1:67 and 1:80. The maximum projection of overbreak is variously stated as between 75 mm and 100 mm. Only the FPS Specification makes the relevant point that this limit is related to the soil conditions.

41. Insofar as the finished head of wall level as cast is concerned, the FPS Specification gives a tolerance related to depth, and one consultant's specification gives a two stage tolerance for trimmed levels either above or below a datum 2 m from the top of the guide wall.

42. *Safety.* This is mentioned only in the FPS document.

Performance

43. *Watertightness.* This is generally required, the consulting engineers stating that the wall 'shall be watertight'. The FPS Specification points out that seepages can arise from wall deflexions, but requires the wall to be free from visible water leaks.

Other factors

44. The above points are of course not the only ones which show some difference of opinion, but they are the main ones.

45. None of the specifications deal with such items as wall thickness limits, permeability and durability of concrete, and noise.

46. The specifications do not set out to cover such items as specification for the site investigation where a diaphragm wall is to be used, nor the use of ground anchors. These are items for which particular specifications might well be developed, in view of the important effect on wall design of having both a really reliable knowledge of soil characteristics and a sound anchor system.

47. The FPS is proposing to review its Specification for cast in place diaphragm walling towards the end of 1974, basing the review on experience of use and on comment received, as well as taking into account points raised at this Conference.

REFERENCES

1. FEDERATION OF PILING SPECIALISTS. *Specification for cast in place concrete diaphragm walling.* FPS, London, 1973. (Reprinted from *Ground Engng*, 1973, **6**, July, No. 4.)
2. OSTERREICHISCHES NORMUNGSINSTITUT. *Slurry trench walls.* Osterreichisches Normungsinstitut, Vienna, 1973, Draft ONORM B4450.
3. CONSTRUCTION INDUSTRY RESEARCH AND INFORMATION ASSOCIATION. *The effect of bentonite on the bond between steel reinforcement and concrete.* CIRIA, London, 1967, Interim Research Report 9.
4. BRITISH STANDARDS INSTITUTION. *Code of practice for foundations.* BSI, London, 1972, CP 2004.
5. BRITISH STANDARDS INSTITUTION. *The structural use of reinforced concrete in buildings.* BSI, London, 1957, CP 114.
6. BRITISH STANDARDS INSTITUTION. *Code of practice for the structural use of concrete.* BSI, London, 1972, CP 110, Part 1.

APPENDIX. COMPARISON OF SPECIFICATIONS

Item	FPS	Draft Austrian ONORM B4450	Consultants A	Consultants B
DESIGN				
Standards to be referred to in preparing design	•	•	N/A*	—
Basic design data to be used (inc. site investigation)	•	•	N/A	•
Stress limitations for concrete	•	—	N/A	•
Wall thickness limits	—	—	N/A	—
Reinforcement stresses, including bond	•	—	N/A	—
Cover to reinforcement	•	•	•	•
Spacing between bars	•	•	—	—
Factors of safety	—	—	N/A	•
Vertical loading and combined horizontal, vertical and bending forces	•	—	N/A	—
Treatment of anchorages and struts	•	—	N/A	—
Information to be submitted with design	•	—	N/A	•
Deflexion limitations or conditions	•	—	N/A	—
Special design limitations	•	—	N/A	—
Consideration of adjacent structures	•	—	•	•
Detailing of wall (i.e. panel lengths, etc.)	•	—	•	•
Design of guide walls	•	—	N/A	•
Specification requires statement of construction procedure—excavation, anchorage, etc.	•	—	•	—
Method of forming recesses	—	—	•	•
Design of post-tensioned prestressed walls	—	—	—	—
MATERIALS				
Concrete				
Details of cement	•	—	•	•
Details of aggregates: general	•	—	•	•
shell content	•	—	•	—
salt content	—	—	•	—
Additives	•	—	•	•
Workability	•	—	—	•
Water/cement ratio	•	—	—	—
Minimum cement content	•	—	—	—
Mixing procedure	•	—	•	•
Transporting concrete	—	—	•	—
Cold weather working	•	—	•	—
Control tests	•	—	•	•
Trial mixes	—	—	•	—
Cement storage	—	—	•	—
Reinforcement				
Quality of steel (also deformed or non-deformed bars)	•	•	•	•
Storage of steel	—	—	•	—
Fixing of steel: general	•	—	•	•
cage stiffness	•	—	•	—
Welding	•	—	•	—
Bending	—	—	•	—
Bentonite suspension				
Sources, type and quality of bentonite	•	—	•	•
Storage of bentonite	—	—	•	—
Mixing of bentonite suspensions	•	—	—	—

Item	FPS	Draft Austrian ONORM B4450	Consultants A	Consultants B
% bentonite in suspension to be stated and consistent	•	—	•	—
Control tests on freshly mixed suspension	•	•	•	•
Control tests on suspension supplied to trench	•	—	•	•
Control tests on suspension in the trench	•	•	•	•
Frequency of suspension control tests	•	—	•	—
Procedures for correction of suspension to fall within specified limits	•	—	•	•
Temperature effects and control	•	—	—	—
Level of bentonite in excavation	•	•	—	•
Remedial action in case of loss of suspension to underground cavities	•	—	—	—
Saline conditions	•	—	—	—
Site cleanliness	•	—	—	•
Bentonite disposal	•	—	•	•
WORKMANSHIP				
Types of equipment required to excavate strata	•	—	•	—
Avoidance of damage to adjacent panels	•	—	•	—
Tolerances: guide walls	•	—	•	—
verticality	•	•	•	•
twist	—	—	•	—
recess and inserts	•	—	•	—
reinforcement	•	—	—	—
cast concrete wall top level	•	—	•	—
protrusions	•	•	•	•
wall thickness	—	—	•	—
Fixing of reinforcement—preventing movement during concreting	•	—	•	•
Handling reinforcement	—	•	—	—
Type and use of stop ends	•	—	—	—
Safety precautions	•	—	—	—
Concrete placing procedure	•	•	•	•
Continuity of construction	•	•	•	•
Trimming of wall top	•	•	•	—
Extraction of stop ends	•	—	—	—
Cleanliness of joints	•	—	•	—
Filling above cast level of concrete	•	—	•	•
Inspection	—	—	•	—
PERFORMANCE				
Stability under various states of load and excavation	•	—	—	—
Deflexion	•	—	—	—
Strength	•	—	—	—
Durability	—	—	—	—
Watertightness	•	—	•	•
Internal finish	•	—	•	—
Records	•	—	•	•
Noise	—	—	—	—
Vibration	—	•	—	—
Adjacent structures	—	•	—	•
Compliance with Local Authority requirements	—	—	—	•

* Not applicable

Discussion on Paper 26

Reported by D. DENNINGTON

Mr Kipps said that the Authors had deliberately been diffident in their approach in the hope of stimulating discussion. He warned against over-emphasis of detail, as may occur in specifications when people concentrate entirely on the bits they know about; an example was the attention given to the testing of bentonite and its suspension, which was only a temporary element in a process and only indirectly affected the quality of the finished work. He considered that the FPS specification, which he and his co-authors helped to draft, was the most comprehensive document of its kind so far: it was concerned with general principles and tried to resolve matters of control that arose with diaphragm wall work; special aspects, particularly on an individual project, could always be covered by additional clauses.

2. The specifications studied by the Authors made no mention of filling materials other than concrete; examples of omissions were the plastic mixes used in cut-off walls and the techniques described by Mr Little in Paper 4. Mr Kipps considered that closer attention should be given to the matter of preliminary site investigations where diaphragm walls were concerned, especially in view of points raised by Professor Veder and Dr Jefferis (Discussion on Papers 10–13, §§ 2 and 9–11). Watertightness was unnecessarily emphasized: a good diaphragm wall was substantially watertight if not completely damp-proof.

3. He pointed out that the Authors of Paper 26 were all primarily concerned with the business of diaphragm wall construction rather than design, and he hoped that the discussion would concentrate on the latter aspect.

SPEAKERS FROM THE FLOOR

Objectives

4. Mr D. J. Palmer, Chairman of the Piling Specification Committee, disagreed with the suggestion (§ 8 of the Paper) of a more general approach in respect of specifying concrete workability; he thought that matters were not so simple as the Authors suggested, and preferred the type of clause which they suggested should be removed. He did not think notes for guidance should be needed, as this suggested that the specification was otherwise inadequate, but he agreed with the Authors that the conditions of contract must not be confused with the specification. He suggested that the Association of Consulting Engineers should be invited to join in any redrafting of the specification, although he foresaw an inherent danger with regard to delay when more people were brought into the matter.

Bentonite suspension

5. Mr K. W. Cole said that he had been associated with design and construction of diaphragm walls since 1962. He recalled a contractor who was incredulous that the engineers were specifying the type of bentonite and its control limits, but who reacted by exceeding the specification and producing very good work. Mr Cole was surprised that Papers 25 and 26 made no reference to the type of bentonite, i.e. its make, origin and properties; he considered this to be a serious omission since there was no standard bentonite and a specification which quoted a particular concentration was not only unhelpful but could be dangerous. Rather than specify mud limits his firm made the practice of requiring contractors to set their own limits.

6. Mr Cole spoke of the properties of poorly controlled slurry and well controlled slurry on two sites in London, both having fill and alluvium overlying London Clay. In Figs. 1–3 the results from the poorly controlled site are given as crosses and those from the well controlled site are given in groups.

7. Figure 1 shows the significant effect of pH on fluid loss. The slurry from the bad site was quickly contaminated, whereas that from the controlled site was kept within limits.

8. From Fig. 2, an upper limit to density much lower than the 1·21 suggested would appear to be desirable, and 1080 g/l seemed to be appropriate on the basis of the experience at these sites. The limit should be flexible to take account of the varying concentrations of bentonite and the desirability of heavier slurry in certain circumstances.

9. In recent years a lower limit of gel strength of 20 lb/100 sq. ft. had been acceptable, as illustrated in Fig. 3 by the progress of slurry quality on the good site. The fresh slurry was good; the first batches of reconstituted slurry were relatively poor (denoted by 'mid period') and after some tightening of control the gel strength fell within the 'later period' envelope. This improvement was also reflected in other properties and the resulting wall looked good with tight joints and little overbreak (Fig. 4).

10. The bad site slurry resulted in a thick cake of London Clay and irregular concrete surface (Fig. 5). Heavy contamination around the wheel spacers occurred (Fig. 6), as the concrete was unable to displace the slurry.

11. Table 1 shows the limits of the various slurry properties found to be acceptable for excavations in London using slurries made from Fulbent 570 or Berkbent. The first group of tests could be conveniently carried out near to the trench, whereas the lower group required site laboratory conditions. Although the Marsh cone test was a qualitative one, Mr Cole thought that it was convenient and could give early warning of slurry problems.

12. Mr Hutchinson said that in Paper 5 he was referring to converted bentonites only and agreed that others

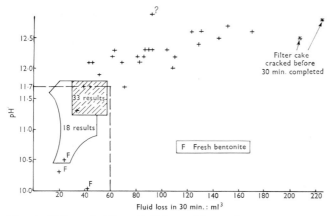

Fig. 1. *Effect of pH on fluid loss of bentonite*

Table 1. *Acceptable limits for properties of diaphragm wall slurry for excavations in London using Fulbent 570 or Berkbent*

	Property	Limits used	Possible revisions
Tests giving immediate field results	Density	1030–1070 g/l³	1035–1080 g/l³
	pH	<11·5	<11·7 (meter)
			<11·0 (paper)
	Marsh cone	38–55	—
	Sand content	<5%	1–8% (used slurry)
Tests requiring laboratory work	Fluid loss	<60 ml (in 30 min)	—
	10 min gel	20–55 lb/100 ft²	—
	Apparent viscosity	3–28 cP	Delete
	Plastic viscosity	2–8 cP	3–10 cP
	Bingham yield	2–38 lb/100 ft²	Delete

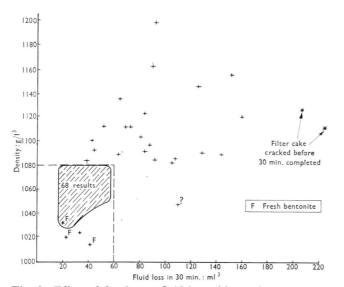

Fig. 2. *Effect of density on fluid loss of bentonite*

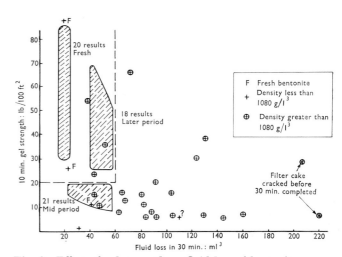

Fig. 3. *Effect of gel strength on fluid loss of bentonite*

Fig. 4. *Diaphragm wall on good site*

Fig. 5. Wall surface on bad site

Fig. 6. Wall surface on bad site

needed specifying. He thought a density of 1.08 g/cm^3 was low compared with his own value of 1.21 g/cm^3, and wished to compare Mr Cole's data with other information. Mr Hodgson noted that the proposed specification did not refer to fluid loss; he thought the standards in it should be as high as was reasonable.

Environmental and other factors

13. Mr W. G. Ascough referred to a project in a shopping precinct in Southend where diaphragm walling had been chosen to avoid noise; purpose-made screens had also been erected to reduce the intrusion of the work. He considered that any future specification should deal with the problem of disposal of used bentonite; in Southend an attempt had been made to dump it on the foreshore but tidal action would not disperse it.

14. As the ground consisted of weathered clay separated from a lower blue clay by 9 m of sand and gravel, there was a certain amount of overbreak; in his opinion this gave rise to the need for a specification to refer to the subsequent treatment of the surface of cast walls. In Paper 3 Mr Kienberger had discussed research in the differential settlement between piled and diaphragm wall foundations; this prompted Mr Ascough to ask what should be the specific procedure for testing diaphragm walls, kentledge tests appearing to be inappropriate.

REPLY AND SUMMARY

15. Dr Fleming accepted many of the points made by Mr Palmer. The notes for guidance were originally included in the text mainly because many specification users would not have direct experience of the technique. It might eventually be decided to banish the notes.

16. It was understood that the document might be regarded with suspicion because it is contractor biased, but at the time it was prepared there was a great need for guidance, and there was no sign of any progress on the matter on the part of any other body.

17. Concerning specification of the type of bentonite, Dr Fleming said that the specification does require details of the proposed bentonite to be given. He said that

bentonite performance depended on a combination of engineering properties and physical chemistry. The FPS specification was based on the material generally used in the UK; for other materials and for use in other parts of the world, the parameters could be amended.

18. Fluid loss was left out of the FPS document because of the dissimilarity between site and test conditions; however, Professor Veder's suggestion for an amended test (Discussion on Papers 5–9, § 15) might offer a means of making the test more realistic. Tests should not only give consistent results; they also should be as meaningful as possible. A consistent control of suspensions, however, led to a better end product.

19. Testing of wall panel units was not covered by the FPS specification, but—as single piles do not settle as do pile groups—the behaviour of diaphragm wall load bearing units in groups might need study.

20. Mr Fuchsberger added that, in his opinion, overbreak was not a matter for the specification and that disposal of used bentonite should be the contractor's liability.

21. Mr Dennington thought that the Authors would not have spoilt their cause by giving their own views. It was disappointing that the Paper had attracted so few speakers and such little fervour.

22. The contractors' bias was referred to by some, almost apologetically. Many would naturally agree with bringing in other sides of the profession, but Mr Palmer's warning on undesirable delay should be heeded.

23. Remarking on the contentious nature of the subject, Mr Dennington said that the Authors obviously agreed with the excellent statement in a Royal Commission report: 'a good specification substitutes objective tests of quality for pious demands for the unattainable'.

CONTRIBUTORS

Mr W. G. Ascough, Borough of Southend.
Mr K. W. Cole, Ove Arup and Partners.
Mr D. Dennington, Bullen and Partners.
Mr F. T. Hodgson, Clarke, Nicholls and Marcel.
Mr M. T. Hutchinson, Cementation Research Ltd.
Mr D. J. Palmer, Lind Piling Ltd.

General discussion

Reported by D. DENNINGTON

Professor J. E. Prentice illustrated the watersealing properties of bentonite with an example of polluted groundwater from an agricultural factory near Cambridge escaping into a tributary of the river Cam. The need was to provide a complete isolation of the polluted ground from the river—to find a material which would not only prevent movement of groundwater, but also to prevent capillarity and diffusive movement. A single trench filled with bentonite–cement mixture was constructed along one side of the site. This proved completely successful. Pollution levels dropped from 4–5 p.p.m. before construction to 0·001 p.p.m. a few months afterwards. The success of this application points to the possibility of a wide range of uses in the control of groundwater movement and pollution.

RESPONSIBILITIES OF CONSULTANTS AND CONTRACTORS

2. Dr R. Bassett discussed the extent of a consultant's involvement in the detailing of bentonite wall specifications. He referred to two philosophies:

(a) the specification requiring an end product of a certain structural standard and finished quality;

(b) the specification in which the consultant fixed many of the details of the construction technique and materials (in particular for the intermediate stage, the bentonite mix).

He firmly supported approach (a) as by everyday experience the specialist contractor must be a better judge of construction details.

3. Mr Mitchell (Paper 17, §§ 30–33) suggested situation (b) in consideration of ground anchor systems. Dr Bassett agreed with him regarding the consultant's responsibility for the wall–anchor system as a complete structural arrangement, but strongly opposed the detailing of construction methods and materials.

4. He suggested that the consultant (or main contractor for temporary works) should be responsible for analysis of the site investigation and decision on site features and the soil parameters, and also for overall analysis of the wall's stability. The specialist anchorage contractors should, in early consultation with the engineers, advise on the site investigation data required in order to make provisional estimates of the anchor size, length and diameter; and they should choose their own construction methods to achieve the consultant's loads, advising the consultant in a method statement.

5. In an experienced firm the construction techniques, including such variables as water/cement ratio, grout pressure, grout take, air or water drilling, rate of case withdrawal etc., were in the hands of the site supervisors, drillers and grout crews. Their integrity and experience must be accepted; their reliability is what a client pays for. Dr Bassett quoted the experience, in apparently identical conditions, of one anchor shaft accepting over 4 t of grout with the pressure never exceeding 50 lb/sq. in., and the adjacent anchors (constructed previously) all accepting some 1½ t at pressures exceeding 120 lb/sq. in.; all anchors tested out satisfactorily, changes being made by the construction crews and site engineer.

6. Certain areas of responsibility and detail could not be clearly divided in the manner suggested; these included the preparation and cleanliness of tendon and strand and their protection against corrosion. In these details the specialist contractor had a superior working knowledge and should be called in as an advisor to the engineers.

ANCHORS

7. Dr Bassett referred to an example[1] of anchor capacity estimation. The basic constructional and soils data for anchors constructed on four separate sites were supplied to five experienced engineers. They used theoretical formulae or contractural experience to estimate ultimate capacities. The actual test loads were then presented for discussion. The results are given in Table 1.

8. The ranges indicated the lack of precision in the available site investigation data. The values showed the real significance of an estimated load capacity. This calculation was usually requested by consultants and main contractors at tender stage and the figure thereafter

Table 1. Estimated anchor capacities (kips)

		Calumet Harbour	Washington metro	Morristown (NJ)	Parque Central
Dr Costa-Nunes, Tecnosolo	Theory	160–510	125–300	80–260	70–380
Mr Maljian, LeRoy-Crandell	Theory	300–500	250	200–250	400–600
Mr Nelson, Spencer, White and Prentis	Experience	250–300	120–150	100–120	250–300
Dr Murphy	Theory	295	150	120	215
Dr Bassett, King's College	Theory	200–290	130	125	145–205
Actual data		320–450	160–220	150–260	200–280

adhered. Dr Bassett considered that estimated load should lie at the low end of the actual production anchor performances.

Compression anchors

9. Mr N. B. Hobbs referred to tension anchors and their poor load carrying capacities compared with compression ones; he considered the Poisson effect of lateral shrinkage was the cause. He referred to the vulnerability of compression anchors to corrosion for the opposite reason, there being more of the anchor in contact with the soil.

10. Dr M. F. Stocker commented on the fear that from the viewpoint of corrosion protection a compression-tube anchor might be worse than an anchor that transferred its load directly from the tendon to the grout body.

11. One reason was that excessive hoop strains in the grout body, caused by an increase in diameter of the compression tube due to Poisson's ratio, might produce some longitudinal cracks. When the Olympic tent roof in Munich was to be erected and stressed with the aid of permanent compression-tube anchors in 1970/1971, some people were concerned and an additional spiral reinforcement was placed around the compression tube. Since then a lot of experience had been gained with compression-tube anchors without spiral reinforcement. More than 30 anchors of this type had been excavated and studied carefully after they had been stressed several times to the yield point of the steel (yield loads 65–112 t). No longitudinal cracks were found whether the anchors had been carried out in gravel, sand or clay. It could be shown theoretically that the hoop strains remained below the failure strain of concrete if the compression tubes were designed correctly.

12. Secondly, due to the greater weight of a compression-tube anchor compared to a normal tension anchor, the anchor might sink to the bottom of the bore hole and lie eccentrically within the grout body, and thus provide insufficient corrosion protection. Since the unit weight of steel was greater than that of cement grout, this fact applied to both anchor types independent of their absolute weight. In soils which did not absorb excessive water from the cement grout during the grouting procedure (e.g. clay), spacers were required to guarantee the central position of the anchor tendon within the bore hole or grout body. In non-cohesive soils spacers were not necessary due to the fact that excessive water was absorbed immediately by the surrounding soil because of the grouting pressure. The resulting grout was stiff enough to hold the anchor tendon or compression tube in the same position as it left the end of the drilling tubes while these were withdrawn. A sufficient concrete cover was thus guaranteed.

13. Apart from the advantage of an uncracked grout body, a compression-tube anchor distinguished itself through its relatively simple corrosion protection system, which could be applied, controlled and tested completely in the factory, and the ruggedness with which it could be handled on site without danger of hurting the corrosion protection.

MUD STABILIZATION

14. Referring principally to Paper 5, Mr M. J. Puller said that the use of stabilizing muds with reverse circulation equipment required a mud with large mud re-use.

The philosophy for mud design was

    (*a*) relatively thin mud;
    (*b*) sufficient mud to ensure trench stability;
    (*c*) minimum fluid loss and cake thickness.

15. These properties were obtainable by using sufficient bentonite for weight purposes, with thinners and a fluid loss inhibitor. The use of the API filter press equipment had contributed considerably to this mud design development. The fluid loss using this test with a 4% bentonite was typically 20 cm³; with 0·2% cement addition this fluid loss increased to 50 cm³. However, the addition of 0·2% Dextrid reduced the fluid loss in the test back to 20 cm³.

16. Whilst it was appreciated that the test provided only indications of cake thickness and fluid loss, practical experience with the muds had shown direct influence on the quality of finished walls.

17. This philosophy had enabled use of the same mud for more than three months with reverse circulation equipment. These same muds were now being used by his firm with Kelly type equipment.

MODEL TESTS

18. Mr James dealt with a few aspects of the model tests described in his Paper 6. Whilst there were severe limitations which might be reflected in the results, he wished to expand on the influence of anchor stressing on some measurements. The model tests formed part of a large investigation including the field instrumentation described in Paper 15. Figure 10 of Paper 6 showed the influence on an already stressed anchor of the stressing of a subsequent layer. This effect was not recorded in the field measurements. Similarly, and possibly more relevant in the light of some of the earlier discussions, Fig. 11 showed the influence of anchor stressing on ground movements at the rear of the wall.

19. If these phenomena occurred then they should be taken into account in the economic considerations of increasing the number of anchor layers and the consequent reduction in bending moments.

REVIEW

20. Two aspects of the Conference were selected for comment by Professor T. H. Hanna.

Bearing capacity

21. Paper 23 by Mr White on the bearing capacity of retaining walls in rock was very timely and it should be appreciated that there was a similar type of problem in soil—perhaps a greater problem.

22. Professor Hanna questioned the ability to predict the bearing capacity of walls and piles in the excavation environment, where the walls are supported either by anchors or by struts. He doubted that such a position had been reached, although the finite element method of excavation modelling would undoubtedly help. In trying to apply such a method the question arose as to whether the excavation and wall supporting processes could be simulated reliably.

23. Work was in progress at the University of Sheffield on development of techniques for the design of load bearing basement type walls. This work involved sophisticated large-scale laboratory tests and also a finite element idealization of the excavation process.

Anchors

24. During the Conference a considerable amount of discussion had centered around ground anchors and their construction in soft soils such as sands and clays. Very positive contributions were made in Paper 18 by Mr Ostermayer and in discussions which it drew. It was appreciated that ground anchoring was a difficult subject.

25. On going through records of different sites one found that currently many anchors failed. Contractors and consultants often blamed the ground conditions, stating that the ground had changed from what they were led to believe existed, based on the site investigation report. Saying that this was not a very positive approach to take, Professor Hanna pointed out that every anchor was drilled and the ground recovered from that drill hole. Also anchors might be formed with or without an underream or series of underreams. He asked how this drilling information could be used to ensure more reliable formation of anchors.

26. Professor Hanna was particularly sorry to find few anchoring contractors telling officially what they were doing in this field. Until anchoring techniques and the philosophy of anchoring improved and until the profession was prepared to make use of the information obtained from each anchor drill hole he felt that the art and perhaps the science of anchor use would not advance very rapidly.

27. Because of these uncertainties in the use of ground anchors the Construction Industry Research and Information Association had set up a small working party to guide study under his direction into ground anchor uses in civil engineering, particularly anchors in soft ground such as sands, clays, marls and chalks. The object of this study was to produce a manual on ground anchors to cover anchoring processes and specifications for anchor construction and testing.

CONTRIBUTORS

Dr R. Bassett, King's College, University of London.
Mr D. Dennington, Bullen and Partners.
Professor T. H. Hanna, University of Sheffield.
Mr N. B. Hobbs, Soil Mechanics Ltd.
Mr E. L. James, Cementation Ground Engineering Ltd.
Professor J. E. Prentice, King's College.
Mr M. J. Puller, A. Waddington and Son Ltd.
Dr M. F. Stocker, Karl Bauer KG.

REFERENCE

1. AMERICAN SOCIETY OF CIVIL ENGINEERS. *ASCE Geotechnical Division Speciality Conference, Austin, Texas, 1974*. Session 5 discussion.

Closing address

C. VEDER, DiplIng, DrTechn, DrIngHC, *Professor, Technische Hochschule, Graz, Austria*

In the still peaceful days of the first half of 1939, I was engaged to study the feasibility of a hydro-electric power plant (Oetz Kraftwerk) in Tyrol, Austria. The main problem was to build a cut-off for a dam near Laengenfeld where the subsoil, to a depth of 100 m, consisted of a rock slide with granite blocks from 0·1 m × 0·1 m to 10 m × 10 m in size, totally below groundwater level. Only freezing of the subsoil seemed to be possible but this method was very expensive. Test borings on the site were carried out using bentonite. While observing the work, I had the idea of forming an impermeable wall by making holes which would be kept open by the bentonite suspension; the holes would then be filled with concrete and new holes would be bored which would overlap the first by 5–10 cm. The customer was very enthusiastic about the idea . . . the only problem was the war.

2. The idea remained dormant until 1950, when it was developed with the help of my friends of ICOS and sponsored by Mr Semensa, director of the SADE firm in Venice. The method was applied for the first time in 1951 at Venafro, near Naples, where a completely watertight wall, 1500 m long and 8–35 m deep, had to be built in and under a compensation basin through very permeable coarse sand and gravel, and embedded in the underlying clay.

3. A test pit of diameter 5 m was constructed 8 m below groundwater level. It showed that the overlapping cylindrical elements constructed with a chisel bit were watertight, and also proved that it was possible to build a rigid, load bearing wall.

4. During this project we soon found out how hard a mixture of bentonite and fine sand can be, and how this can completely hinder boring at the bottom of the borehole. We also realized that fresh bentonite suspension could easily loosen up the hard mixture; we thus discovered one of the most important characteristics of bentonite for construction engineering.

5. At this time, the special clamshell designed by Mr Brunner for constructing elongated diaphragm wall elements was used exclusively in constructing the 450 000 m² diaphragm wall on the first line of the Milan subway, 16 km long.

6. The diaphragm wall is embedded in the soil and can be loaded horizontally and vertically. For this reason, in construction of the subway in Milan, the previous method of cut and cover could be converted into the method of cover and cut, known now as the Milan method, with great economic and transportational advantages. After completion of the diaphragm wall on both sides of the tunnel route, the final roof could be installed and traffic could resume undisturbed while, below the roof, the floor was being excavated to the necessary depth.

7. Numerous subways the world over were and still are being built by this method. Further, the diaphragm wall system has been developed by many famous researchers and builders, with or without valuable changes.

PROPERTIES OF BENTONITE

8. When bentonite is used for construction, the importance of knowing the properties of bentonite is comparable with the importance of knowing the properties of cement such as heat development, time of hardening, and shrinking. More investigation is required concerning the properties of bentonite and the effect of the bentonite on the wall. For instance, the ideas on adhesion and the importance of electrical forces, brought out at the conference in Mexico,[1] should be developed further.

9. Concerning the observations made by Dr DiBiagio of the Norwegian Geotechnical Institute (discussion on Papers 1–4, § 4), I had the same experience in Milan, where the soil was composed of gravel, sand and silt above groundwater level. The bentonite suspension disappeared into a hole below the excavation and the trench remained standing for a time (10–20 min), held in place by only the bentonite cake.

10. The report by Dr James (discussion on Papers 5–9, §§ 16–18) regarding the edge to face structure of bentonite is important and the ideas ought to be followed up. Using an electron microscope I have observed clearly this edge to face structure in a photograph of a bentonite cake taken from an excavated trench.

11. A content of 25% of sand in the bentonite, as indicated by Mr Hutchinson (Paper 5, § 35), seems to me a theoretical maximum value. In practice the value has to be determined case by case and, according to my experience, should not exceed 10%. Figure 1 shows an example of a wall where some elements were built with a sandy suspension (they look rough) and the two elements on the edge of the construction were built with a clean suspension (they look smooth).

FURTHER DEVELOPMENT

12. Further studies have been carried out on the action of the bentonite cake, and the international bentonite specifications and test procedures have been improved. There has been increasing application of the finite element method to calculate the behaviour of the wall in combined action with anchored or non-anchored soil. The deep cutoff has also been developed further; a depth of 131 m was achieved at Manicouagan, and 200 m or more seems to be quite feasible.[2]

13. The use of prefabricated elements should be developed further, with special attention paid to avoid distressing the soil when introducing the prefabricated elements into the trench, and bearing in mind that the cement will usually 'spoil' the bentonite suspension to a certain degree.

14. Prestressing the wall leads to a very elegant and thin construction.

15. Also of interest are composite constructions, as designed for example for the quay of Redcar, described

Diaphragm walls and anchorages. Institution of Civil Engineers, London, 1975, 221–222.

221

Fig. 1. Effect of sand in bentonite suspension: clean suspension (centre) gives smooth surface; sandy suspension (right and left) gives rough surface

Fig. 2. Fish-eye view of the circular car park in Bloomsbury Square (dia. 52 m, excavated depth 20 m, thickness of wall 80 cm)

in Paper 2. The tensile joint, designed by Irwin-Child, enabled achievement of a real combined action of the diaphragm wall with the included soil.

16. The circular protection wall for excavations, as built, for example, for the car park of Bloomsbury Square (Fig. 2) and at Lille (Fig. 3), has undergone development. Interesting laboratory tests were carried out by Ridgen and Rowe with the centrifuge (Paper 9). Because of these tests, a relatively thin shell could be built with little or no reinforcement.

17. Development in the practice of building the diaphragm wall means better joints, and also faster work, especially when going through boulders and hard rock strata. Much is being done in this field but I have the impression that still more could be done.

18. A special problem is the confiding of important and difficult projects only to firms who have the necessary staff and experience to carry them out. Our real enemies are not our capable competitors but those who fail. The method seems easy but there are many details and special problems which must be attended to.

CONCLUSION

19. The bentonite diaphragm wall, in my opinion, is ahead of other construction methods when it represents a finished structure in the subsoil without any need of further supplementary constructions; it is not an environmental pollutant as it produces, for example, little or no noise.

Fig. 3. Aerial view of the circular car park in Lille, France (dia. 80 m, depth 20 m, thickness of wall 80 cm)

20. I believe that this method of construction has enormous possibilities in the future, requiring a great deal of work but also producing a great deal of satisfaction.

REFERENCES

1. KIENBERGER H. Testing results on the behaviour of bentonite suspensions in trenches. *Proc. 7th Int. Conf. Soil Mech., Mexico, 1969*, Speciality sessions 14 and 15.
2. DREVILLE J. *et al.* Diaphragme en béton moulé pour l'étanchéité des fondations du barrage. *Proc. 10th Int. Congr. Large Dams, Montreal, 1970*, **2**, 607–630.

Conversion table

Plane angle	degree minute second	$1^\circ = \dfrac{\pi}{180}$ rad $1' = \dfrac{1^\circ}{60}$ $1'' = \dfrac{1'}{60}$
Length	inch foot mile	1 in. $= 25\cdot40$ mm 1 ft $= 0\cdot3048$ m 1 mile $= 1\cdot609$ km
Area	hectare	1 ha $= 10^4$ m^2
Volume	gallon litre millilitre	1 gal $= 4\cdot546 \times 10^{-3}$ m^3 1 l $= 10^{-3}$ m^3 1 ml $= 1$ cm^3
Mass	pound kip ton tonne	1 lb $= 0\cdot4536$ kg 1 kip $= 453\cdot6$ kg 1 ton $= 1016$ kg 1 t $= 10^3$ kg
Mass per unit length	pound per foot	1 lb/ft $= 1\cdot488$ kg/m
Density	pound per cubic foot gramme per millilitre	1 lb/cu. ft $= 16\cdot02$ kg/m^3 1 g/ml $= 10^3$ kg/m^3
Moment of inertia	kip ft^2	1 kip ft^2 $= 42\cdot14$ kg m^2
Force	pound force kip force ton force dyne	1 lbf $= 4\cdot448$ N 1 kipf $= 4\cdot448 \times 10^3$ N 1 tonf $= 9\cdot964 \times 10^3$ N 1 dyn $= 10^{-5}$ N
Moment of force	pound force foot kip force foot	1 lbf ft $= 1\cdot356$ N m 1 kipf ft $= 1\cdot356 \times 10^3$ N m
Pressure and stress	megapascal	1 MPa $= 1$ N/mm^2 $= 10^6$ N/m^2
Dynamic viscosity	centipoise	1 cP $= 10^{-3}$ N s/m^2